THE I TATTI
RENAISSANCE LIBRARY

James Hankins, General Editor

ZABARELLA

ON METHODS
VOLUME 2

ON REGRESSUS

ITRL 59

JACOPO ZABARELLA

✦ ✦ ✦

ON METHODS

VOLUME 2 ✦ BOOKS III–IV

ON REGRESSUS

EDITED AND TRANSLATED BY

JOHN P. McCASKEY

THE I TATTI RENAISSANCE LIBRARY
HARVARD UNIVERSITY PRESS
CAMBRIDGE, MASSACHUSETTS
LONDON, ENGLAND
2013

Series design by Dean Bornstein

Library of Congress Cataloging-in-Publication Data

Zabarella, Jacopo, 1533–1589.
[Opera logica. Selections. English]
On methods / Jacopo Zabarella ; edited and translated by John P. McCaskey.
volumes cm. — (The I Tatti Renaissance library ; 58–59)
Latin text with English translation.
Includes bibliographical references and index.
ISBN 978-0-674-72479-2 (v. 1) — ISBN 978-0-674-72480-8 (v. 2)
1. Logic — Early works to 1800.   2. Methodology — Early works to 1800.
I. Zabarella, Jacopo, 1533–1589. Opera logica. Selections.   II. Title.
III. Title: De methodis.   IV. Title: De regressu.   V. Title: On regressus.
B785.Z23O613 2013
160 — dc23   2013017883

Contents

ॐ९९ॐ

ON METHODS

ON REGRESSUS   356

DE METHODIS

ON METHODS

LIBER TERTIUS

: I :

De differentia ordinis et methodi.

1   Quo discrimine methodus ab ordine discrepet iam in praecedentibus declaratum est, cùm enim ambo sint instrumenta logica et processus à noto ad ignotum, ordo tamen quatenus ordo est, vim colligendi non habet, sed disponendi solùm. Methodus verò vim habet illatricem et hoc ex illo colligit.

2   Ad hoc discrimen declarandum videntur illa aptari posse, quae dicuntur ab Averroe in primo commentario primi libri Posteriorum Analyticorum de praecognitionibus, inquit enim praecognitionum demonstrationi necessariarum duo esse genera, alia namque est praecognitio dirigens, alia verò agens. Dirigens illa dicitur, sine qua conclusio non colligeretur, sola tamen ad collectionem faciendam non sufficit, veluti praecognoscere quòd subiectum sit et quid eius nomen et quid nomen affectionis significent, nam si una aliqua istarum desit, nihil potest demonstrari, ex his tamen solis nihil demonstratur.

3   Praecognitio autem veritatis propositionum est praecognitio agens, quia iam agit et infert necessariò conclusionem.

4   Dicere igitur in praesentia possumus tam ordinem quàm methodum esse progressum à notioribus, ordinem quidem à notioribus cognitione tantùm dirigente ad aliorum notitiam,° methodum verò à notioribus etiam cognitione agente, quae videtur esse propria methodi differentia, qua methodus ab ordine separatur, in qua

BOOK THREE

: I :

On the differentia of order and method.

The discriminating difference by which method is different from 1
order has now, in the preceding, been made clear. For although
both are logical instruments and proceedings from known to unknown, nevertheless order, insofar as it is order, does not have the
power to gather,[1] only to dispose. Method, however, has inferential power and gathers this from that.

To make this discriminating difference clear, those things that 2
are said about prior knowledge by Averroës in the first commentary to the first book of the *Posterior Analytics*[2] appear to be appli-
cable. For he says that there are two kinds of prior knowledge
necessary for demonstration. One is prior knowledge directing and
the other acting. The one said to be directing—as knowing beforehand that the subject is, and what its name and what the name of
[its] affection signify—without which no conclusion is gathered, is
nevertheless not alone sufficient to make the gathering.[3] For if any
one of these is absent, nothing can be demonstrated, but nothing
is demonstrated from these alone.

On the other hand, prior knowledge of the truth of the prem- 3
ises is prior knowledge acting, because it acts now and infers a
conclusion necessarily.

At the present time, therefore, we can say that order as well as 4
method is a progression from [things] more known—order being
from [things] more known, by knowledge just directing [us] to
knowledge° of other [things], while method is from [things] more
known, by knowledge acting too. This appears to be method's

significatione methodum accepit Aristoteles in contextu 4 et 5
primi libri de Anima.[1]

5   Propterea rectè illi dixere, qui statuerunt methodum esse instrumentum notificans, id est iam faciens cognitionem ignoti, ordinem verò esse instrumentum disponens, partes enim disciplinae
ita disponit, ut ea anteponat, quorum cognitio nos dirigat et nos
iuvet ad ea, quae posterius tractantur, cognoscenda.

6   De hoc cùm abundè locuti fuerimus et de methodo agendum
relinquatur, propriè sumendo nomen methodi prout ab ordine
distinguitur, primo loco ipsam methodi naturam universè conside-
rabimus, deinde ad species descendemus.

: II :

In quo definitio methodi ponitur et declaratur.

1   Ex iis, quae dicta sunt de methodi atque ordinis differentia, haec
methodi definitio colligi posse videtur, Methodus est intellectuale
instrumentum faciens ex notis cognitionem ignoti. Intellectuale
instrumentum genus methodi est, quod ordinem quoque complectitur. Facere autem ex notis cognitionem ignoti est differentia, qua
methodus ab ordine separatur, quemadmodum alii quoque dicunt
proprium esse methodi notificare. Hanc differentiam si benè

proper differentia; by it method is separated from order. Aristotle accepted method with this signification in texts no. 4 and no. 5 of the first book of *On the Soul*.[4]

Therefore those who thought that method is an instrument 5 making [something] known, that is, bringing about knowledge of [what is] now unknown, spoke correctly. Order on the other hand is a disposing instrument. For it so disposes parts of a discipline that it places those things first, knowledge of which directs us and helps us to know those things that are treated later.

Since we have spoken about this abundantly and method re- 6 mains to be dealt with, by taking the name of method properly, in that it is distinguished from order, we will consider in the first place the very nature of method universally, and then descend to the species.

: II :

In which the definition of method is posited and made clear.

From the things that have been said about the differentia of 1 method and order, it appears that this definition of method can be gathered: method is an instrument of [the] understanding, bringing about knowledge of [what is now] unknown from [things now] known. The genus of method is *instrument of [the] understanding*, and this encompasses order also. But to bring about knowledge of [what is now] unknown from [things now] known is the differentia by which method is separated from order, just as others too say that to make known is proper to method. If we consider

consideremus, dicere cogimur, eam nil aliud significare, quàm illationis necessitatem, ut quando talis est progressus, quo aliquid ex aliquo necessariè colligatur, ea propriè methodus appelletur. Hoc namque nisi dicamus, nullum amplius inter ordinem et methodum discrimen remanet, cùm uterque sit progressus quidam à notis ad ignota. Nam si ab eiusmodi notis progrediamur ex quibus ignota non inferantur, sed quorum cognitio praeparet solùm nos ad rerum ignotarum cognitionem, et hanc methodum esse dicamus, methodum cum ordine confundimus; quoniam ordo quoque est progressus talis. Methodus igitur, ut ab ordine distinguatur, vim illativam habeat necesse est, qua aliquid ex aliquibus per necessariam consecutionem colligatur.

2   Praeterea quisquis hoc negat, negat methodum esse methodum, nam vox haec, $\mu\acute{\epsilon}\theta o\delta o\varsigma$, significat viam ex aliquo in aliquod, duos igitur terminos habeat omnis methodus necesse est, unum, à quo, alterum, ad quem; et terminum ad quem, esse ignotum, fieri autem notum beneficio termini à quo, ex quo in huius cognitionem ducimur, nam si terminus ignotus per methodum non innotesceret, quomodo methodus esset instrumentum notificans, ut omnes concedere videntur? Notificare autem rem ignotam non potest sine illationis necessitate, diximus enim aliud esse dirigere nos ad alicuius rei notitiam,° id est nobis auxiliari et prodesse ut aliquid cognoscamus, aliud esse eius rei cognitionem facere.° Potest enim aliquid nobis prodesse ad rem aliquam cognoscendam, quod tamen pro eius notitia° parienda non sufficiat, ideoque eius cognitionem

this differentia well, we are forced to say that it signifies nothing other than the necessity of inference; when progression is such that by it something is gathered from something [else] necessarily, this is properly called method. Now unless we say this, no discriminating difference any longer remains between order and method, since each is a sort of progression from knowns to unknowns. For if we progress from knowns of the type that from them the unknowns are not inferred but knowledge of them merely prepares us for knowledge of things unknown, and we say that this is method, then we conflate method with order, since order too is such a progression. It is necessary, therefore, that method, so as to be distinguished from order, have inferential power by which something is gathered from some other things by means of a necessary consequence.

Moreover, whoever denies this denies that method is method, 2 for this [Greek] word, *methodos*, signifies a way from something to something else — it is necessary, therefore, that every method have two termini, one *a quo* (from which), the other *ad quem* (to which) — and [denies that][5] the terminus *ad quem* is unknown but becomes known thanks to the terminus *a quo*, from which we are led to knowledge of the former. For if the unknown terminus were not made known by means of method, how would method be an instrument making [something] known, as everyone appears to concede? It is not possible to make known something unknown without the necessity of inference. For we said that it is one thing to direct us to knowledge° of some thing, that is, to aid and profit us so that we may know something; it is something else to produce° knowledge of this thing. For something can be of profit to us for knowing some other thing, that is nevertheless insufficient

non faciat.° At facere alicuius rei cognitionem illud dicitur, cuius notitiam° cognitio illius rei necessariò consequatur.° Itaque si natura methodi ea est ut sit via notificans id, quod erat ignotum, necesse est ut habeat illationis necessitatem, quia sine hac non diceretur notificans.

3   Nec possumus de hoc dubitare dum methodum cum ordine conferimus, et utriusque officium consideramus, ordo enim totam scientiam respicit, methodus verò non totam, sed problemata ipsius singula. Singulum enim problema dum ignoramus et inquirimus, sumimus aliqua nota, ex quibus in eius cognitionem ducamur. Propterea suprà dicebamus ordinem non tradere nobis aliquam rei ignotae cognitionem, sed solùm facere, ut melius ac facilius eorum, quae inquirimus, notitiam° consequamur, sed hanc sola methodus nobis tradit. Hinc factum est ut illi, qui res logicas tractant, dum de instrumentis cognoscendi[2] sermonem faciunt, solas methodos complecti videantur, nulla facta de ordinibus mentione, nam Aristoteles quoque in logicae artis traditione solas methodos consideravit, de ordinibus autem nihil docuit. Quocirca ab eius consilio recessisse non videntur, dum in logicis eius libris versantes, et de instrumentis sciendi disputationem instituentes methodos tantùm propriè acceptas instrumenti nomine comprehenderunt.

4   Quae igitur sit methodi natura, manifestum est.

for producing° knowledge° of it, and as a result cannot bring about knowledge of it.[6] But something is said to bring about knowledge of some other thing [when] knowledge° of it necessarily is followed° by knowledge of that other thing. And so if it is the nature of method that it be a way of making known that which was unknown, then it is necessary that it have the necessity of inference, because without that it would not be said to be making [something] known.

Neither can we have doubts about this when we compare method with order and consider the function of each. For order regards the whole science, and method [regards] not the whole but each of its topics of dispute. For when we are ignorant of each topic of dispute and ask about it, we take some known things, from which we are led to knowledge of it. Accordingly, we said above that order does not convey to us some knowledge of something unknown, but only makes it that we gain better and more easily the knowledge° of what we ask about; only method conveys this [knowledge] to us. And thus it is that those who treat issues of logic, when they discourse on instruments for knowing, appear to cover only methods, no mention being made of orders. And Aristotle too, in conveying logical art, considered only methods, and taught nothing about orders. Hence they [i.e., those writers on logic] appear not to have deviated from his intent, when, concerning themselves with the books on logic and putting together a debate about instruments for knowing scientifically, they comprehended only methods, accepted properly, under the name of instrument.

What, therefore, the nature of method is, is manifest.   4

3

: III :

Quòd inter methodum et syllogismum, proinde inter definitionem methodi ac definitionem syllogismi vel nulla vel parva est differentia.

1 Si ea, quae modò diximus, vera sunt, nil aliud videtur esse methodus, quàm syllogismus et definitio methodi à definitione syllogismi non differt.[3] Dixit enim Aristoteles, Syllogismum esse orationem, in qua quibusdam positis necesse est aliud quiddam sequi eò quòd illa sunt, volens per haec necessitatem consequentiae significare, quae est à propositionibus ad conclusionem, haec enim est propria syllogismi conditio. Sed in hoc methodi quoque naturam constitutam esse diximus, quid igitur prohibet ne methodum dicamus esse syllogismum ipsum ab Aristotele definitum? Nam si methodus est progressio cum illationis necessitate, sequitur methodum esse orationem, in qua quibusdam positis necesse est aliud quiddam sequi eò quòd illa sunt; in eo tantùm videtur differre methodus à syllogismo, quòd in definitione methodi hae duae dictiones ponuntur, notum et ignotum, quas Aristoteles in definitione syllogismi non protulit. Ratio autem est quoniam quilibet syllogismus duos ex necessitate fines habet, unum generalem, qui nullam materiam requirit, alterum particularem, qui non est sine propria materia; finis quidem generalis est ipsa necessaria collectio conclusionis ex positis, quem Aristoteles in eius definitione expressit. Finis autem alter particularis est vel scientiae adeptio, quae materiam necessariam postulat; vel opinio sive victoria, quae probabili materia satis habet; vel persuasio, quae fit in materia civili; vel deceptio, quae fit in materia falsa; vel aliquis fortasse alius. De syllogismo quidem ut generalem finem respiciente Aristoteles egit in

: III :

That there is either little or no difference between method
and syllogism, and accordingly between the definition
of method and the definition of syllogism.

If what we have just said is true, then method appears to be noth- 1
ing other than syllogism, and the definition of method does not
differ from the definition of syllogism. For Aristotle said, "A syllogism is speech in which, some things having been posited, it is
necessary that something else follow, because the former are [so],"[7]
wanting to indicate by means of this the necessity of consequence,
which is from premises to a conclusion; for this is a characteristic
proper to the syllogism. But we said that the nature of method,
too, is constituted in this. What, therefore, prohibits us from saying that method is syllogism itself, [as] defined by Aristotle? For if
method is a progression with the necessity of inference, it follows
that method is speech in which some things having been posited,
it is necessary that something else follow, because the former are
[so]. Method appears to differ from syllogism only in that these
two locutions, "known" and "unknown," are put in the definition of
method, and Aristotle did not advance these in the definition of
syllogism. The reason, however, is that, since any syllogism has
out of necessity two ends — one general, which requires no matter,
the other particular, which does not exist without proper matter — the general end is, of course, the necessary gathering itself of
the conclusion from what was posited; Aristotle expressed this in
its [i.e., syllogism's] definition. And the other, particular end is
either the obtaining of scientific knowledge, which demands necessary matter, or opinion or victory, which has enough in probable
matter, or persuasion, which happens in civil matters, or decep-
tion, which happens in false matters, or something else perhaps.

Prioribus Analyticis, ubi aliorum particularium finium nulla mentio fit, quamvis enim Aristoteles[4] de syllogismo loquatur praecipuè propter scientiam, quam parit in materia necessaria, de qua in Posterioribus Analyticis docturus[5] erat, attamen noluit ipsum ita restringere, ut ad alios quoque fines accommodari non posset, sed ipsum in sua amplitudine considerare voluit, nulli adhuc fini particulari, nullique materiae addictum. Hoc igitur cùm ipsius consilium fuerit, nullam debuit in definitione syllogismi mentionem facere noti et ignoti, haec enim cognitionis adeptionem respiciunt et syllogismi considerationem ita coarctant, ut libros tantùm Posteriores Analyticos respiciat. Ob hanc causam maluit in ea definitione dicere ex positis, quàm ex notis, quia ad conclusionem colligendam satis est si propositiones positae et concessae sint.

2   Nomen autem methodi aliquanto arctius est syllogismo, significat enim syllogismum ut ad cognitionis adeptionem directum et est commune genus omnium instrumentorum sciendi, de quibus agit Aristoteles in Posterioribus Analyticis, idem enim penitus esse arbitramur instrumentum sciendi et methodum, proinde apertissimè deceptos esse quàm plurimos, qui putarunt Aristotelem nullibi in logica de methodis locutum esse, cùm nullus alius sit eius scopus in Posterioribus Analyticis, quàm de methodis agere.

3   Syllogismi autem Dialectici et oratorii non propriè dicuntur methodi, quia solam habent methodi formam, at materiam convenientem pro scientiae adeptione non habent neque cognitionem ut finem respiciunt, qui omnium methodorum finis est. Ideò ratione suae formae possunt appellari methodi, sed non ratione finis

In the *Prior Analytics* Aristotle dealt, of course, with syllogism as regarding the general end; no mention is made there of the other, particular ends. For even though Aristotle speaks about syllogism principally for the sake of scientific knowledge, which is productive in necessary matter, and about which he was about to teach in the *Posterior Analytics*, he nevertheless did not want to so restrict it [i.e., syllogism] that it could not be accommodated to other ends also. He wanted instead to consider it in its wide sense, committed to no particular end and to no matter. Since this, therefore, was his intent, he ought to have made no mention of known and unknown in the definition of syllogism. For these regard the obtaining of knowledge, and so narrow down the consideration of the syllogism such that it regards only the *Posterior Analytics* books. Because of this, he preferred to say in the definition "from what are posited" rather than "from what are known." Because for gathering a conclusion, it is enough if the premises are posited and conceded.

The name "method," however, is somewhat narrower than syllogism,[8] for it signifies syllogism as directed to the obtaining of knowledge and as it is the common genus of all instruments for knowing scientifically—what Aristotle deals with in the *Posterior Analytics*. For we think that method and the instrument for knowing scientifically are completely the same, and accordingly that many were very plainly deceived, those who held that Aristotle spoke nowhere in [his] logic about methods, though his goal in the *Posterior Analytics* is nothing other than to deal with methods. 2

Now the dialectical and oratorical syllogisms are not properly said to be methods, because they only have the form of method but do not have matter appropriate for the obtaining of scientific knowledge, nor do they regard knowledge as an end; that [i.e., knowledge] is the end of all methods. Therefore [these] can be called methods by reason of their form but not by reason of the 3

neque ratione materiae. Quemadmodum ensis ligneus potest ratione formae ensis vocari, at si materiam, et finem ensis spectemus, non est ensis, quia in illa materia finem et operam ensis non praestat.

4 Si quo igitur discrimine methodus à syllogismo dissidet, id non aliud est, quàm illud, quod modò exposuimus. Quare manifestum est syllogismum esse commune genus et communem formam onmium methodorum, proinde posse vocari methodum, id est methodum formalem, in quo tota illationis necessitas posita est. Nam quatuor dicuntur esse logica instrumenta, de quibus loquitur Aristoteles in Prioribus Analyticis, syllogismus, enthymema, inductio et exemplum, enthymema quidem truncatus syllogismus est, qui nisi perficiatur, nihil ex necessitate colligit, si verò perfectus fiat, iam est syllogismus, propterea tunc vim habet illativam. Exemplum est inductio imperfecta, quae nihil ex necessitate concludit, ut docet Aristoteles in calce secundi libri Priorum Analyticorum. Inductio autem si perfecta sit, vim habet necessariam colligendi, sed per syllogismum, ea enim ratione concludit, qua vim syllogismi habet et ad bonum syllogismum redigi potest, ut ibidem declarat Aristoteles in capite de inductione. Neque erit ab re si ipsius verba in principio illius capitis posita referamus, quae sunt haec, 'Quoniam autem non solùm Dialectici et Demonstrativi syllogismi per praedictas efficiuntur figuras, sed etiam oratorii et simpliciter quaecumque fides et secundùm quamcumque methodum, nunc dicendum est; omnia enim credimus aut per syllogismum, aut ex inductione,' in quibus verbis duo sunt maximè notanda, unum est, quòd ibi testatur philosophus omnem methodum, quae fidem aliquam faciat, formam habere syllogismi et necessariò ad aliquem modum utilem trium figurarum reduci; et è contrario nil roboris

end or by reason of the matter. Just as a wooden sword can by reason of the form be called a sword, yet if we look at the matter and end of the sword, it is not a sword, because in that matter, it does not really fulfill the end and do the work of a sword.

If method, therefore, is distinguished from syllogism by this 4 discriminating difference, it is nothing other than what we just laid out. And so it is manifest that syllogism is the common genus and common form of all methods, and accordingly can be called method, that is, formal method; in it is located the whole necessity of inference. For there are said to be four logical instruments, about which Aristotle speaks in the *Prior Analytics*: syllogism, enthymeme, induction, and example.[9] An enthymeme, of course, is a truncated syllogism; unless it is perfected, it gathers nothing out of necessity. If it is made perfect, it is then a syllogism, and thereby then has inferential power. An example is an imperfect induction; it concludes nothing out of necessity, as Aristotle teaches at the end of the second book of the *Prior Analytics*.[10] If the induction is perfected, however, it does have the power to gather necessarily, but by means of a syllogism. It is conclusive for the reason that it has the power of a syllogism and can be reduced to a good syllogism, as Aristotle makes clear in the same place, in the chapter on induction. It will not be beside the point if we recount his words, placed in the beginning of the chapter, which are these: "It now has to be said that not only are dialectical and demonstrative syllogisms effected by means of the previously described figures, but oratorical also, and absolutely any belief whatever and according to any method whatever, for we believe everything either by means of syllogism or from induction."[11] In these words, two things especially have to be noted. The first is that the philosopher here attests that every method that brings about some belief has the form of a syllogism and necessarily reduces to some useful mode of the three figures, and on the contrary

habere et nullam facere fidem, si ad formam boni syllogismi reduci nequeat.

5   Alterum est quòd omnia instrumenta logica apud Aristotelem sub syllogismo continentur, à quo habent totam consequentiae necessitatem. Cùm enim enthymema sit syllogismus, exemplum verò sit inductio, illa quatuor ad duo rediguntur, ad syllogismum et inductionem, ideò Aristoteles omnia credi ait vel per syllogismum vel per inductionem, nec vult aliquo alio instrumento fidem fieri posse praeter haec duo. Sed postea statim probat etiam inductionem merito syllogismi habere vim faciendae fidei. Ideò rectè colligit id, quod prius dixerat, nullam esse vim illativam nisi in syllogismis utilibus trium figurarum, ad quos methodi omnes fidem aliquam facientes reducuntur. Quomodo autem inductio perfecta in bonum syllogismum vertatur, ita ostendit, inductio perfecta illa est, in qua sumuntur omnia particularia, sed haec suo universali aequalia sunt et cum eo reciprocantur, ut omnes individui homines cum homine communi, ideò si per illos ostendere velimus omnem hominem esse risibilem, haec erit inductio, hic homo et hic et ille et singulus alius est risibilis, ergo omnis homo est risibilis, quae in syllogismum primae figurae convertitur, si homines individui in medio collocentur, omnes enim simul sumpti de specie sua praedicantur et ei aequales sunt, ut ibi docet Aristoteles, quare inductio perfecta vim syllogismi habet. Itaque omnis necessitas illationis in syllogismo consistit et omnis methodus vim habens inferendi syllogistica est.

6   Eandem sententiam significavit Aristoteles in primo libro Priorum Analyticorum in fine sectionis secundae, ubi loquitur de via divisiva, qua utebatur Plato, inquit enim, 'divisio est parva quaedam particula dictae methodi,' et nomine methodi intelligit syllogismum, ut ibi exponit Alexander, imò et Aristoteles sese ipse

[a method] has no weight and brings about no belief if it cannot be reduced to the form of a good syllogism.

The other is that in Aristotle all logical instruments are con-   5
tained under syllogism; by it they have the whole necessity of consequence. For since enthymeme is a syllogism and example is an induction, these four are reduced to two, to syllogism and induction. Because of this Aristotle says that everything is believed either by means of syllogism or by means of induction. He is determined that belief can be brought about by no instrument other than these two. But he afterward proves straightaway that induction too has the power to produce belief by merit of the syllogism. He therefore correctly gathers what he had said first, that there is no inferential power except in useful syllogisms of the three figures; to these, all methods producing any belief are reduced. And how a perfect induction is turned into a good syllogism, he shows as follows.[12] A perfect induction is that in which all particulars are taken in, but these are equal in total extent to their universal and reciprocate with it, as [for example] "all individual men" is with "man" in general. Therefore if we want to show by means of them that every man is risible, this will be the induction: This man and this one and that one and each of the others is risible; therefore every man is risible. This is converted into a syllogism of the first figure if individual men are included in the middle [term], for all taken together are predicated of their species and are equal in total extent to it, as Aristotle there teaches. Perfect induction, therefore, has the power of a syllogism. And so every necessity of inference consists in a syllogism, and every method having the power to infer is syllogistic.

Aristotle indicated the same position in the first book of the   6
Prior Analytics at the end of the second section, where he speaks about the divisive way that Plato used. For he says, "Division is some tiny part of the said method,"[13] and by the name of method he understands syllogism, as Alexander lays out there;[14] indeed

interpretatur subiungens, 'est enim divisio veluti imbecillus syllogismus,' imbecillum vocat, id est in prava forma constructum, quae
nihil ex necessitate colligit, ut ipse postea declarat. Via igitur syllogistica apud Aristotelem fuit norma, qua methodos utiles ab inuti-
libus internoscere deberemus, nam viae divisivae efficacitatem
cognoscere volens, eam ad syllogisticam formam redigit et videns
pravum syllogismum ex ea fieri, inutilem esse iudicavit viam divisivam ad rem ignotam ex aliquo noto notificandam, propterea ipsam
vocavit parvam particulam methodi syllogisticae et imbecillum syllogismum.

7   Alexander quoque in praefatione sua in primum librum Topicorum tale testimonium profert de Aristotelis opinione, dicit enim
Dialecticam apud Platonem nil aliud fuisse, quàm viam divisivam,
quod unum instrumentum cùm videatur cognovisse Plato, totam
artem logicam in benè dividendo consistere arbitratus est. Aristoteles autem (inquit Alexander) et qui ab eo profecti sunt, non ita
utuntur nomine Dialecticae, sed aiunt eam esse viam et rationem
quandam syllogisticam. Significat itaque Alexander Aristotelem
nullas methodos, nullaque instrumenta tractare in logica voluisse,
nisi quae vim haberent syllogisticam, reliquas vias sprevisse et
tanquàm inutiles reiecisse, cùm non habeant consequentiae necessitatem.

8   Philoponus quoque interpretans contextum 176 primi libri Posteriorum, dicit Aristotelem primum et solum logicas methodos
tradidisse nec de aliqua methodo loquitur, sed de omnibus, quia in
plurali numero et cum articulo inquit τὰς λογικὰς μεθόδους.
Articulus enim eandem vim habet, quam nota universitatis et per-
inde est ac si dixisset, πάσας. Omnes autem logicas methodos
tradidit, et invenit Aristoteles, quoniam syllogisticam formam primus invenit, quae est commune genus et communis forma omnium
methodorum. Nam etiam Alexander in memorato loco primi libri

Aristotle himself also comments on himself, adding, "for division is like a feeble syllogism."[15] He calls feeble what is constructed in a defective form; this gathers nothing out of necessity, as he afterward makes clear. In Aristotle, therefore, the syllogistic way was the norm by which we ought to distinguish useful methods from useless [ones]. For, wanting to know the efficacy of the divisive way, he reduces it to the syllogistic form, and seeing that a defective syllogism results from it, he judged the divisive way to be useless for making an unknown thing known, from something else known. And so he called it a tiny part of syllogistic method and a feeble syllogism.

Alexander, too, in his preface to the first book of the *Topics*,[16] advances a similar testimony to Aristotle's opinion. For he says that in Plato the dialectical was nothing other than the divisive way. Since Plato appears to have known this one instrument, he thought the whole logical art consisted in dividing well. Aristotle, however, and those who followed him (Alexander says) do not use the name "dialectic" in this way. They say it is some sort of syllogistic reasoning and way. Thus Alexander indicates that in logic Aristotle had wanted to treat no methods and no instruments unless they had syllogistic power. He left other ways aside and rejected [them] as useless, since they do not have the necessity of consequence.

Philoponus,[17] too, commenting on text no. 176 of the first book of the *Posterior* [*Analytics*],[18] says that Aristotle had conveyed logical methods first and only, and does not speak of some method, but of all, because he says, in the plural and with an article, *hai logikai methodoi* (the logical methods). For the article has the same power as a mark of universality and is just as if he had said *pasa* (all). Aristotle, moreover, discovered and conveyed all the logical methods, since he was the first to discover the syllogistic form, which is the common genus and common form of all methods. Now Alexander[19] also asserts in the passage referred to, in the first

7

8

Priorum Analyticorum asserit Aristotelem primum syllogismi inventorem fuisse.

9 Sed et ipse Aristoteles hoc profitetur in calce secundi libri Elenchorum sophisticorum, dum inquit se de syllogismo nihil ab aliis traditum habuisse, sed proprium ipsius inventum fuisse postquàm multo tempore laboraveret.

10 Satis igitur demonstratum est syllogismum esse commune genus omnium methodorum et instrumentorum logicorum, proinde methodum appellari posse et instrumentum generaliter acceptum, quod omnia logica instrumenta complectitur, ut vocat° ipsum Alexander dum syllogismi definitionem interpretatur tum in principio primi libri Priorum tum in principio primi Topicorum.

: IV :

In quo vera sententia de numero, ac differentiis
methodorum breviter exponitur.

1 Postquàm de Methodo universè locuti sumus, sequitur ut ipsius divisionem faciamus in species ac de iis singulis disseramus. Qua in re cùm difficultas magna et arduum cum aliis omnibus certamen nobis proponatur, hunc ordinem facilioris doctrinae gratia servare statuimus, primùm quidem ipsam rei veritatem paucis verbis exponemus, deinde aliorum placita ac dubia, quae nos ad ea refutanda traxerunt, referemus. Postmodum ad veram sententiam revertentes eam fusius ac diligentius explanare et comprobare nitemur.

2 Dictum à nobis est methodi nomen scientiam ut finem respicere, est enim via ad scientiae adeptionem ducens, nomine autem

book of the *Prior Analytics*, that Aristotle was the first discoverer of
the syllogism.

But Aristotle himself claimed this, too, at the end of the second   9
book of the *Sophistical Refutations*,[20] where he says he had conveyed
[to him] by others nothing about the syllogism and that it was
instead — after he labored for a long time — his very own discovery.

It has, therefore, been sufficiently demonstrated that syllogism   10
is the common genus of all methods and logical instruments and
accordingly can be called method and instrument, accepted in the
general sense, since it encompasses all the logical instruments, as
Alexander says° when he comments on the definition of syllogism
both in the beginning of the first [book] of the *Prior* [*Analytics*][21]
and in the beginning of the first of the *Topics*.[22]

: IV :

*In which the true position regarding the number and
differentiae of methods is briefly laid out.*

Now that we have spoken about Method universally, it follows   1
that we make a division of it into species and discuss each of them.
Now in this, since a great problem and a tough dispute with all
others have been set out for us, we have decided, for the sake of
easier teaching, to maintain this order: First, of course, we will lay
out in a few words the truth itself of the issue. We will then recount the preferences and doubts of others that drove us to confute them. A little later, returning to the true position, we will
endeavor to explain and confirm it at greater length and more
carefully.

It was said by us that the name of method regards scientific   2
knowledge as an end, for it is a way leading to the obtaining of

scientiae quamlibet certam cognitionem intelligimus. Ita videtur methodum accepisse Eustratius in sua praefatione in libro secundo[6] Posteriorum Analyticorum, quando de methodorum numero, atque utilitate loquens dixit quatuor esse methodos, quibus omnis disciplina scientialis acquiritur.

3   Quoniam igitur methodus est via et forma syllogistica, certum est hanc non posse scientiam alicuius rei parere, nisi materiei necessariae applicetur, hoc est nisi ex propositionibus necessariis constet. Necessarias autem propositiones Aristoteles illas esse docuit, quae sint de omni, per se et universales, ad quarum conditionum declarationem[7] nos librum scripsimus de propositionibus necessariis, in quo ostendimus nullam propositionem hoc modo necessariam esse posse, nisi essentialem praedicati cum subiecto connexionem habeat, et nullam praedicati cum subiecto connexionem essentialem esse posse, nisi alterum alterius causa sit.

4   Hinc fit ut in omni syllogismo sciendi gratia constructo necesse sit vel à causa ad effectum vel contrà ab effectu ad causam progressum fieri, aliud genus scientifici syllogismi et ex propositionibus eo modo necessariis constituti non invenio. Cùm enim in propositione maiore medium et maior extremitas collocentur et necessum sit alterum alterius causam esse, medium quidem maioris causa si fuerit, syllogismus est à causa ad effectum; si verò maior extremitas sit causa medii, est ab effectu ad causam. Duae igitur scientificae methodi oriuntur, non plures, nec pauciores, altera per excellentiam demonstrativa methodus dicitur, quam Graeci κυρίως ἀπόδειξιν vel ἀπόδειξιν τοῦ διότι vocant, nostri potissimam demonstrationem vel demonstrationem propter quid appellare consueverunt. Altera, quae ab effectu ad causam progreditur, resolutiva nominatur, huiusmodi enim progressus resolutio est,

scientific knowledge. By the name "scientific knowledge" (*scientia*) we understand any certain knowledge (*certa cognitio*).[23] It appears Eustratius accepted method this way in his preface to the second book of the *Posterior Analytics*;[24] when speaking about the number and utility of methods, he said that there are four methods by which every scientific discipline is acquired.

Since, therefore, method is a syllogistic way and form, it is certain this cannot bring forth scientific knowledge of any thing unless it is applied to necessary matter, that is, unless it is composed out of necessary premises. Now Aristotle taught that necessary premises are those that are *de omni, per se*, and *universales*. For the clarification of these characteristics, we wrote a book, *On Necessary Premises*,[25] in which we showed that no premise can be necessary in this way unless it has an essential connection of predicate with subject, and no connection of predicate with subject can be essential unless one is the cause of the other. 3

It happens, therefore, that in every syllogism constructed for the sake of knowing scientifically, it is necessary that progression occurs either from cause to effect or, on the contrary, from effect to cause. I do not find another kind of scientific syllogism also constituted from premises [that are] necessary in this way. For since the middle [term] and the major extreme are included in the major premise, and it is necessary that one be the cause of the other, then, of course, if the middle [term] was the cause of the major, the syllogism would be from cause to effect; and if the major extreme were the cause of the middle [term], it would be from effect to cause. Two scientific methods, therefore, arise — neither more nor fewer. One, which the Greeks call *kuriōs apodeixis*[26] (demonstration of the strongest sort) or *apodeixis tou dioti* (demonstration of the on-account-of-which), and we have been accustomed to call demonstration *potissima* or demonstration *propter quid*, is said to be demonstrative method *par excellence*.[27] The other, which progresses from effect to cause, is named resolutive. For a progression of this 4

sicuti à causa ad effectum dicitur compositio, methodum hanc vocant Graeci, συλλογισμὸν τοῦ ὅτι vel διὰ σημείων, nostri demonstrationem quia vel syllogismum à signo vel secundi gradus
demonstrationem.

5   Has duas methodos in Aristotelis disciplina° reperio, demonstrativam et resolutivam, alias[8] nec posuit Aristoteles nec ratio videtur admittere, quod cum ex iis, quae modò dicta sunt, facilè
colligi potest, tum ex iis, quae dicenda manent, fiet manifestissi-
mum. Nos autem fortasse non incongrua appellatione uteremur, si
demonstrativam vocaremus compositivam, cùm enim compositio
sit via contraria resolutioni, necesse est ut quemadmodum progressus ab effectu ad causam dicitur resolutio, ita eum, qui est à causa
ad effectum, liceat appellare compositionem. Sub resolutivam methodum reducitur inductio, ut postea declarabimus.

6   Praeter has nullam dari aliam scientificam methodum, nul-
lumque aliud sciendi instrumentum ego constanter existimo, quod
mihi et ipsa ratio et Aristotelis authoritas persuasit. Omnino enim
credere, atque confiteri debemus praecipuum eius scopum in logica fuisse methodorum traditionem, haec enim instrumenta sciendi sunt, unde logica vocatur instrumentalis. Non est igitur asse-
rendum aliquam methodum ab Aristotele omissam fuisse, de qua
non docuerit. Attamen nullam aliam methodum tradidit, nisi demonstrativam et resolutivam; demonstrativam quidem primariò, cùm
ad eam omnes logici libri dirigantur; resolutivam autem secundariò, cuius rationem in sequentibus declarabimus. Ad maiorem
enim veritatis comprobationem decrevimus huic nostrae de methodis tractationi disputationem annectere de Aristotelis consilio in
Posterioribus Analyticis, in iis enim credimus Aristotelem diligentissimè atque artificiosissimè de methodis disseruisse. Fortasse

type is a resolution, just as composition is said to be from cause to effect. The Greeks call this method *syllogismos tou hoti* (syllogism of the that-it-is-the-case) or *dia sēmeiōn* ([syllogism] by means of signs). We [call it] demonstration *quia* or syllogism *a signo* or demonstration of the second degree.

In Aristotle's teaching,° I find these two methods, demonstrative and resolutive. Aristotle did not posit, nor does reason appear to admit, others. This can easily be gathered from the things that have just been said, and will become most manifest from the things that remain to be said. But now perhaps we would not use an inapt appellation if we were to call the demonstrative compositive, for since composition is the way contrary to resolution, it is necessary that just as resolution is said to be a progression from effect to cause, so that which is from cause to effect may be called composition. Induction is reduced under resolutive method, as we will afterward make clear.[28]

I continue to judge that there is no other scientific method besides these and no other instrument for knowing scientifically. Both reason itself and Aristotle's authority have persuaded me of this. For we ought altogether to believe and confess that his principal goal in logic was the conveying of methods, for these are the instruments for knowing scientifically; from this logic is called instrumental. It is not, therefore, to be asserted that there was any method omitted by Aristotle, [any] about which he did not teach. Yet he conveyed no method other than demonstrative and resolutive — the demonstrative primarily, of course, since all the books on logic are directed toward it, and the resolutive secondarily, the reasoning for which we will clarify in what follows. For greater confirmation of the truth, we have decided to annex to this tract of ours on methods a debate about Aristotle's intent in the *Posterior Analytics*.[29] For we believe that Aristotle discussed methods most carefully and skillfully in these [books]. And perhaps our

enim et nostra contemplatio non parvum lumen libris illis intelli-
gendis allatura est et vicissim ipsa Aristotelis sententia dogma
nostrum mirificè confirmabit.

7   Nunc quid alii de numero methodorum sentiant exponamus.

: V :

In quo aliorum de methodis sententia declaratur.

1 Est hac tempestate communis omnium sententia, quatuor esse
methodos, demonstrativam et resolutivam, quas diximus, et praeter has etiam definitivam ac divisivam. Videtur autem et antiquorum fuisse haec opinio cùm apud Graecos Aristotelis interpretes
eam[9] saepe legamus. Sed eam apertè ponit ac fusè declarat Eustratius in sua praefatione in secundum librum Posteriorum Analyticorum, ubi et numeri necessitatem et singularum methodorum
utilitatem ostendit. Scopum quidem methodi definitivae dicit esse
rei essentiam declarare, demonstrativae accidentia, quae per se insunt, nota facere; divisivae autem utilitatem esse ut methodo definitivae materiam subministret, genera enim et differentiae sunt
materia, ex qua definitio constituitur, quae materia per divisionem
suppeditatur. Tandem methodi resolutivae scopus est à particularibus ad universalia progredi et à posterioribus ad priora.

2   Has methodos inter se conferens Eustratius dicit definitivam ac
demonstrativam principem locum tenere, et magis definitivam,
quae omnium praestantissima est, cùm substantiam rei declaret.
Duas autem reliquas secundarias esse et illis duabus ministrantes

contemplation will bring no little light to the understanding of these books, and Aristotle's position itself will in turn confirm our doctrine wonderfully well.

Let us now lay out the position others take about the number 7 of methods.

: V :

In which others' position on methods is made clear.

At this time it is the common position of everyone that there 1 are four methods, demonstrative and resolutive, which we talked about, and besides these also definitive and divisive. This also appears to have been the opinion of the ancients, since we often read it in the Greek commentators on Aristotle. But Eustratius posits it plainly and clarifies it at length in his preface to the second book of the *Posterior Analytics*,[30] where he shows both the necessity of the number and the utility of each of the methods. He says that the goal of definitive method, of course, is to make the essence of something clear and of demonstrative to make known the accidents that belong *per se*, and that the utility of divisive [method] is that it furnishes the matter for definitive method. For the genera and differentiae are the matter from which the definition is constituted; this matter is supplied by means of division. The goal of resolutive method, finally, is to progress from particulars to universals and from posteriors to priors.

Comparing these methods to each other, Eustratius[31] says that 2 the definitive and demonstrative hold the foremost place and more so the definitive, which is the most excellent of all, since it makes the substance of something clear. The remaining two, on the other hand, are secondary, ministering to and serving the former two.

ac servientes. Nam divisiva materiam methodo definitivae suppeditat, genera ac differentias. Resolutiva quoque est ministra definitivae, quoniam à particularibus ad universalia, ex quibus definitio omnis est constituenda, progreditur. Eaedem inserviunt etiam methodo demonstrativa, quia demonstrativa sine definitiva esse non potest, cùm absque definitione non possit esse demonstratio. Nec tamen definitiva dicitur ministra demonstrativae, quamvis ei medium terminum suppeditet, quia definitiva est etiam sibi sufficiens absque demonstrativa, cùm ad substantiam declarandam à demonstrativa non pendeat et nulla alia methodo egeat. Sed post substantiae declarationem prodest etiam methodo demonstrativae, quod quidem est potius imperare, quàm ministrare, at divisiva et resolutiva, si earum ad definitivam utilitas amoveatur, redduntur prorsus inutiles, tota enim earum utilitas in hac ministratione consistit.

3   Principem igitur locum tenet methodus definitiva, secundum demonstrativa, postea duae reliquae.

4   Quòd autem plures his quatuor non dentur neque pauciores, ita probat Eustratius, duo ad summum sunt illa, quae scire volumus, substantia et accidentia, quae ipsam consecuntur, ad substantiae cognitionem necessaria est methodus definitiva, siquidem absque definitione substantia non cognoscitur. Ad accidentia verò cognoscenda utimur methodo demonstrativa. At definitio constitui non potest, nisi partes, ex quibus constituenda est, nempè genera et differentiae in promptu habeantur, has autem venari possumus vel à prioribus progrediendo vel à posterioribus; à prioribus quidem per divisionem, dum à summo genere ordientes ad propinquiora genera differentiasque descendimus; à posterioribus verò per methodum resolutivam, quae à posterioribus ad priora, ex quibus definitio constare debet, progreditur.

For the divisive supplies the matter, that is, the genera and differentiae, for definitive method. And the resolutive is minister to the definitive, since it progresses from particulars to universals, from which every definition has to be constituted. The same [two] also serve demonstrative method, because there cannot be demonstrative without definitive, given that there cannot be demonstration without definition. Nevertheless, the definitive is not said to be minister to the demonstrative, even though it supplies a middle term to it, because the definitive is sufficient by itself even without the demonstrative, since it does not depend on the demonstrative and needs no other method to clarify the substance. But after the clarification of a substance, it is still useful to demonstrative method — that is, of course, to command rather than to minister to. But divisive and resolutive are rendered utterly useless if their utility to definitive is removed, for their whole utility consists in that ministering.

Definitive method, therefore, holds the foremost position, demonstrative second; the remaining two [follow] afterward. 3

Moreover, that there are neither more nor fewer than these 4 four, Eustratius proves as follows. There are above all two things that we want to know scientifically: substance and the accidents that ensue from it. Definitive method is necessary for knowledge of substance, since without definition substance is not known. But for knowing accidents we use demonstrative method. Yet a definition cannot be constituted unless the parts from which it is to be constituted, namely genera and differentiae, are right at hand. And we cannot search for these by progressing either from priors or posteriors — from priors by means of division, when, beginning from the highest genus we descend to closer genera and differentiae, or on the other hand from posteriors by means of resolutive method, which progresses from posteriors to priors, out of which the definition ought to be composed.

5   Quatuor igitur methodi necessariae sunt, duae quidem[10] ut primariae, duae verò ut secundariae et illis duabus inservientes.

6   Hanc eandem de methodis sententiam apud Ammonium legimus in suis commentariis in praefationem Porphyrii.

7   Posteriores quoque omnes, qui hac de re scripserunt, hanc opinionem secuti sunt, nec de methodorum numero aliquis unquàm dubitare visus est, quanquam complures, ut aliquid è proprio penu[11] depromere et proprio ingenio invenisse viderentur, nonnulla adiecerunt ad harum quatuor methodorum necessitatem atque utilitatem pertinentia, nec sine magna inter ipsos altercatione, quam quoties considero, non possum equidem pugnae caecorum, quam aliquando in foro spectavi, non recordari, non est autem hîc omnium, quae ab eis dicuntur, mentio facienda, multa enim puerilia sunt et prorsus indigna, quae in medium afferantur. Multa etiam satis erit postea inter contemplandum considerare et singulatim expendere. Illud in praesentia omittendum non est, eos non modò divisivam ac definitivam praeter demonstrativam et resolutivam, quas nos posuimus, in methodorum numerum retulisse; sed etiam resolutivam methodum longè aliter accepisse, quàm nos suprà declaraverimus,° nam sub demonstrativam methodum omnes demonstrationis species reduxerunt, illam quoque, quae à signo ad causam progreditur, quam nos diximus esse methodum resolutivam; nomine autem methodi resolutivae aliud quiddam intellexere, nempè processum divisioni contrarium, ut quemadmodum à summo genere ad inferiora descendere dividere est, et methodus divisiva, ita ab infima specie ad superiora genera, ac differentias ascendere sit resolvere et methodus resolutiva. Ideò aliqui declarare volentes id, quod Eustratius dixit, divisivam ac resolutivam methodos servas esse methodi definitivae, dixerunt hoc ita esse intelligendum, ut quando alicuius superioris generis definitionem venari volumus, methodo utamur resolutiva ab inferioribus ad superiora;

Four methods, therefore, are necessary, two as primary, two as 5
secondary and serving the former two.

We read this same position regarding methods in Ammonius, 6
in his commentary on the preface to Porphyry.[32]

All those who later wrote on this issue followed this opinion. 7
Nor does anyone ever appear to have doubts about the number of
methods, although some, so as to appear to draw from their own
store and to have discovered something by their own wit, added
some things pertaining to the necessity and utility of these four
methods, though not without great wrangling among themselves.
(Whenever I consider this, I cannot but remember a fight I once
saw in the market between blind men.) Mention does not have to
be made here of all that is said by them. For many things that
were brought up are puerile and utterly worthless. It will be
enough to consider many [of the others] below and weigh each
one when we come to contemplate it. At the present time, it is not
to be omitted, that not only did they count in the number of
methods the divisive and definitive, besides the demonstrative and
resolutive, which we too posited, but they also accepted resolutive
method much differently than how we explained° it above. For
they reduced all species of demonstration under demonstrative
method, including that which progresses from sign to cause—
what we said is resolutive method—and understood something
else by the name of resolutive method, namely the procedure contrary to division: just as to descend from the highest genus to
lower ones is to divide and is divisive method, so to ascend from
the lowest species to higher genera and differentiae is to resolve
and is resolutive method. And so some, wanting to clarify that
which Eustratius said, that divisive and resolutive methods are
servants of definitive method, said this has to be understood in
this way: that when we want to search for the definition of some
higher genus, we use resolutive method, from the lower to the

cùm verò infimae alicuius speciei definitionem inquirimus, tunc divisiva methodus partes definitionis nobis ordinatim suppeditet.

8   Haec de aliorum sententia breviter retulimus, nunc ad eam diligenter examinandam accedamus, à divisiva methodo exordientes.

: VI :

Quid sit divisio, et cur sit appellanda methodus secundùm alios.

1 Ante omnia non est ignorandum, cùm divisionis multa genera sint, non omnem divisionem appellari methodum divisivam, fit enim saepe divisio tum vocis tum orationis ambiguae in multas significationes. Dividitur etiam totum quantum in partes, quae integrantes dicuntur. Dividitur liber aliquis in capita vel tractationes. Dividitur species aliqua in partes suas essentiales, ut homo in corpus et animam, et quodlibet naturale corpus in materiam et formam, quae divisio vocatur etiam resolutio. Solet etiam dividi res aliqua per accidentales differentias, ut si quis dicat equos alios esse nigros, alios albos, alios rubeos et alios misti coloris. Nulla tamen harum divisionum vocatur ab aliquo methodus divisiva; sed omnes, qui de hac methodo locuti sunt, nil aliud intellexere, quàm divisionem generis in species per essentiales differentias, ut quando animal per rationale et irrationale in hominem et bruta dividimus.

2 Huius methodi duae feruntur esse utilitates, una, quam solam tetigit Eustratius, ad venandas partes definitionis ignotas et colligendam definitionem. Altera, quam alii adiiciunt, ad cognoscendum

higher, but when we ask about the definition of any of the lowest species, then divisive method supplies us with the parts of the definition in order.

We have briefly recounted this position of others. Let us now 8 move on to carefully examine it, beginning from divisive method.

: VI :

What division is, and why, according to others,
it has to be called a method.

Before all else, it should not be ignored that although there are 1 many kinds of division, not every division is called a divisive method. For often there is a division of both an ambiguous word and also [ambiguous] speech into many significations. Also, a whole of any size is divided into parts that are said to be constitu- ent. A book is divided into chapters or tracts. A species is divided into its essential parts, as [for example] man into body and soul; and any natural body into matter and form, a division that is also called resolution. Now normally some thing is divided by means of accidental differentiae, as [for example] when someone says that some horses are black, some white, some red, and some of mixed colors. But none of these divisions is called divisive method by anyone; everyone who spoke about this method understood noth- ing other than the division of genera into species by means of es- sential differentiae, as [for example] when, by means of rational and irrational, we divide animal into man and brutes.

The utilities of this method are taken to be two. One, which is 2 the only one Eustratius touched on, is for searching for the un- known parts of the definition and gathering the definition. The

numerum specierum, qui antea ignorabatur, nam si numerum specierum animalis cognoscere voluerimus, eum certè inveniemus, si animal per omnes suas differentias diviserimus donec ad ultimas pervenerimus. Quam utilitatem veluti finalem divisionis causam aliqui in definitione methodi divisivae expresserunt dicentes eam esse methodum, qua per notas differentias progredimur ad cognoscendum quotuplex res sit; sed cùm methodum hanc ita definiverint, non video cur postea eandem in duas species diviserint ratione duorum memoratorum finium, quando dixerunt aliam ducere ad cognoscendum numerum specierum, aliam dirigi ad invenienda genera ac differentias, ex quibus definitio constituatur. Videtur enim vel utrumque finem in ea definitione exprimendum fuisse vel neutrum, sed aliquid commune sumendum, quod utrosque complecteretur; vel saltem si alterius tantùm finis expressione methodum divisivam ad alteram solam speciem restrinxere, non potuerunt postea methodum divisivam in duas illas species partiri, cùm altera sub illa definitione non contineatur, quando enim per divisionem partes definitionis venamur, non quaerimus quotuplex res sit, sed solùm quid sit, partes namque definitionis in definitione non sumuntur ut plures, sed ut ad unam rei essentiam constituendam sive declarandam concurrunt.

3   Attamen literatos atque eruditos viros eos esse constat. Aliud fortasse intellexerunt, quod ego animadvertere non potui. Mihi quidem videtur satis explicatam fuisse divisionis naturam; si dixissent divisionem esse processum à genere noto ad species ignotas per differentias notas, species namque ignorari possunt vel quot sint[12] vel quid sint, ideò rectè postea dividitur haec methodus in duas illas species habita duorum finium ratione.

other, which others add, is for knowing the number of species, which was unknown earlier. For if we want to know the number of species of animal, we will certainly discover it if we divide animal by means of all its differentiae until we come through to the ultimate [ones]. Some expressed this utility as the final cause of division in the definition of divisive method, saying that it is the method by which we progress, by means of known differentiae, to knowing how many types of something there are; but since they so defined this method, I do not see why they afterward divided the same [method] into two species by reason of the two ends referred to, when they said one leads to knowing the number of species, and the other is directed to discovering the genera and differentiae from which the definition is constituted. For it appears that either each end would have had to be expressed in the definition, or neither would have; instead, something common that would have encompassed both had to be taken up, or at least, if they restricted divisive method to just the one species by expressing just the one end, they could not afterward partition divisive method into those two species, since the other would not be contained under that definition. For when we search for the parts of a definition by means of division, we do not inquire how many types of something there are, but only what it is. For the parts of a definition are not taken in the definition as many things, but as they concur in constituting or clarifying the one essence of the thing.

But nevertheless, it is evident that those men are lettered and 3 learned. Perhaps they understood something that I have not been able to notice. It appears to me, of course, that the nature of division would have been sufficiently explicated if they had said division is a proceeding from a known genus to an unknown species by means of known differentiae. For the species — either how many there are or what they are — can be unknown. And so this method is correctly divided afterward by taking account of the two ends.

4   Sed utcumque res sese habeat et quicumque methodi divisivae finis esse statuatur, illud certum esse debet, quod etiam illi, qui talem methodum ponunt, negasse non videntur, aliquem esse in ipsa divisione terminum, à quo, notum, et aliquem ignotum, ad quem, et demum illationem aliquam huius ex illo, hoc enim omnino dicere cogitur quisquis divisionem methodum esse asserit, quandoquidem methodus omnis est processus à noto ad cognitionem ignoti cum illationis necessitate. Haec igitur quomodo in divisione locum habeant considerandum est.

: VII :

*In quo eorum sententia refellitur, qui dicunt
divisionem esse methodum utilem ad
venandas partes definitionis ignotas.*

1   Cùm duo fines methodi divisivae ab aliis statuantur, si respectu utriusque ostenderimus divisionem esse prorsus inutilem ad aliquid ignotum notificandum, satis erit demonstratum ipsam inter methodos non esse collocandam. Priorem igitur illam utilitatem consideremus, qua dicitur divisio ad partes definitionis notificandas conferre et videamus quisnam sit terminus notus, à quo et quis terminus ignotus, ad quem, et denique qualis via ab illo ad hunc. Certè nil aliud dicere possumus, nisi quòd summum genus, à quo divisionis exordium sumitur, est terminus notus, à quo; differentiae verò sunt[13] terminus ignotus, ad quem. Nam hae quaeruntur pro definitionis constitutione; processus verò à genere ad differentias est methodus ipsa divisiva, sic enim videtur Aristoteles in primo

But however that may be, and whatever it is decided the end of   4
divisive method is, it ought to be certain that those who posit such
a method do not appear to have denied that there is in division
itself some known terminus *a quo* (from which), and some unknown *ad quem* (to which), and lastly some inference of the latter
from the former. For whoever asserts that division is a method is
altogether forced to say this, since every method is a proceeding
from the known to knowledge of the unknown with the necessity
of inference. In what way, therefore, these have a place in division,
has to be considered.

: VII :

In which is refuted the position of those who say
division is a method useful in searching for the
unknown parts of a definition.

Although two ends of divisive method are fixed by others, if we   1
show, with regard to each, that division is utterly useless for making known something unknown, that will be enough to have demonstrated that it is not to be included among methods. First,
therefore, let us consider that utility by which division is said to
contribute to making the parts of a definition known, and let us
see what the known terminus *a quo* (from which) is, and what the
unknown terminus *ad quem* (to which) is, and lastly what sort of
way there is from the former to the latter. Certainly we can say
nothing else but that the highest genus from which the beginning
of a division is taken is the known terminus *a quo* and the differentiae are the unknown terminus *ad quem*. For these are inquired after for the sake of constituting the definition, and proceeding from
genus to differentiae is the divisive method itself. For this is how

libro Priorum Analyticorum in calce secundae sectionis viam divisivam Platonis intellexisse.

2   Ut itaque rem melius explicemus et veritatem claram reddamus, tali exemplo utamur, sit hominis definitio investiganda per methodum divisivam, accipiendum est in primis aliquod superius genus notum, sit illud corpus, etenim notum est hominem esse corpus, quod deinde dividendum est per proximas eius[14] differentias, ut per animatum et inanimum; deinde corpus animatum per sensibile et insensibile; mox sensibile, quod animal est, per rationale et irrationale, unde accepto rationali colligitur hominis definitio, quae est animal rationale vel, corpus animatum sensibile rationale, quod idem est, si hoc modo intelligatur via divisiva, nullam habet illationis necessitatem, neque ex noto notificat aliquid ignotum, quare methodus appellanda non est.

3   Ad hoc ostendendum eo ipso argumento utor, quo usus est Aristoteles in memorato loco primi libri Priorum Analyticorum, ubi deridet ac spernit tanquàm inutilem viam hanc divisivam inquiens, 'Divisio est quaedam parva particula dictae methodi, est enim divisio imbecillus syllogismus,' quod ipse postea declarat, videns enim nullam posse fieri illationem nisi per syllogisticam formam, hanc divisioni applicat et facit ex divisione syllogismum et ostendit eum vel pravum esse et nihil concludere vel si bonus et concludens fiat, non colligere id, quod erat colligendum. Sic enim est debilis syllogismus, quia vel est pravus et nihil concludens, vel non concludens propositum et ita inutilis.

4   Haec omnia cum Aristotele sic ostendo. Si ab ipso genere fit processus illativus ad differentias, quae homini competant, investigandas, est ergo corpus medius terminus notus et differentiae sunt maior extremitas ignota, quam quaerimus, homo verò minor extremitas, cui ostendere volumus eas differentias inesse. Sumatur igitur

Aristotle in the first book of the *Prior Analytics*, at the end of the second section, appears to understand Plato's divisive way.[33]

So that we may explicate the issue better and render the truth 2 clear, let us use this example. Let the definition of man be investigated by means of divisive method. In the first place, some known higher genus has to be grasped. Let it be body, for indeed it is known that man is a body; this then has to be divided by means of its proximate differentiae, as by means of animate and inanimate; then animate body by means of sentient and insentient; next sentient, which animal is, by means of rational and irrational. From this, rational having been grasped, the definition of man is gathered, and it is rational animal, or rational, sentient, animate body, which is the same thing. If the divisive way is understood in this way, it has no necessity of inference; nor does it make known something unknown from the known. Therefore it is not to be called a method.

To show this, I use the very argument that Aristotle used in the 3 passage in the first book of the *Prior Analytics* [that I] referred to, where he derides this divisive way and leaves it aside as useless, saying, "Division is a tiny part of the said method, for division is a feeble syllogism";[34] he clarifies this afterward. For seeing that no inference can occur except by means of the syllogistic form, he applies this to division and makes a syllogism out of division, and shows that it is either defective and not conclusive, or if it is made good and conclusive, does not gather what was to be gathered. And so the syllogism is weak, because either it is defective and not conclusive or it does not conclude what was set out and is thus useless.

I show all this, with Aristotle, as follows. If an inferential pro- 4 ceeding occurs from the genus itself to the investigated differentiae that appertain to man, then body is the known middle term, and the differentiae are the unknown major extreme that we inquire after, and man is then the minor extreme, to which we want to

in primis haec minor propositio nota, homo est corpus, et quia
primae differentiae, per quas dividitur corpus, sunt animatum et
inanimum, nos autem eligere volumus animatum ut homini com-
petens et reiicere inanimum, ideò animatum est maior extremitas,
quam quaerimus, ergo si maiorem propositionem ex corpore et
animato constituere volumus, maior erit una ex tribus vel, omne
corpus est animatum vel, aliquod corpus est animatum vel, omne
animatum est corpus. Si primam sumamus, ea falsa est. Si secundam, erit syllogismus ex maiore particulari, qui pravus est et nihil
concludit in prima figura, si tertiam, erit syllogismus in secunda
figura ex duabus affirmantibus, qui similiter pravus est et nihil
concludens. Nulla igitur ratione per eam divisionem colligere possumus hanc differentiam homini competere, vanaque est omnium
sententia, qui putant, divisionem nobis talem notitiam° tradere.

5   Quòd si in iis terminis bonum syllogismum facere volumus,
debemus in maiore extremitate utramque simul differentiam sumere, non alteram solam et dicere, omne corpus est animatum vel
inanimum, at homo est corpus, unde in prima figura sequitur,
ergo omnis homo vel animatus est vel inanimus, qui bonus quidem
syllogismus est, sed non colligit propositum, volumus enim colligere solum animatum homini competere reiecto inanimato.

6   Hac ratione utitur ibi Aristoteles ad ostendendam divisionis
inutilitatem ad differentiarum, si latuerint, inventionem. Idque
revera cuilibet manifestum esse deberet, quando enim hominis
definitionem, id est differentias in ipsius definitione accipiendas
inquirimus, et aliquod superius genus notum accipimus, ut corpus,
et ipsum dividimus per animatum, et inanimatum, ut altera differentia reiecta reliquam homini tribuamus, tunc vel est per se

show that these differentiae belong. This minor premise, "Man is body," is taken, therefore, in the first place as known, and because the first differentiae by means of which body is divided are animate and inanimate, we want to choose animate as coinciding with man and to reject inanimate. So animate is the major extreme that we inquire after. If, therefore, we want to constitute the major premise from body and animate, the major [premise] will be one of these three, either "Every body is animate," or "Some body is animate," or "Every animate thing is [a] body." If we take the first, it is false. If the second, the syllogism will be from a particular major; it is defective and, in the first figure, not conclusive. If the third, the syllogism will be [made] in the second figure out of two affirmations; this again is defective and not conclusive. By no reasoning, therefore, can we gather by means of a division that this differentia appertains to man, and vain is the position of all who hold that division conveys to us such knowledge.°

Now if we want to make a good syllogism in these terms, we 5 ought to take both differentia together in the major extreme, not just one, and say, "Every body is animate or inanimate," and "Man is [a] body." From these it follows in the first figure that, therefore, "Every man is animate or inanimate." This is, of course, a good syllogism, but it does not gather what was set out. For we want to gather that only animate appertains to man, inanimate being rejected.

Aristotle uses this reasoning there to show the uselessness of 6 division for discovery of differentiae, if they were hidden. It ought to be manifest to anyone that, in truth, this is so. For when we ask about the definition of man, that is, the differentiae to be accepted into the definition of it [i.e., of man], and we accept some higher known genus, such as body, and we divide it by means of animate and inanimate, so that, the one differentia being rejected, we

notum quòd homo sit animatus vel ignotum. Si per se notum, accipitur statim animatum relicto inanimato, non tamen quòd ex
ipsa divisionis vi illud elucescat, sed quia est per se notum. Si verò
sit per se ignotum, divisio ad id notificandum non sufficit et facta
divisione corporis per animatum et inanimum adhuc ignoratur
utra differentia homini tribuenda sit. Verùm quia res per se nota
est hominem esse animatum, ideò data est multis occasio errandi
et existimandi eam per divisionem fieri notam, cùm tamen non
divisionis beneficio, sed per seipsa sit nota.

7   Quod quidem manifestè inspiciemus, si eo exemplo dimisso
aliud aliquod sumamus definitionis per se ignotae et in controversia positae, veluti si quaeramus definitionem Zoophyti, quod non
satis est clarum, an vitam animalis vivat an stirpis; quia an sentiat,
necne, non omnino cognoscitur. Sumamus genus[15] notum corpus,
certum est enim Zoophytum esse corpus, primùm quidem si divi-
damus corpus per animatum et inanimum, his differentiis propo-
sitis apparet illico Zoophytum esse animatum, non inanimum,
tamen non quòd id nobis divisio praestet, sed quia per se notum
est Zoophytum sumere alimentum et nutriri. Postea verò quando
accepto animato corpore ipsum dividimus per sensibile et insensibile, per hanc divisionem non apparet magis Zoophytum esse
sensibile, quàm insensibile, quia ubi in proposita re neutra differentia per se cognoscitur, divisio nullam prorsus notificandi vim
habet. Tunc igitur nisi ad aliquam methodum confugeremus, quae
illatricem vim haberet, nullam nobis cognitionem divisio ipsa prae-
staret. Confugere autem possemus ad aliquod accidens sensum
consequens et methodo uti resolutiva, veluti si ostenderemus Zoophytum dolere quando pungitur, ex eo enim dolore colligeremus

ascribe the remaining [one] to man, then either it is known *per se* that man is animate or it is unknown. If it is known *per se*, then animate is straightaway accepted, inanimate left to the side, not because it shines out from the very power of the division, but because it is known *per se*. If, on the other hand, it is unknown *per se*, the division is not sufficient to make it known, and the division of body having been made by means of animate and inanimate, it is still unknown which differentia of the two is to be ascribed to man. But because the fact that man is animate is known *per se*, an occasion for erring and judging that it became known by means of division was given to many. Nevertheless, it is not known thanks to the division, but by means of the things themselves.

Of course, we will observe this manifestly, if, putting this example aside, we take up another one, of a definition unknown *per se* and posited in controversy, such as if we inquire after the definition of Zoophyte; it is not clear enough whether this lives the life of an animal or of a plant, because whether it senses or not is not at all known. Let us take body as the known genus, for certainly Zoophyte is a body. Of course, if we first divide body by means of animate and inanimate, these differentiae having been set out, it is immediately apparent that Zoophyte is animate, not inanimate. Nevertheless, it is not that the division really does this for us, but because it is known *per se* that Zoophyte takes nourishment and is nourished. But then afterward, when animate body has been accepted, we divide that by means of sentient and insentient. By means of this division it is not apparent that Zoophyte is more sentient than insentient, because where in a given thing neither differentia is known *per se*, division has no power at all to make [something] known. At that point, therefore, unless we take refuge in some method that has inferential power, division itself furnishes us no knowledge. We could, however, following sense, take refuge in some accident and use resolutive method—just as if we showed that Zoophyte feels pain when pricked—for from this pain we

Zoophyto sensum inesse, quod ipsa divisio ostendere haudqua-
quàm potest, idque mihi videtur ita esse manifestum, ut nulla
probatione indigeat.

: VIII :

In quo dubium quoddam solvitur et declaratur quidnam
sit per se notum, et quid per se ignotum.

1 Caeterum adversus hoc obiicere aliquis posset, videntur enim plurimae definitiones per solam divisionem notificari, quae ante divisionem ignorabantur, veluti definitio hominis, definitio circuli,
nam si rusticum percontemur, quid est homo? quid est circulus?
nullam sciet definitionem assignare, sed facta divisione, et nulla
alia methodo adhibita, utramque cognoscet. Divisio itaque notificat illud, quod per se ignotum erat, nam si omnes partes definitionis hominis fuissent per se notae, rusticus certè etiam ante divisionem eas cognovisset. Declaratio igitur definitionis ignotae tota
divisioni attribuenda est.

2 Hanc difficultatem facilè solvemus, si intelligamus quidnam dicatur per se notum, et quid non per se notum, id est per se ignotum. Nam, ut docet Averroes in commentario 25 secundi libri
Posteriorum Analyticorum, illud dicitur per se ignotum sive naturaliter ignotum, quod lumine proprio et propria evidentia à nobis
cognosci non potest, sed opus habet ostendi per aliud, proinde per
syllogismum notificatur, ut haec propositio, triangulum habet
tres angulos aequales duobus rectis, ipsa enim per se nunquàm
nota fieret neque per longam mentis considerationem neque per
diligentem trianguli inspectionem, quoniam illa trianguli affectio

could gather that sense belongs to Zoophyte, something that division itself can in no way show. This appears to me to be so manifest that it needs no proof.

: VIII :

In which a doubt is done away with and what is known per se and what is unknown per se are clarified.

But now someone could object to this. For many definitions that were unknown before a division appear to be made known by means of division alone, such as the definition of man [or] the definition of circle. For if we probe a peasant, "What is a man?" "What is a circle?" he will not know, scientifically, how to assign a definition, but once a division is made, he will know both, even when no other method has been applied. And so division makes known that which was unknown *per se;* for if all parts of the definition of man had been known *per se,* the peasant would certainly have known them before the division. Clarification of an unknown definition, therefore, has to be attributed wholly to division. 1

We will easily do away with this problem if we understand what is said to be known *per se* and what is not known *per se,* that is, unknown *per se.* For, as Averroës teaches in commentary 25 to the second book of the *Posterior Analytics,*[35] that which cannot be known to us by its own light and its own evidentness is said to be unknown *per se* or naturally unknown; it needs to be shown by means of something else and is accordingly made known by means of a syllogism—as this premise, "A triangle has three angles equal in total magnitude to two right angles." This never becomes known *per se,* either by means of long consideration by the mind or by means of careful inspection of a triangle, since that affection of 2

minimè sensilis est. Propterea ad aliud notius confugiendum est illudque est medium[16] in syllogismo, per quod eam demonstramus. De hoc per se ignoto si loquamur, certissimum est id per divisionem minimè notificari, ut ait eo in loco Averroes, imò et Aristoteles.

3   Per se autem notum seu naturaliter notum duplex est, ut ex ipso Aristotele passim in Posterioribus Analyticis colligimus, non solùm enim illud, quod omnibus evidentissimum est, dicitur per se notum, sed illud quoque, quod etiam si non sit cognitum, tamen, si proponatur vel statim vel post aliquam eiusdem considerationem absque ullius medii ope cognoscitur. Aristoteles enim in secundo capite primi libri Posteriorum Analyticorum inter principia immediata et indemonstrabilia non sola ponit axiomata, quae per se nota sunt discipulo antequàm ea audiat ex ore magistri, sed etiam definitiones ac suppositiones, quae non sunt discipulo cognitae antequàm à doctore exprimantur. Ut enim per se nota dicantur, satis est si non probentur per aliud, sed ipsa per se, quando proponuntur, credantur. Postea verò in tertio eiusdem libri capite postquàm in secundo species principii immediati distinxerat, incipit eorum sententiam refellere, qui nihil sciri posse asseverabant, necnon eorum, qui omnia per demonstrationem sciri posse arbitrabantur et ostendit utrumque errorem ex eo processisse, quòd qualisnam sit primorum principiorum cognitio non adverterunt, putarunt enim principia quoque, si sciri debeant, per aliud esse demonstranda, quod omnino negavit[17] Aristoteles, sed dixit, prima principia non ex alio, sed ex seipsis nota fieri. Nec de solis

triangle is not at all sensible. Therefore one must take refuge in something else more known, and that is the middle [term] in a syllogism; by means of it [i.e., the middle term] we demonstrate the [affection of the triangle]. If we speak about what is unknown *per se*, it is most certain that it is not at all made known by means of division, as Averroës says in this passage and Aristotle [says] too.

But now [what is] known *per se*, or known naturally, is of two   3 types, as we gather from Aristotle himself throughout the *Posterior Analytics*. For not only is that which is most evident to everyone said to be known *per se*, but also that which, even if it is not known, nevertheless, if it is set out, is known without need of any middle [term], either at once or after some consideration of it. For Aristotle in the second chapter of the first book of the *Posterior Analytics*[36] locates among the immediate and indemonstrable beginning-principles not only axioms, which are known *per se* by the student before he hears them from the mouth of the master, but also definitions and suppositions, which are not known to the student before they are expressed by the teacher. For, for it to be said they are known *per se*, it is enough if they are not proved by means of something else but are themselves believed *per se* when they are set out. But then afterward, in the third chapter of the same book,[37] after he had distinguished the species of immediate beginning-principles in the second, he starts to refute the position of those who averred that nothing could be known scientifically and of those who thought that everything could be known scientifically by means of demonstration. He shows that each error proceeded from this: They did not notice of what sort the knowledge of first beginning-principles is. For they also held that beginning-principles, if they ought to be known scientifically, have to be demonstrated by means of something else. Aristotle altogether denied this and said that first beginning-principles become known not from something else but from themselves. He says° this not only about axioms, but about all the beginning-

dignitatibus loquitur,° sed de omnibus principiis, quae prius inter species principii indemonstrabilis numeraverat. Omnia igitur vocat per se nota.

4   Unde colligimus, duplex esse per se notum, aliqua enim dicuntur per se nota, quia sunt omnibus actu per se cognita; aliqua verò quia, licèt non sint actu cognita, sunt tamen per se cognoscibilia, si proponantur, veluti definitio circuli rustico non est actu cognita, si tamen, ipsi proponatur, statim absque ullo medio eam intelliget. Sic definitio hominis, etsi non est actu cognita rustico, tamen, simulatque ipsi proponitur, ab eo cognoscitur per se, non per aliud. Hoc autem facilè videre possumus, si singulas definitionis partes hominis[18] rustico sine ulla divisione, imò etiam sine ullo ordine proponamus. Nam si rusticum percontemur, est ne homo rationis particeps? annuet proculdubio. Similiter si interrogemus an sit sensibilis an sit animatus an sit corpus; haec namque omnia rustico notissima sunt. Quoniam igitur ea, dum sine divisione proponuntur alicui, statim cognoscuntur, sequitur quòd etiam quando cum divisione dicuntur, non per divisionem nota fiunt, sed per se ipsa cognoscuntur. Illa verò, quae per se non cognoscuntur, ignota manent, sive cum divisione sive absque divisione proferantur, nam divisio ad ea notificanda est prorsus inefficax.

principles that he had first counted among the species of indemonstrable beginning-principles. He calls them all, therefore, known
per se.

From this we gather that [what is] known *per se* is of two types.   4
For some things are said to be known *per se* because they are actually known *per se* by everyone, and others because, granted that
they are not actually [so] known, nevertheless, if they are set out,
they are knowable *per se.*[38] For example, the definition of a circle is
not actually known to the peasant, but nevertheless, if it is set out
for him, he understands it at once, without any middle [term]. So
also the definition of man, although it is not actually known to the
peasant, nevertheless, as soon as it is set out to him, it is known by
him *per se,* not by means of something else. Now we can easily see
this if we set out to the peasant each of the parts of the definition
of man without any division, even without any order. For if we
probe the peasant, "Does man not partake of reason?" he will
without doubt agree. Similarly if we were to ask whether [man] is
sentient or is animate or is a body, for these are all most known to
the peasant. And since these, therefore, when they are set out to
someone without division, are known at once, it follows that even
when they are said *with* division, they are not made known *by
means of* division, but are known by means of the things themselves. And those that are not known *per se* remain unknown, be
they advanced either with division or without division, for division
is utterly ineffectual for making them known.

: IX :

Adversus eos, qui dicunt divisionem esse methodum utilem
ad notificandum numerum specierum ignotum.

1 Relinquitur alter error confutandus eorum, qui dicunt per methodum divisivam inveniri numerum specierum ignotum. Primùm
quidem hi in eo mihi videntur errasse, quòd numerus specierum in
quaesitis scientialibus non numeratur, nam Aristoteles in principio
secundi libri Posteriorum Analyticorum dixit quatuor ad summum esse illa, quae quaeri ac sciri possunt, an sit, quid sit, an insit, cur insit, non adiecit quot sint species sub genere. Si verò totam Aristotelis philosophiam legamus, videbimus ipsum nunquàm
proponere numerum specierum investigandum; vult quidem in
philosophia naturali cognitionem nobis tradere specierum corporis
naturalis, scilicet ut earum naturas et accidentia cognoscamus, at
quot illae sint nunquàm declaravit, parum enim refert cognoscere
an species animalis sint mille an bis mille, sed sat est si ipsarum
specierum naturas et accidentia noscamus.° Hanc notitiam° adepti
possumus species, si volumus, numerare. Quòd si numerare eas
nolimus, non propterea imperfecta vel diminuta scientia nostra
nuncupanda est.

2   Videtur autem haec cognitio numeri esse quiddam consequens
ipsarum specierum cognitionem sive distinctam sive confusam,
nam ex cognitione quid sint numerum ipsarum colligere possumus, quem nunquàm perfectè cognoscimus dum earum naturas
ignoramus. Sed cognitionem etiam specierum imperfectam consequitur numeri quoque cognitio imperfecta, quando enim solùm

: IX :

Against those who say that division is a method useful for making known the unknown number of species.

Another error remains to be confuted, that of those who say that   1
the unknown number of species can be discovered by means of
divisive method. To me, of course, they appear to have erred first
in this, that in scientific inquiries the number of species is not
counted. For Aristotle said in the beginning of the second book of
the *Posterior Analytics*[39] that there are at most four things that can
be inquired after and known scientifically: whether it is, what it is,
whether it belongs, why it belongs. He did not add: how many
species are under the genus. And if we read the whole philosophy
of Aristotle, we will see that he never sets out to investigate the
number of species. In natural philosophy, of course, he wants to
convey to us knowledge of the species of natural body, evidently so
that we may know their natures and accidents. But how many
there are, he never makes clear. For it matters little to know
whether there are one thousand or two thousand species of animal. It is enough if instead we know° the natures and accidents of
the species themselves. Having obtained this knowledge,° we can,
if we want, count the species. But if we do not want to count
them, our scientific knowledge is not on that account to be called
imperfect or wanting.

It appears, however, that this knowledge of the number is a   2
consequence of knowledge of the species themselves, whether distinct or confused. For from knowledge of what they are, we can
gather their number; we can never know this perfectly when we
are ignorant of their natures. But even imperfect knowledge of
the number ensues from imperfect knowledge, also, of the species.

quòd sint, non quid sint, cognoscimus, debilem quoque numeri notitiam° habemus, quia dum earum naturas ignoramus, possumus saepe dubitare an duae aliquae species sint revera duae, an una et eadem, ut asinus et mulus, quod in multis avibus, atque in multis piscibus contingere facilè potest. Cognitio igitur numeri specierum non est seiuncta à cognitione an sint et à cognitione quid sint, sed cum eis ex necessitate est coniuncta et eas consequitur.

3   Propterea huiusmodi quaestionis nullam fecit mentionem Aristoteles et in scientiis nunquàm numerus specierum quaeritur vel aliarum rerum, nisi dum quaeritur an sint et quid sint, quemadmodum in primo libro Physicorum notare possumus, etenim quaerit ibi Aristoteles, quot sint principia rerum naturalium, cui quaestioni non per divisionem satisfacit, sed probando quòd sint et quid sint, postquàm enim ostensum est prima omnium principia esse duo contraria et unum subiectum, factus est manifestus principiorum numerus absque ulla divisione. Fit quidem divisio quaedam ab Aristotele in contextu sexto illius libri, sed per eam proponitur solùm quaestio, non explicatur. Quod in aliis quoque illius libri locis notare possumus.

4   Quandoquidem, ut antea dicebamus, divisio nil ignotum notificat quando nullum membrum est per se notum, ideò qui dicit principia vel plura esse vel unum, et, si plura, vel finita vel infinita, et, si finita, vel tria vel quatuor vel in alio aliquo statuto numero, non propterea numerum principiorum declarat, sed solùm quaestionem proponit de numero principiorum, quoniam nullum ex iis membris est per se notum. Facta autem quaestione Aristoteles non eam solvit per divisionem, siquidem vel caecus videre posset

For when we know only that they are and not what they are, we also have weak knowledge° of the number, because when we are ignorant of their natures, we can often doubt whether some two species are in truth two, or one and the same, as [for example] ass and mule. This can easily happen with many birds and many fishes. Knowledge, therefore, of the number of species is not separate from knowledge of whether they are and from knowledge of what they are; it is, rather, out of necessity conjoined with them and ensues from them.

And so Aristotle made no mention of this type of question and in the sciences the number of species or of other things is never inquired after, except when it is inquired whether they are and what they are, as we can note in the first book of the *Physics*.[40] Indeed Aristotle there inquires how many beginning-principles of natural things there are, a question he does not answer by means of division, but by proving that they are and what they are. For after it is shown that the first beginning-principles of everything are two contraries and one subject, the number of beginning-principles is made manifest without any division. A sort of division, of course, is made by Aristotle in text no. 6 of that book,[41] but by means of it the question is only set out, not explicated. We can note this in other passages of that book also.

Now, as we said earlier, division makes nothing unknown known, when no branch [of the division] is known *per se*; therefore whoever says that beginning-principles are either many or one, and if many, either finite or infinite, and if finite, either three or four or in some other fixed number, does not on that account make the number of beginning-principles clear. He only sets out the question about the number of beginning-principles, since none of the branches is known *per se*. Having raised this question, however, Aristotle does not resolve it by means of division, since even

divisionem ad hoc munus idoneam non esse, sed per methodum resolutivam, etenim inductione probat omnium rerum principia esse duo contraria, deinde ostendit dari primam materiam per demonstrationem ab effectu. Itaque demonstrando quòd sint et quid sint principia numerum quoque principiorum consequenter declarat et nulla ibi notari potest ostensio numeri seiuncta à declaratione quòd sint et quid sint. Quare nulla methodo numerus ignotus declarari potest, nisi illa eadem, qua vel essentia vel saltem existentia declaretur.

5   Qui igitur dicunt, methodo divisiva numerum specierum ignotum inveniri, commentum dicunt, quod ne excogitari quidem meo iudicio potest, quandoquidem si earum specierum existentia nota est, numerus quoque cognoscitur. Si verò etiam essentia, adhuc magis cognoscitur numerus neque opus est ut per divisionem investigetur. Quòd si et essentia et existentia ignoretur, vana est numeri indagatio per divisionem, is enim nunquàm invenietur, nisi essentia vel saltem existentia nota fiat, at neutra potest per divisionem ostendi, ergo neque numerus.

6   Duos igitur errores illi commisere, unum quidem, quoniam quaestio numeri seiuncta à quaestione an sit et à quaestione quid sit, non est in usu apud philosophos; alterum verò quia eo modo, quo est in usu, nimirum coniuncta cum altera memoratarum quaestionum, divisio ad eam quaestionem dissolvendam et numerum quaesitum declarandum nihil roboris habet.

7   Verùm ut[19] luce clariorem huius dogmatis falsitatem inspiciamus, consideremus quisnam sit in eo progressu terminus notus et

a blind man could see that division is not fit for this job. [He does it] instead by means of resolutive method. For indeed he proves by induction that the beginning-principles of all things are two contraries and then shows that there is first matter by means of a demonstration *ab effectu*.[42] And so by demonstrating that the beginning-principles are and what they are, he consequently makes the number of beginning-principles clear. And no presentation of the number can be noted there, separated from the clarification that they are and what they are. Therefore no unknown number can be made clear by any method, except that same one by which either the essence or at least the existence is made clear.

Those, therefore, who say that the unknown number of species 5 is discovered by divisive method speak a fiction—what in my judgment cannot, of course, even be imagined, since if the existence of these species is known, [their] number is known, also. And if the essence is also [known], then even more so is the number known, and it does not need to be investigated by means of division. And if both the essence and the existence are unknown, tracking down the number by means of division is vain, for it will never be discovered unless the essence or at least the existence is made known. But neither can be shown by means of division; therefore, the number cannot [be shown] either.

They, therefore, committed two errors: first, since the question 6 of the number, separated from the question of whether it is and from the question of what it is, is not in use in the philosophers; second, because of the way in which it is used, that is, conjoined with one or the other of the questions referred to, division has no weight for resolving this question and for making the number inquired after clear.

Now, so that we may observe the falsity of this doctrine more 7 clearly than light, let us consider which is the known term in this progression and which is the unknown, and what sort of inference

quis ignotus et qualis illatio huius ex illo. Necesse est ut qui numerum specierum per divisionem notificari dicunt, fateantur tres tantummodò terminos hîc posse considerari, genus, differentias, species. Vel igitur processus est à genere ad differentias vel à genere ad species vel à differentiis ad species cognoscendas.

8   Quorum nullum dici posse ostendemus. Non est à genere ad differentias, quoniam ex genere non possunt colligi differentiae, id enim si esset, ultimae naturalium corporum differentiae non ignorarentur, hae tamen sunt ignotae, licèt genera cognoscantur. Quòd si quis dicat hanc consequentiam validam esse, est animal, ergo est rationale vel irrationale, proinde ex genere posse differentias inferri; respondemus, ad differentiarum cognitionem nos non duci ex cognitione generis, nam si ignoraremus differentias animalis esse rationale et irrationale, dicere non possemus, est animal, ergo vel est rationale vel irrationale. Alio igitur medio ducti sumus in harum differentiarum notitiam,° nimirum aliquo signo vel effectu posteriore sive per sensum, at certè non per genus. Differentiis autem iam cognitis eas ex generis positione licet inferre, cùm enim cognoscamus eas esse animalis differentias et inter illas nullum dari medium, certi etiam sumus quicquid est animal, illud vel rationale esse vel irrationale. Igitur ex genere cognito ad differentiarum notitiam° per necessariam illationem duci non possumus; sed neque ex genere in cognitionem specierum, eadem enim ratio viget, quia ubi notae sunt species omnes alicuius generis, valida est consequentia à genere ad eas omnes cum disiunctione prolatas, veluti si dicamus, est triangulum, ergo vel aequilaterum vel aequicrus vel scalenum, quia iam novimus alias trianguli species praeter has tres non dari. Ast ubi species omnes notae non sunt, quamvis

there is of the latter from the former. It is necessary that those who say the number of species is known by means of division acknowledge that only three terms can be considered here: genus, differentiae, species. The proceeding, therefore, is either from the genus to knowing differentiae, or from the genus to knowing the species, or from the differentiae to knowing the species.

We will show that none of these can be said. It is not from the genus to the differentiae, since differentiae cannot be gathered from a genus. For if that were the case, the ultimate differentiae of natural bodies would not be unknown. But they are unknown, even granted that the genera be known. Now if anyone were to say that this consequence is valid, "It is an animal, therefore it is rational or irrational, and accordingly the differentiae can be inferred from the genus," we would respond that we are not led from knowledge of the genus to knowledge of the differentiae; for if we did not know that the differentiae of animal are rational and irrational, we could not say, "It is an animal, therefore it is either rational or irrational." We were led, therefore, to knowledge° of these differentiae by another middle [term], namely of course by some sign or posterior effect or by means of sense, but certainly not by means of the genus. Once the differentiae are known, however, we may infer them from positing the genus. For since we know that they are the differentiae of animal and that there is no middle between them, we are then certain that whatever is an animal, it is either rational or irrational. We cannot, therefore, be led by means of a necessary inference from the known genus to knowledge° of the differentiae, but also not from the genus into knowledge of the species, for the same reasoning applies: where all the species of some genus are known, the consequence is valid from the genus to all those given in the disjunction, as [for example] if we were to say, "It is a triangle, therefore it is either equilateral or isosceles or scalene," because we already knew that there are no species of triangle other than these three. But where not all the species are

8

notum sit genus, ea illatione uti non possumus, notum enim nobis est genus animal, tamen ex eo numerus specierum inferri non potest, cùm plurimae sint nobis ignotae. Quare si dicamus, est animal, ergo vel homo vel bos vel equus vel aliquod aliud nobis notum, ratio est prorsus inefficax, quia esse potest aliqua alia animalis species ex earum numero, quas ignoramus. Genus igitur non est tale notum, ex quo numerus differentiarum, vel specierum ignotus colligi ac notificari possit.

9   Superest ut dicant in methodo divisiva processum fieri à differentiis ad species tanquàm à noto ad ignotum, ut ex differentiis notis ducamur in cognitionem numeri specierum ignoti. Sed neque hoc dici potest, in tali enim processu fieret petitio principii manifesta, quoniam talis illatio fieri non potest nisi hoc fundamento constituto quòd numerus specierum ignotus inferatur ex numero differentiarum noto, hoc autem est inferre idem ex seipso, quia differentia est ipsamet essentia speciei, nec potest unquàm esse cognita dum species ignoratur.[20] Propterea Aristoteles in secundo libro Posteriorum Analyticorum, ubi docet è demonstratione definitionem extrahere, inquit illa tantùm posse demonstrari, quae causam habent externam, sic enim aliud ex alio demonstratur; at non illa, quorum causa interna est et in ipsorum essentia comprehenditur. Ideò sola accidentia demonstrantur, quoniam ipsorum causae extra eorum essentiam sunt. At substantiarum non est demonstratio, quia causa, qua sunt, est ipsa earum essentia, causa enim, qua homo est homo, est ipsa forma, ac differentia propria hominis, quae non est extra hominis essentiam, sed ipsa essentia et natura hominis est. Quare si hominem demonstraremus per rationale, peteremus illud idem, quod ostendere vellemus, rationale namque nil aliud est, quàm homo ipse neque cognosci unquàm

known, even though the genus is known, we cannot use this inference. For the genus animal is known to us, but from it the number
of species cannot be inferred, since many are unknown to us.
Therefore if we were to say, "This is an animal, therefore it is a
man or an ox or a horse or something else known to us," the reasoning is utterly ineffectual, because there can be some other species of animal outside their number, of which we are ignorant. The
genus, therefore, is not so known that from it the unknown number of differentiae or species can be gathered and made known.

It remains for them to say that in the divisive method, proceeding occurs from differentiae to species as from known to unknown,
so that we are led from the known differentiae into knowledge of
the unknown number of species. But this cannot be said either.
For in such a procedure, a *petitio principii* becomes manifest, since
such an inference cannot occur unless this foundation is established, that an unknown number of species is inferred from a
known number of differentiae; this, however, is to infer the same
thing from itself, because the differentia is itself the very essence of
the species and cannot ever be known when the species is unknown. And so Aristotle, in the second book of the *Posterior Analytics*,[43] where he teaches the extracting of a definition from a
demonstration, says that only those things that have an external
cause can be demonstrated, for thus is something demonstrated
from something else, but not those whose cause is internal and
included in their essence. Because of this, only accidents are demonstrated, since their causes are outside their essence. But demonstration is not of substances, because the cause by which they are
is their very essence. For the cause by which man is man is the
very form and proper differentia of man; this is not outside man's
essence but is the very essence and nature of man. Therefore if we
demonstrated man by means of rational, we would presume the
same thing that we wanted to show, for the rational is nothing
other than man himself, and it cannot ever be known, if man is

9

potest dum homo ignoratur, eaque est sententia clara Averrois et
Eustratii in eodem loco et Alexandri referente Eustratio, asserunt
enim omnes quòd demonstrando speciem per propriam differentiam petitur id, quod in principio quaerebatur.

10   Praeterea species propriis differentiis semper notiores sunt, quia
cuiusque rei confusa cognitio distinctam ex necessitate vel praece-
dit vel saltem subsequi nullo modo potest. Cognitio autem differentiae nil aliud est, quàm cognitio distincta speciei. Ergo si
species omnes sunt suis differentiis notiores et prius cognitae,
nunquàm evenire potest ut ex numero differentiarum cognito ducamur in cognitionem numeri specierum ignoti, sed prius species
ipsas novimus,° postea differentias, quibus discrepant, indagamus,
dum ipsarum essentiam noscere° contendimus. Idque nemo sanae
mentis inficiari deberet, quia res ipsa loquitur et in ipso vocabulo
veritas apertè conspicitur; differentia enim non est aliquid per se
ipsum consistens, sed alteri inexistens, nempè id, quo res differt ab
aliis, quomodo igitur cognoscere vel cogitare differentiam aliquam
possumus sine specie ipsa differente? Idcirco nemo unquàm diceret animal aliud rationale esse, aliud irrationale, nisi prius aliquam
nosceret° animalis speciem rationis compotem et aliquam rationis
expertem. Sic dicimus animalium alia terrestria esse, alia volatilia,
alia aquatilia, quia iam novimus has species, quibus hae differentiae competunt; nullam enim harum differentiarum pronunciare-
mus, nisi speciem, cui inest, nosceremus,° quia differentia extra
speciem neque est neque cognosci potest. Quòd si fieri posset
ut speciebus omnibus ignoratis has differentias cognosceremus et
ipsis uteremur, cur non etiam quartam differentiam adiiceremus

unknown. This is the clear position of Averroës and Eustratius in that same passage, and of Alexander referring to Eustratius. For all assert that, in demonstrating a species by means of its own proper differentia, that is presupposed which was in the beginning inquired after.

Species, moreover, are always more known than their own 10 proper differentiae, because confused knowledge of any thing, out of necessity, either precedes or at least can in no way follow distinct [knowledge]. Knowledge of the differentia, however, is nothing other than distinct knowledge of the species. If, therefore, all species are more known than their own differentiae, and known first, then it can never happen that from the known number of differentiae we are led into knowledge of the unknown number of species. Rather, we first came to know° the species themselves; afterward we track down the differentiae by which they are different, when we venture to know° the essence of the [species] themselves. No one of sound mind ought to have denied this, because the issue speaks for itself,[11] and the truth is plainly seen in the very vocabulary. For a differentia is not something enduring by means of the thing itself but is existing in something else, namely that by which a thing differs from other things. How, therefore, can we know or ponder some differentia without the species itself differing? And so no one could ever say that some animal is rational, another irrational, unless he first knew° some species of animal possessing reason and another not having reason. Thus we say of animals that some are terrestrial, some winged, some aquatic, because we have already come to know the species to which these differentiae appertain. Indeed we would have stated none of these differentiae unless we should know° the species to which they belong, because the differentia is not outside the species and cannot be known [outside the species]. Now if it could happen that, all the species being unknown, we could know these differentiae and could use them, why would we not then add a fourth differentia,

dicentes alia animalia in terra, alia in aquis, alia in aere, alia in igne degere? Attamen hanc non addimus, quia prius novimus nullam esse animalis speciem, quae in igne vivat. Nunquàm igitur differentias, per quas genus dividitur, pronunciamus, nisi species, quibus competant et in quibus existant, saltem aliquas cognoscamus.

11   Non est igitur verum id, quod illi dicunt, ex numero differentiarum nos duci posse in cognitionem numeri specierum ignoti; quod in ipsa quoque divisione manifestum est, quando enim genus per differentias dividimus, nullam illationem facimus, nullo utimur ratiocinio à differentiis ad species procedendo tanquàm à notis ad ignotas, non enim dicimus rationale, ergo homo, irrationale, ergo bruta, sed propositas differentias speciebus ut prius cognitis applicamus, dicimus enim, animal aliud est rationale, aliud irrationale, rationale quidem est homo, irrationale verò per alias differentias dividitur, quae singulae speciebus notis tribuuntur, nunquàm species ut ignotae ex eis deducuntur. Nullus igitur processus illativus in divisione apparet, neque enim à genere ad differentias neque à genere ad species neque à differentiis ad species. Quare nulla ratione eorum sententia admitti potest, qui dicunt methodo divisiva nos duci ad cognoscendum numerum specierum ignotum.

12   Sed huius erroris causa[21] fuit quòd non cognoverunt, non posse numerum specierum ignotum declarari, nisi ipsarum essentia vel saltem existentia nota fiat. Divisio autem cùm nullam habeat vim declarandi essentiam vel existentiam ignotam, numerum quoque specierum notificare non potest.[22] Quòd autem divisio essentiam ignotam non declaret, prius ostendimus dum de definitionis venatione per divisionem locuti sumus. Quòd verò neque existentiam, nunc demonstravimus, quoniam ex necessitate prius est cognita speciei existentia, quàm differentiae.

saying that some animals spend their lives on earth, some in water, some in air, and some in fire? But we do not add this, because we first come to know that there is no species of animal that lives in fire. We never, therefore, state the differentiae by means of which a genus is divided, unless we know at least some species to which they appertain and in which they exist.

What they say, therefore, is not true, that we can be led from the number of differentiae into knowledge of the unknown number of species; this is also manifest in the division itself. For when we divide a genus by means of differentiae, we make no inference, we use no ratiocination in proceeding from differentiae to species as from knowns to unknowns. For we do not say, "Rational, therefore man; irrational, therefore brute." Instead we apply the given differentiae to the species as known first. For we say, "One animal is rational, another irrational." The rational is man, and the irrational is divided by means of other differentiae, each of which is ascribed to known species. The species, as unknown, are never deduced from them. No inferential procedure, therefore, appears in division. For neither [is there one] from the genus to the differentiae, nor from the genus to the species, nor from the differentiae to the species. And so in no way of reasoning can the position be admitted of those who say that we are led by divisive method to knowing the unknown number of species.

But the cause of this error was that they did not know that the unknown number of species cannot be made clear unless their essence or at least existence becomes known. Division, however, since it has no power to make an unknown essence or existence clear, also cannot make known the number of species. And that division does not make an unknown essence clear, we showed earlier, when we spoke about the search for a definition by means of division. And that [it does not make clear] the existence either, we have now demonstrated, since out of necessity the existence of the species is known prior to the [existence of] the differentia.

13   Haec igitur sententia rationi consentanea non est, sed et antiquorum philosophorum authoritati adversatur, Aristoteles enim
quando Platonis de divisione opinionem recenset, nil aliud ipsi
attribuit, nisi quòd dixerit eam esse idoneam methodum, per
quam partes definitionis ignotas investigemus. De numero autem
specierum per divisionem indagando nihil prorsus dixit Aristoteles
neque ab aliis dictum commemoravit.

: X :

De utilitate divisionis.

1   In divisione nullam esse notificandi vim satis est in praecedentibus
demonstratum, nunc declarandum est quaenam sit divisionis utilitas, ne quis fortè suspicetur nos eam ut prorsus inutilem reiecisse.

2   Dicimus igitur divisionis utilitatem non in notificando, sed in
disponendo potius esse constitutam, proinde ipsam ordinis potius,
quàm methodi conditiones et naturam prae se ferre. Nam ad
partes totius disciplinae rectè disponendas non parùm iuvat divisio
rerum considerandarum, ut in naturali philosophia credendum est
Aristotelem divisione usum esse ad eius scientiae partes benè ordi-
nandas, cùm enim subiectum ipsi proponeretur considerandum
corpus naturale generaliter acceptum, huius divisionem fecit in
simplex et mistum latè sumptum, deinde misti in imperfectum et
perfectum, deinde misti perfecti in homogeneum, et heteroge-
neum, postea homogenei in lapides et metalla et heterogenei in
animalia et stirpes et horum omnium in species ultimas, ut primo

This position, therefore, does not agree with reason, but it is 13
also opposed to the authority of ancient philosophers. For when
Aristotle recalls Plato's opinion on division,[45] he attributes nothing
else to him except that he said it is a fitting method by means of
which we may investigate the unknown parts of a definition. But
Aristotle said nothing at all about tracking down the number of
species by means of division, and neither did he rehearse anything
said by others.

: X :

On the Utility of Division.

In what preceded, it was demonstrated enough that in division 1
there is no power to make [something] known. Now it has to be
made clear what the utility of division is, lest anyone by chance
suspect that we have rejected it as utterly useless.

We say, therefore, that the utility of division is constituted not 2
in making [something] known, but rather in disposing. Accordingly, it exhibits the characteristics and nature of order rather than
of method. For a division of things being considered helps not a
little for correctly disposing the parts of the whole discipline, as
[for example] in natural philosophy it has to be believed that Aristotle used division for well ordering the parts of this science. For
although the subject he set out to be considered [was] natural
body accepted in the general sense, he made a division of it into
simple and mixed taken in a broad sense, and then of mixed into
imperfect and perfect, and then of perfect mixed into homogeneous and heterogeneous, and afterward of homogeneous into
stones and metals and of heterogeneous into animals and plants
and into the ultimate species of all these, so as to make in the first

loco de corpore naturali communi tractationem faceret, postea de simplici, postea de misto communi, deinde de speciebus iam memoratis usque ad infimas. Qualem divisionem videtur etiam tetigisse Aristoteles in principio primi libri de Coelo dum corpus naturale partitur in simplex et mistum, ut de simplici corpore tractationem aggrederetur; necnon in primo capite primi libri Meteorologicorum, ubi omnia tum dicta tum dicenda commemorat. In hac autem divisione illud inspicere possumus, quod antea de ordine declaravimus, quemadmodum enim ordo nihil ignotum notificat, sed solùm disponit, ita dum divisionem facimus corporum naturalium, nullam rem ignotam declaramus, sed dividendo res naturales disponimus. Mox autem methodis uti incipimus in singularum partium tractatione, quibus à noto ducimur ad cognitionem ignoti.

3   Eadem est divisionis utilitas in definitionum assignatione, sive definitionis partes per se notae sive ignotae fuerint. Quando enim sunt per se notae, id est saltem per se absque alio medio cognoscibiles, ut partes definitionis hominis respectu rustici, tunc divisio facit ut ordinatè omnes[23] proferantur et genus differentiis omnibus anteponatur et differentiae superiores inferioribus. Quod nisi fieret, nugatio committeretur. Si verò vel omnes partes vel aliquae sint naturaliter ignotae et investigatione atque probatione indigeant, divisio facit ut ordinatè investigentur, ut si assignanda esset haec hominis definitio, corpus animatum sensibile rationale et nullam harum esset per se notum homini inesse, proinde singulae demonstrandae essent, primùm quidem demonstrandum esset hominem esse corpus, deinde esse animatum, et ita deinceps, quae partium ordinatio commodè per divisionem fit, sine qua possemus

place a treatment on common natural body, afterward on simple, and after that on common mixed, and then on the species just referred to, all the way to the lowest. Aristotle also appears to have touched on such a division in the beginning of the first book of *On the Heavens*,[46] where he partitioned natural body into simple and mixed, so that he could undertake a treatment on simple body; and also in the first chapter of the first book of the *Meteorology*,[47] where he rehearses everything both said and to be said. In that division, however, we can observe what we made clear earlier about order. For just as order makes nothing unknown known and instead only disposes it, so when we make a division of natural bodies, we clarify no unknown thing; instead, by dividing, we dispose natural things. Next, however, in a treatment of each of the parts, we start to use methods, by which we are led from the known to knowledge of the unknown.

The utility of division is the same in assigning definitions, 3 whether the parts of the definition were *per se* known or unknown. For when they are known *per se*, that is, at least knowable *per se* without some middle [term], as the parts of the definition of man with regard to the peasant, then the division makes it that they are all advanced in order and that the genus is placed before all the differentiae and the higher differentiae before the lower. Unless this happens, a *nugatio*[48] is committed. Indeed, if either all the parts or some are naturally unknown and in need of investigation and proof, division makes it that they are investigated in order, so that if this definition, "a body animate, sentient, [and] rational," were assigned to man and it were known *per se* that none of these belongs to man, and accordingly each had to be demonstrated, then first, of course, it would have to be demonstrated that man is a body, then is animate, and so on; this ordering of the parts by means of division would be advantageous; without it we could

nugationem committere demonstrando prius hominem esse rationalem, deinde esse animatum. Sed cùm divisionem summi generis per differentias facimus, ordinatè procedimus et quando in aliquam partem non per se cognoscibilem incidimus, ad methodum aliquam confugimus et probationem adhibemus, quoniam divisio vim probandi non habet, ut suprà dicebamus. Methodus igitur partes definitionis ignotas notificat, divisio verò non notificat, sed solùm disponit.

4   Est igitur divisio ordo potius, quàm methodus et ordo quidem modò universalis, modò particularis; universalis quidem in totius disciplinae dispositione facienda; particularis verò in partibus alicuius definitionis in aliqua disciplinae parte benè ordinandis.

5   Sic etiam dicere possumus, divisionem ad numerum specierum ignotum notificandum nihil prodesse, tamen conferre ad species iam cognitas ordinatè et distinctè numerandas, ut quaecumque dicatur esse divisionis utilitas, eam non in notificando, sed solùm in disponendo constituamus.

6   Si verò cuipiam videatur non propriè divisionem vocari ordinem, id saltem ille inficiari non debet, divisionem ministram quandam ordinis esse, cùm enim duo sint praecipua instrumenta logica, ordo et methodus, non est necessarium ut omne, quod logicus docet et quo philosophus utitur, ordo sit vel methodus, sed satis est si ad haec duo, reliqua omnia vel tanquàm partes vel tanquàm principia vel tanquàm ministrantia redigantur. Nam praeter hanc viam divisivam, de qua nunc locuti sumus, nonne apud philosophos in usu est etiam divisio quanti in partes, quae integrantes dicuntur? Nonne dividunt saepe philosophi vocem ambiguam in suas significationes? Ea tamen divisio neque ordo est, neque methodus, sed horum ministra. Nam de aliqua re sermonem habituri vel aliquam rem demonstraturi debemus ante omnia vocis

commit a *nugatio* by first demonstrating that man is rational and then that [man] is animate. But since we make the division of the highest genus by means of the differentiae, we proceed in order, and when we come upon some part not knowable *per se*, we take refuge in another method and apply the proof, since division does not have the power to prove, as we said above. Method, therefore, makes known the unknown parts of a definition, and division does not make known; instead it only disposes.

Division, therefore, is an order rather than a method, and, of 4 course, an order now universal, now particular — universal in making a disposition of the whole discipline, particular in ordering well the parts of any definition in any part of the discipline.

And so we can say that division profits us nothing for making 5 known the unknown number of species. Nevertheless it does contribute to counting through the species already known in order and distinctly, so that whatever the utility of division may be said to be, we established it not in making [something] known, but only in disposing.

If indeed it appears to someone that division is not properly 6 called an order, at least he ought not to deny that division is some sort of minister to order. For although there are two principal logical instruments — order and method — it is not necessary that everything that the logician teaches and the philosopher uses be an order or a method. It is instead enough if all the rest, either as parts or as beginning-principles or as things ministering, are reduced to these two. For besides this divisive way, about which we have now spoken, is there not also in use in the philosophers division of a quantum into the parts that are said to be constituent? Do philosophers not often divide an ambiguous word into its significations? Nevertheless, this division is neither an order nor a method, but a minister to them. For in preparing discourse about some thing or in demonstrating some thing, we ought before all

ambiguitatem tollere, et eius significationes distinguere. Tamen
huiusmodi divisionem indignam esse arbitramur, quae instrumentum logicum appelletur et malumus ministram et servam logicorum instrumentorum nominare.

7   Nos igitur talem putamus esse divisionis naturam, ut ordinis
potius conditiones, quàm methodi prae se ferat, proinde vel ordo,
vel ordinis pars vel saltem ministra dicenda sit.

8   Hanc sententiam Aristotelis fuisse, non est dubitandum, ipse
namque in logica sibi proposuit de methodis atque instrumentis
notificantibus, non de ordinibus agere, propterea de divisione
nusquàm locutus est, nisi eam reiiciens et eius inutilitatem
ostendens, ut legere possumus in libris Analyticis tum Prioribus
tum Posterioribus. At si divisio instrumentum notificandi esset,
certè reprehendendus esset Aristoteles, qui de huiusmodi instrumentis diligentissimè in logica locuturus viam[24] divisivam omisis-
set, cùm de ratione benè dividendi aliqua omnino dicere debuisset.

9   Quòd igitur de via divisiva non sit locutus Aristoteles, nisi eam
refellendo, ea[25] causa° fuit, quia ipse de ordinibus nihil docere in
logica voluit, sed de methodis solùm, divisio autem ordo quidam
est, non methodus et quando Aristoteles eam impugnat, non alia
utitur ratione, quàm quòd vim syllogisticam non habet. Qua in re
videtur carpere Platonem, qui divisionem ut methodum accipere
volens, non solùm ut ordinem, ipsam sufficere arbitratus est ad
venandas partes definitionis ignotas, quod omnino negavit Aristoteles. Sic autem manifestum est ordinem quoque omnem inutilem
esse, quatenus ordo,[26] nempè ad notificandum, utilis tamen est ad
disponendum, qualem ipsam quoque divisionem esse diximus.

10   De hac re potest quisque Aristotelem legere in primo libro
Priorum Analyticorum in calce sectionis secundae, et in secundo

else to get rid of ambiguity in the word and distinguish its significations. But still, we do not think a division of this type is worthy of being called a logical instrument, and we prefer to name [it] a minister or a handmaid of logical instruments.

We hold, therefore, that the nature of division is such that it   7
exhibits the characteristics of order rather than of method, and accordingly has to be said to be an order, or part of an order, or at least a minister.

That this was Aristotle's position is not to be doubted. For in   8
logic he intended to deal with methods and instruments that make [something] known, not with orders. Nowhere, therefore, did he speak about division except rejecting it and showing its useless-ness, as we can read in both books, the *Prior* and *Posterior* [*Analyt-ics*]. But if division were an instrument for making [something] known, Aristotle would certainly have to be censured, for he, about to speak most carefully in the logic about instruments of this type, omitted the divisive way, when he altogether ought to have spoken somehow about this way of dividing well.

The reason,° therefore, that Aristotle did not speak about the   9
divisive way, except in refuting it, was this: because he wanted in the logic to teach nothing about orders and instead only about methods. Division, however, is a sort of order, not a method, and when Aristotle attacks it, he uses no other reason than that it does not have syllogistic power. On this issue he appears to criticize Plato, who, wanting to accept division as a method, not just as an order, thought that it was sufficient for searching for the unknown parts of a definition; Aristotle altogether denied this. So now it is manifest that every order is also useless — insofar as it is order — for making [something] known. But it is nevertheless useful for disposing, just as we said division itself is.

Anyone can read Aristotle on this issue in the first book of the   10
Prior Analytics at the end of the second section[49] and in the second

Posteriorum in 23 et in 73 contextu aliisque sequentibus[27] et ipsius dicta benè considerando ex eis hanc, quam exposuimus, sententiam desumere.

: XI :

In quo methodus definitiva refellitur.

1   Reiecta methodo divisiva sequitur ut de definitiva sermonem faciamus, quam refellere multò facilius nobis erit, quandoquidem in divisione via quaedam inest et processus ab hoc ad illud, ut non absque ratione plurimos haec similitudo decipere potuerit, ut crediderint divisionem methodum esse, cùm revera non sit.

2   At in definitione nihil tale conspicimus. Definitio enim est simplex quaedam essentiae expressio, una et individua, in qua nullum processum ab aliqua re ad aliam rem notare queamus. Ideò per magnum errorem ab omnibus fuit inter methodos collocata, et summoperè semper miratus sum quo modo in mentem tot clarissimorum virorum cadere tantus error potuerit, quòd definitio sit methodus, quae ex noto ducat in cognitionem ignoti et dicatur methodus definitiva ab aliis methodis distincta.

3   Huius sententiae falsitas satis demonstrari potest per ea, quae antea dicta sunt de natura methodi ac de eius definitione, diximus enim methodum esse processum syllogisticum et habere illationis necessitatem, cuiusmodi certè non est definitio, quare methodus vocari non potest, cùm formam illam, quae in omni methodo requiritur, non habeat.

4   Sed ut clarius veritatem demonstremus et omnem fugae locum adversariis auferamus, singillatim omnia considerare oportet, quae

[book] of the *Posterior* [*Analytics*] in texts no. 23 and no. 73[50] and in what follows, and by considering well what is said can draw from them this position that we have laid out.

: XI :

In which definitive method is refuted.

The divisive method having been rejected, it follows that we dis- 1
course on the definitive, which will be much easier for us to refute.
Seeing that in division there is a sort of way and a proceeding
from this to that, it is not without reason that this resemblance
was able to deceive many, such that they believed that division is a
method, although in truth it is not.

But in definition we see nothing of this sort. For definition is 2
some sort of simple, single, and individual expression of essence, in
which we can note no proceeding from some thing to some other
thing. And so by means of a great error was it included by everyone among methods. And I always greatly wondered how such an
error—that definition is a method that leads from the known into
knowledge of the unknown and is said to be the definitive method,
distinct from the other methods—could have occurred to the
mind of so many illustrious men.

The falsity of this position can be sufficiently demonstrated by 3
means of that which was said earlier about the nature of method
and about its definition. For we said that method is a syllogistic
procedure and has the necessity of inference. Definition is certainly not of this sort. It cannot, therefore, be called a method,
since it does not have the form that is required in every method.

But so that we may demonstrate the truth more clearly and take 4
away from our opponents every place of refuge, everything that

ipsi hac in re dicere aut imaginari possunt, cùm enim omnes usque ad hodiernam diem sententiam hanc secuti sint, eaque in eruditorum virorum mentibus ita sit radicata, ut opinio nostra primo ipso aspectu fortasse παράδοξος esse[28] videatur, necessarium esse duximus ad inveteratum aliorum errorem eradicandum et novum dogma introducendum, omnia diligentissimè expendere. Praestat enim ut multi dicant librum nostrum prolixitate orationis laborare, quàm ut verborum inopia et obscuritate fiat ut nostrae sententiae veritas et alterius falsitas non benè à multis percipiatur.

: XII :

Quòd in definitione neque ab una parte ad aliam processus fiat neque à definito ad definitionis cognitionem ducamur.

1   Qui igitur dicunt definitionem esse methodum ducentem nos à noto ad cognoscendum ignotum, ostendere debent quisnam sit in hac definitiva methodo terminus ignotus, ad quem; et quisnam terminus notus, à quo; ut qualisnam ab hoc ad illud via sit consideremus. Quatuor ad summum invenio, quae responderi possint, eaque omnia vana et absurda esse demonstrabimus. Vel enim dicunt in methodo definitiva processum fieri à parte definitionis nota ad aliam partem ignotam; vel à re definita ut notiore ad definitionem ut ignotiorem, cùm Aristoteles dicat in prooemio primi libri Physicorum nomen rei definitae esse nobis notius definitione;

they can say or imagine about this issue must be considered one by one. Now since everyone down to this very day has followed this position and it is so rooted in the minds of learned men that our opinion perhaps appears on first glance to be *paradoxos*, we have come to think it necessary to weigh everything most carefully so as to eradicate the long-standing error of others and introduce a new doctrine. For it really is better if many say that our book labors with a prolixity of speech, than if it happens that, by obscurity and a shortage of words, the truth of our position and the falsity of the others is not well grasped by many.

∻ XII ∻

*That in definition neither does proceeding occur
from one part to another nor are we led from
the defined to knowledge of the definition.*

Those who say, therefore, that definition is a method leading us 1
from the known to knowing the unknown ought to show what the
unknown terminus *ad quem* (to which) is in this definitive method
and what the known terminus *a quo* (from which) is, so that we
may consider what sort of way there is from the latter to the former. I find at most four things that could be said in response; we
will demonstrate that all of them are vain and absurd. For either
they say [(a)] that in definitive method proceeding occurs from the
known part of the definition to the other, unknown part; or [(b)]
from the thing defined as more known to the definition as more
unknown, since Aristotle says in the proem of the first book of the
Physics[51] that the name of the thing defined is more known to us

vel dicunt (quae potius videtur eorum esse sententia) methodum definitivam esse processum ab ipsa definitione ad essentiam sive (ut aiunt) quidditatem rei definitae cognoscendam; vel demum dicunt, ut aliqui dicere visi sunt, methodum definitivam esse processum à partibus definitionis praecognitis ad totam definitionem, quae ex earum compositione colligitur.

2 Primum quidem, minimè dici potest, cùm manifestum sit nullum esse in definitione processum illativum à parte ad partem, sed totam definitionem continuata locutione pronunciari tanquàm unum, non tanquàm multa. Quamvis enim in definitione multae sint partes, tamen non ut plures proferuntur, sed ut unum, quemadmodum admonuit nos Aristoteles in libello de Interpretatione. Ideò naturae definitionis repugnat dictio coniunctiva, quae multitudinem prae se fert, non enim coniungimus nisi ea, quae ut plura proponuntur. Ob eandem rationem illatio in partibus definitionis nulla esse potest, quia non minùs multitudinem significat, quemadmodum enim idem cum seipso non coniungitur, ita neque ex seipso infertur. Partes autem definitionis, etsi plures sunt, tamen non ponuntur ut plures, sed ut ad constituendam naturam et essentiam unam conveniunt, quatenus alia pars potestatem dicit et determinatur, alia verò determinat et significat actum. Sed rem hanc diligentius explicare ad nos non pertinet. Satis est per haec ostendisse in definitione nullum fieri discursum à parte definitionis ad partem.

3 Praeterea genus in definitione differentiis semper anteponitur, dicimus enim animal rationale esse hominis definitionem, ergo si processus aliquis illativus fieret à parte praecedente ad subsequentem, esset utique à genere ut noto ad differentias ut ignotas, tamen

than the definition; or [(c)] they say (this appears rather to be their position) that definitive method is a proceeding from the definition itself to the essence or to knowing the quiddity (as they say) of the thing defined; or lastly [(d)] they say, as some have appeared to say, that definitive method is a proceeding from the parts of a definition known beforehand to the whole definition, which is gathered from the composition of them.

The first [(a)], of course, can not be said at all, since it is mani- 2 fest that in definition there is no inferential proceeding from part to part; instead the whole definition is said in a continuous utterance as one [thing], not as many. And even though there are many parts in a definition, nevertheless they are not advanced as many but as one, as Aristotle points out in the little book *On Interpretation*.[52] And so a conjunctive locution that exhibits a multitude is incompatible with the nature of a definition. For we do not conjoin [things] except those that are set out as many. By the same reasoning, there can be no inference in the parts of a definition, because it no less signifies a multitude. For just as the same thing is not conjoined to itself, so neither is it inferred from that thing itself. Now the parts of a definition, although they are many, are nevertheless not posited as many, but as they come together to establish one nature and essence, insofar as one part says the potential and is determined, and the other part determines and signifies the actual. But it does not pertain to us to explicate this issue more carefully. It is enough to have shown by means of this that in definition no discursive movement occurs from part of the definition to [another] part.

Moreover, in a definition the genus is always placed before the 3 differentiae; for we say that "rational animal" (*animal rationale*) is the definition of man. If therefore some inferential proceeding occurred from the preceding part to the subsequent, then by all means it would be from the genus as known to the differentiae as

haec illatio nihil roboris habet et est penitus inefficax, quia non possumus dicere, est animal, ergo est rationale, quare nunquàm possumus ex parte praecedente inferre partem sequentem in definitione, cùm pars praecedens sit semper universalior sequente. Fatua igitur sententia est, si intelligant definitionem ita esse methodum, quia in ea fiat processus à parte praecedente ad sequentem cum illationis necessitate.

4   Si verò dicant in definitione progressum fieri à re definita ut nota ad definitionem ut ignotam, multa secuntur absurda, primùm quidem erit progressus ab eodem ad idem, definitio namque idem est ac definitum; sequeretur etiam definitum esse definitionis partem, quod quidem minimè dicendum est, nam si ipsa definitio est via et methodus quaedam, oportet in ipsa definitione comprehendi terminum notum, à quo, et terminum ignotum,[29] ad quem, ut in ipsa methodo transitus ille consideretur nullo extrinsecus assumpto. Itaque res definita pars erit methodi definitivae, proinde etiam definitionis, cùm apud illos definitio sit methodus ipsa definitiva. Erit igitur repugnantia magna in hac positione, quia dum dicitur hanc viam esse à definito ad definitionem, definitum extra definitionem ponitur. Quatenus verò definitum dicitur esse terminus, à quo, in methodo definitiva, sic definitionis pars esse dicitur.

5   Sed quid in re manifesta pluribus opus est verbis? Certum est ex cognitione rei definitae nos non duci per necessariam illationem in cognitionem definitionis ignotae, hoc enim si ita se haberet, facile esset omnia definire, qualibet enim re proposita, statim ex ea inferretur et in lucem prodiret ipsius definitio, quemadmodum ex fumo, quem intuemur, inferimus ibi ignem esse. Attamen id non

unknown. But this inference has no weight and is completely ineffectual, because we cannot say, "It is an animal, therefore it is rational." And so in a definition we can never infer the subsequent part
from the preceding part, since the preceding part is always more
universal than the subsequent. The position, therefore, is foolish,
if they understand definition to be a method, because in it a proceeding occurs from preceding part to subsequent with the necessity of inference.

But if they say [(b)] that in definition a progression occurs   4
from the thing defined as known to the definition as unknown,
many absurdities follow. First, of course, progression will be from
one thing to the selfsame thing, for the definition and what is defined are the same. And it would also follow that what is defined
is part of the definition; this, of course, cannot be said at all. For
if definition itself is some sort of way and method, then in the
definition itself must be comprehended a known terminus *a quo*
(from which) and an unknown terminus *ad quem* (to which), so
that this passage may be considered in the method itself, with
nothing assumed from outside. And so the thing defined will be
part of the definitive method, and accordingly also of the definition, since for them [i.e., the proponents of this position], definition is definitive method itself. In this position, therefore, there
will be great inconsistency. Because when it is said that this way
is from the defined to the definition, what is defined is located
outside the definition. But insofar as the defined is said to be the
terminus *a quo* (from which) in definitive method, it is thus said to
be part of the definition.

But what need is there for more words in a manifest issue? It is   5
certain that from knowledge of the thing defined we are not led by
means of necessary inference into knowledge of the unknown definition. For if this were so, it would be easy to define everything.
For with any thing set out, at once its definition would be inferred
and come to light, just as from smoke that we see, we infer there is

contingit, multas enim res cognoscimus, quarum definitiones ignoramus. Quoniam igitur ex sola cognitione rei definitae non ducimur in cognitionem definitionis ignotae, sequitur processum hunc à definito ad definitionem non esse methodum, cùm methodus sit via syllogistica cum necessitate illationis ignoti ex noto, et ipsius ignoti notificatione.

6   Neque nobis obest dictum Aristotelis in prooemio primi libri Physicorum quòd nomen rei definitae est nobis notius definitione, notius enim est, non tamen tale notius, ex quo possimus duci in cognitionem definitionis ignotae, quia iam diximus duas esse praecognitiones, quibus iuvamur ad cognitionem rerum ignotarum adipiscendam, una dicitur praecognitio dirigens, altera verò agens. Dirigens dicitur illa praecognitio, quae est quidem necessaria ad illius rei notitiam° comparandam, tamen ad eam notificandam non sufficit, ut praecognitio significationis nominis eclipsis est necessaria, si ipsius eclipsis cognitionem consequi debeamus, attamen ex ea sola in cognitionem eclipsis non ducimur; quia haec non habet processum syllogisticum. At praecognitio agens illa est, quae per necessariam illationem potest in nobis illius rei notitiam° parere, ut cognitio fumi facit ignem cognosci, quia ex hac praecognitione fit syllogisticus processus, quem possumus methodum appellare. Concedimus igitur definitum notius esse definitione, hoc enim nil aliud significat, quàm quòd cognitio cuiusque rei confusa prius et facilius acquiritur, quàm distincta, nam cognitio nominis rei definitae est cognitio confusa, cognitio autem definitionis non est

fire there. But nevertheless this does not happen, for we know many things whose definitions we do not know. Since, therefore, we are not led solely from knowledge of the thing defined into knowledge of the unknown definition, it follows that this proceeding from the defined to the definition is not a method, since method is a syllogistic way with necessity of inference of the unknown from the known and a making known of the very [thing] unknown.

Aristotle's dictum in the proem of the first book of the *Physics*,[53] 6 that the name of the thing defined is more known to us than the definition, is no obstacle to us. It is more known, but not more known in such a way that from it we can be led into knowledge of the unknown definition, because we already said[54] that there are two sorts of prior knowledge by which we are helped in obtaining knowledge of things unknown. One is said to be prior knowledge directing, and the other acting. The prior knowledge said to be directing is that which is, of course, necessary for procuring knowledge° of a thing. It is, nevertheless, not sufficient for making the [thing] known, as [for example] prior knowledge of the signification of the name "eclipse" is necessary if we are going to gain knowledge of eclipse itself, but, nevertheless, we are not led from it alone into knowledge of eclipse, because this does not have a syllogistic procedure.[55] But prior knowledge acting is that which, by means of necessary inference, can bring forth knowledge° in us of a thing—as knowledge of smoke makes it that fire is known—because from this prior knowledge a syllogistic procedure occurs that we can call a method. We concede, therefore, that the defined is more known than the definition, for this indicates nothing other than that confused knowledge of some thing is acquired earlier and more easily than distinct [knowledge]. For knowledge of the name of the thing defined is confused knowledge. And knowledge

alterius rei cognitio, sed est eiusdem rei cognitio distincta. Et potest quidem cognitio rei confusa nos dirigere ac iuvare ad acquirendam distinctam, quam non habemus, tamen eam in nobis parere potens non est.

: XIII :

Qùod processus à definitione ad essentiam definiti declarandum non sit methodus definitiva.

1 Qui verò dicunt in eo esse constitutam methodi definitivae naturam, ut sit processus à definitione ad essentiam et quidditatem rei definitae ignotam declarandam, in eorum sententiam incidere videntur, qui dicunt definitionem esse instrumentum sciendi quid est, idem enim est methodus atque instrumentum sciendi. Cùm igitur utrumque dogma in unum et eundem sensum cadat, eadem erit utriusque confutatio.

2 In primis cognoscendum nobis est quid sit definitio et quid sit id, quod Graeci τὸ τί ἐστι, nostri quid est seu quidditatem appellare consueverunt, ut videamus quid intersit inter definitionem et quidditatem, hoc enim cognito[30] non difficile erit videre qualisnam sit ille processus, quem aliqui imaginantur, à definitione ad quidditatem et quem putant esse methodum definitivam.

3 Discrimen inter definitionem et quidditatem sumi potest ex iis, quae dicuntur ab Aristotele in primo capite libri de Interpretatione de nomine, conceptu et re, dicit enim, conceptum esse imaginem

of the definition is not knowledge of another thing, but is distinct knowledge of the same thing. And confused knowledge of something can, of course, direct and help us in acquiring distinct [knowledge] that we do not have, but it is not able to bring it forth in us.

: XIII :

That proceeding from a definition to clarifying the essence of the defined is not definitive method.

But some say [(c)] that the nature of definitive method is consti- 1 tuted in this, that it is a proceeding from a definition to making the unknown essence and quiddity of the thing defined clear. Those who say that definition is the instrument for scientifically knowing what [something] is, appear to fall into their position. For method is the same thing as an instrument for knowing scientifically. Since, therefore, each doctrine would fall under one and the same sense, the refutation of both will be the same.

It has to be known to us in the first place what a definition is 2 and what that is that the Greeks were accustomed to call *to ti esti* and we "what [something] is" (*quid est*) or quiddity, so that we may see the difference between definition and quiddity. For once this is known, it will not be difficult to see what sort of thing is this proceeding that some imagine to be from definition to quiddity and that they hold to be definitive method.

The discriminating difference between definition and quiddity 3 can be taken from the things that are said by Aristotle about name, concept, and thing in the first chapter of the book *On Interpretation*.[56] For he says that a concept is an image[57] existing in

in animo existentem rei extra animum positae, nomen verò esse signum conceptus sive signum rei per medium° conceptum, postquàm enim imaginem rei mente concepimus, eam aliis per vocem significamus. Definitio igitur ad quidditatem habet eandem rationem,° quam nomen habet ad rem.

4   Res enim nobis proponitur cognoscenda et concipienda vel ut totum quiddam confusum vel ut in essentiales partes distributa ut omnes distinctè conspici possint. Illud quidem dicitur res ipsa, haec verò dicitur essentia seu quidditas rei. Ideò conceptum rei possumus utroque modo formare. Aut enim totam confusam concipimus et ita dicimur rei conceptum in animo habere; aut in essentiales partes divisam et ita dicimur conceptum habere quidditatis rei, quidditas enim ab eo, cuius est quidditas, reipsa non differt, sed sola ratione, ut dictum est. Possumus igitur etiam in voce hoc idem discrimen notare, nam si rem ipsam confusè conceptam significare volumus, nomine utimur et dicimus, homo. Si verò eandem distinctè conceptam, id est ipsius quidditatem, utimur definitione et dicimus, animal rationale. Ideò Aristoteles in prooemio primi libri Physicorum dicit definitionem[31] in singulas essentiales partes rem distribuere, quae per nomen tota confusa significabatur. Nomen igitur à definitione non differt, nisi ut idem confusè acceptum à se distinctè sumpto, quale est discrimen inter illa, quae ab utrisque significantur, res enim ut totum quoddam confusum significatur à nomine, eadem ut distinctè sumpta est id, quod per definitionem significatur.

5   Hinc sit ut definitio nomen unum esse non possit, sed semper sit oratio, ideò Aristoteles in definitione definitionis eam dicit esse

man's soul, of something located outside man's soul. But a name is the sign of a concept or the sign of the thing by means of a mediating° concept. For after we have conceived an image of something in the mind, we indicate it to others by means of a word. Definition, therefore, has the same relationship° to quiddity that name has to thing.

Something to be known and conceived is set out to us either as some sort of confused whole or as subdivided into essential parts, so that all can be seen distinctly. The former is said to be the thing itself, but the latter is said to be the essence or the quiddity of the thing. We can, therefore, form the concept of a thing in either way. For we conceive either a confused whole and are thus said to have in our soul a concept of the thing, or [we conceive what is] divided into essential parts and are thus said to have a concept of the quiddity of the thing. For not in reality, but only in reasoning, does the quiddity differ from that of which it is the quiddity, as was said. We can, therefore, also note this same discriminating difference in the words. For if we want to indicate the thing itself conceived confusedly, we use the name and say, "man." But if [we want to indicate] the same thing conceived distinctly, that is, its quiddity, we use the definition and say, "rational animal." Aristotle, accordingly, says in the proem of the first book of the *Physics*[58] that a definition subdivides into each of the essential parts something that was, by means of a name, signified [as] a confused whole. The name, therefore, does not differ from the definition except as the same thing accepted confusedly [differs] from itself taken distinctly. Such is the discriminating difference between those things that are signified by both. For a thing, as some confused whole, is signified by the name. The same thing, as taken distinctly, is that which is signified by means of the definition.

Thus it is that a definition cannot be one name; it is instead always speech. Aristotle, therefore, in the definition of definition[59]

orationem, est enim oratio veluti genus definitionis, propria autem differentia, qua ab aliis orationibus definitio separatur, est significatio quidditatis rei, omnis enim oratio est alicuius rei vel aliquarum rerum significatrix, at sola definitio est significatrix quidditatis.

6 Haec omnia ita sunt manifesta, ut à nemine negari possint.

7 Ex his colligamus magnam propositi dogmatis absurditatem, dicentes enim definitionem esse instrumentum sciendi quidditatem rei, sive methodum, qua ad quidditatis cognitionem ducimur, dicunt processum esse à definitione tanquàm nota et termino, à quo, ad quidditatem tanquàm terminum ignotum, ad quem. Qua quidem sententia nil potest esse absurdius, quandoquidem definitio et quidditas non discrepant nisi ut significans et significatum, non potest igitur definitio esse nota, dum quidditas ignoratur, quomodo enim notam habere possumus hominis definitionem et quidditatem hominis ignorare? Tamen si dogma illud verum esset, oporteret definitionem esse notam dum est ignota quidditas, ut à definitionis cognitione ad cognoscendam quidditatem duceremur. Quod an dici, imò an excogitari possit, quisque rationis compos consideret, sic enim dicitur idem notum esse et ignotum et progressum fieri ab eodem in idem.

8 At rei veritas haec est, quòd definitio est ipsa quidditas, quia significat quidditatem, ideò definitionem intelligere est quidditatem intelligere; et quidditatem ignotam dicere est dicere definitionem ignorari. Quidditatem quaeri est definitionem quaeri; et inventa definitione nil aliud nos assecutos esse dicimus, quàm cognitionem quidditatis. Nulla igitur methodus est à definitione ad quidditatem tanquàm à noto ad ignotum, et vana est eorum sententia, qui

says that it is speech. For speech is as it were the genus of definition, and the proper differentia by which definition is separated
from other speech is the signification of the quiddity of the thing.
For all speech is a signifier of some thing or some things; but the
definition alone is a signifier of the quiddity.

All this is so manifest that it cannot be denied by anyone.   6

From all this we may gather the great absurdity of the proposed   7
doctrine. For those saying that definition is an instrument for scientifically knowing the quiddity of something or a method by
which we are led to knowledge of the quiddity, say that it is a
proceeding from the definition as known and as the terminus *a quo*
(from which) to the quiddity as the unknown terminus *ad quem*
(to which). There can be, of course, nothing more absurd than
this position, since the definition and the quiddity are not different except as signifying and signified. The definition, therefore,
cannot be known when the quiddity is unknown. For how could
we have a known definition of man and not know the quiddity of
man? Now if, nevertheless, that doctrine were true, it would have
to be that the definition is known while the quiddity is unknown,
so that we might be led from knowledge of the definition to knowing the quiddity. Whether this can be said, or rather even imagined, let anyone in possession of reason consider. For here the
same thing is said to be known and unknown, and progression
occurs from one thing to the selfsame thing.

But the truth of the issue is this, that the definition is the quid-   8
dity itself, because it signifies the quiddity. And so to understand
the definition is to understand the quiddity, and to say that the
quiddity is unknown is to say that the definition is unknown. To
inquire after the quiddity is to inquire after the definition, and
once the definition has been discovered, we say that we have se-
cured nothing other than knowledge of the quiddity. There is no
method, therefore, from definition to quiddity as from known to
unknown. And vain is the position of those who hold in this way

hoc modò putant definitionem esse instrumentum ducens ad cognitionem quid est.

9   Quod autem eos fefellit, fuit ambiguitas huius vocis, instrumentum, ignorarunt enim quomodo definitio sit instrumentum et quaenam sint illa instrumenta, in quorum traditione logicus versatur et quae possunt appellari methodi, est enim definitio instrumentum quidem, non tamen notificandi quidditatem ignotam, sed solùm significandi. Non enim per discursum à noto ad ignotum dicitur definitio notam facere quidditatem rei, sed quia eam significat eodem modo, quo nomen significat rem per conceptum medium, neque ullum inter haec discrimen assignari potest, si enim res ignota sit, nomen quoque eam ignotam significabit, non eam notificabit et ita ipsum quoque ignotum dicitur. Sic quando quidditas est ignota, definitio eam significat ignotum, proinde ignota ipsa definitio dicitur neque rem notificabit. Est igitur definitio instrumentum significandi quidditatem rei, sicuti nomen est instrumentum significandi rem confusè acceptam.

10   At non huiusmodi sunt instrumenta logica, quorum fabricatio° in logica quaeritur, sed sunt instrumenta notificandi ignotum ex noto per discursum à noto ad ignotum et illationem necessariam huius ex illo, quae dicitur tertia mentis nostrae operatio. At conceptio definitionis ad primam pertinet operationem, quae simplicium apprehensio dicitur. Ideò Aristoteles in contextu 14 et 79 primi libri Posteriorum negat definitionem enunciare, quare si secundam operationem non habet, multo minùs habet tertiam,

that definition is an instrument leading to knowledge of what [something] is.

Now what deceived them was the ambiguity of this word, "instrument." For they did not know in what way definition is an instrument, and which are the instruments the conveying of which the logician is concerned with, and which can be called methods. For definition is an instrument, of course, but nevertheless not [one] for making known the unknown quiddity, but only for signifying [it]. For the definition is not said to make the quiddity of something known by means of discursive movement from known to unknown, but because it signifies it in the same way that a name signifies the thing, by means of a mediating concept, and no discriminating difference can be assigned between these. For if the thing is unknown, then the name too will signify it is as unknown and will not make it known, and it too is thus said to be unknown. Thus when the quiddity is unknown, the definition signifies it as unknown, and accordingly the definition itself is said to be unknown and will not make the thing known. Definition, therefore, is an instrument for signifying the quiddity of something, just as a name is an instrument for signifying the thing accepted confusedly.

But the logical instruments whose construction° is inquired after in logic are not of this type. They are instead instruments for making known the unknown from the known by means of discursive movement from known to unknown and of necessary inference of the latter from the former; this is said to be the third activity of our mind.[60] But the conception of definition pertains to the first activity, which is said to be the apprehension of simples. And so Aristotle in texts no. 14 and no. 79 of the first book of the *Posterior* [*Analytics*][61] denies that a definition forms a proposition. And so if it does not have the second activity, much less does it have the

quae non est sine secunda; non est igitur Logicum instrumentum, omnia namque instrumenta logica sunt discursus à noto ad ignotum.

11   Aliqui verò, qui dicunt aliqua dari instrumenta logica ad primam, non[32] ad tertiam mentis operationem pertinentia et eiusmodi esse ipsam definitionem, quid sit instrumentum logicum penitus ignorare videntur. Cùm enim logica ἀπὸ τοῦ λόγου dicta sit non tanquàm ab oratione, sed tanquàm à ratione et discursu, dicere instrumentum logicum sine discursu, est dicere λογικὸν ἄνευ λογου et calidum sine calore.[33] Logicae igitur[34] artis officium, ut nomen ipsum logicae significat, nullum aliud est quàm docere modum ratiocinandi et transeundi à noto ad ignotum, hic autem discursus vocatur et tertia mentis operatio.

12   Quod ipsa quoque Aristotelis tractatio declarat, de simplicibus enim terminis, nomine et verbo, quibus prima mentis operatio significatur, brevissimè locutus est in logica, neque ut de instrumentis cognoscendi, nihil enim ab Aristotelis mente et à veritate alienius est, quàm aliquod eiusmodi simplex vocare instrumentum logicum; sed ut de partibus, ex quibus enunciatio postea constituenda erat. Nam haec quoque non ut instrumentum cognoscendi consideratur in logica, sed ut principium instrumenti, nempè ut materia, ex qua syllogismus constare debet. Instrumenta verò sciendi illa tantùm sunt, quae sub syllogismo tanquàm sub communi genere continentur. Nec ob aliam causam Aristoteles in Posterioribus Analyticis de methodis atque instrumentis sciendi locuturus constituit Priores Analyticos iis anteponere et in iis agere de syllogismo latissimè sumpto, quàm ut prius de genere omnium methodorum, quàm de speciebus ipsis loqueretur, est enim syllogismus

third, which is not [done] without the second; it is, therefore, not a *logical* instrument. For all logical instruments are discursive movements from known to unknown.

But others, who say that there are some logical instruments ⅠⅠ pertaining to the first, not to the third activity of the mind, and that definition itself, which is a logical instrument, is of this type, appear to be completely ignorant. For since logic is said to be *apo tou logou* (from the word), not in that it is from speech but in that it is from reason and discursive movement, to say "logical instrument without discursive movement" is to say *logikon aneu logou* ("wordy without word") and "warm without warmth." And so the function of the art of logic, as the very name of logic indicates, is nothing other than to teach the way of ratiocinating and of moving on from known to unknown. This, however, is called discursive movement and the third activity of the mind.

Aristotle's treatment itself also makes this clear. For in the logic, ⅠⅠⅠ he spoke most briefly about the simple terms, noun and verb, by which the first activity of the mind is signified, and not about them as instruments for knowing (for nothing is more alien to the mind of Aristotle and to the truth than to call any simple thing of this type a logical instrument) but about them as the parts from which a proposition has to be afterward constituted. For in logic this too is not considered as an instrument of knowing, but as the beginning-principle of an instrument, that is, as the matter out of which a syllogism ought to be composed. Indeed the instruments for knowing scientifically are only those things that are contained under syllogism as under a common genus. Aristotle, planning to speak in the *Posterior Analytics* about the methods and instruments for knowing scientifically, decided to place the *Prior Analytics* before it and in this to deal with syllogism taken in the broadest sense, for no other reason than to speak about the genus of all methods prior to the species themselves. For *syllogism* is the common genus

commune genus et communis forma omnium logicorum instrumentorum, ut antea declaravimus.

13   Hinc fit ut, cùm omne instrumentum logicum sit progressus à
termino ad terminum, hi termini eiusdem ordinis et eiusdem rationis esse debeant, nempè vel ambo in voce vel ambo in re, vel
ambo in conceptu animi sumantur, sic enim dicimur ab uno duci
in cognitionem alterius, qui est verè discursus logicus. At à voce ad
conceptum vel à voce ad rem nullus est logicus discursus, quia
statim sine ullo discursu, sine ullo ratiocinio vox significat rem per
medium conceptum.

14   Definitio igitur, cùm sit vox, non potest eatenus dici methodus
et instrumentum logicum, quatenus significat conceptum quidditatis rei. Et qui hoc dicunt, debeant eadem ratione confiteri omnem orationem et omnem vocem esse methodum et instrumentum logicum, omnis enim oratio et omne nomem aliquam rem
significat. Id tamen asserere absurdissimum est, ut adversarii quoque ipsi concederent. Ideò si adversus eos tali argumento uteremur, quidnam ad illud responsuri essent, non video, eadem est
ratio° definitionis ad quidditatem ab ea significatam et nominis ad
rem, quae ab ipso significatur, ergo si definitio est methodus et instrumentum logicum notificans quidditatem, nomen quoque methodus erit et instrumentum rem ipsam notificans, hoc tamen
nemo assereret, quia est manifestè falsum, ergo neque definitio est
logicum instrumentum, sed instrumentum tantùm significandi.
Omnis enim vox significativa sive simplex sive complexa, instrumentum est ab hominibus inventum ad rem aliquam sive conceptum significandum, in quarum vocum numero est ipsa definitio.
Ideò definiens eam Aristoteles dixit esse orationem significantem
quid est, rectè enim dixit significantem, quia non est instrumentum, nisi significandi. Similiter primae propositiones et dignitates,

and common form of all logical instruments, as we made clear earlier.

And thus it happens that since every logical instrument is a 13 progression from terminus to terminus, these termini ought to be of the same order and of the same [sort of] reasoning, namely either both are taken in words or both in things or both in concepts in our soul. For thus we are said to be led from one into knowledge of another. This truly is logical discursive movement. But from word to concept or from word to thing is no logical discursive movement, because a word at once signifies something by means of a mediating concept, without any discursive movement, without any ratiocination.

Definition, therefore, since it is word, cannot be said to be 14 method and logical instrument, insofar as it signifies the concept of the quiddity of something. And those who say this by the same reasoning ought to confess that all speech and every word is a method and a logical instrument, for all speech and every name signifies some thing. But to assert this is so very absurd, as even our opponents themselves would concede. And so if we used such an argument against them, I do not see how they would respond to it. The relationship° of the definition to the quiddity signified by it is the same as [that] of the name to the thing that is signified by it. Therefore if definition is a method and a logical instrument making the quiddity known, then a name will be a method and an instrument making the thing itself known. But no one would assert this since it is manifestly false. Definition, therefore, is not a logical instrument but only an instrument for signifying. For every significative word, whether simple or compound, is an instrument discovered by men for signifying some thing or concept. Definition itself is to be counted among such words. And so, defining it, Aristotle said that it is speech signifying what [something] is, and he correctly said "signifying," because it is not an instrument except [one] for signifying. Similarly the first premises and axioms,[62]

dum ore proferuntur, sunt voces significantes rem per se notam, nec tamen dicuntur esse logica instrumenta aliquid notificantia, idque[35] si admitteremus, res omnes dicerentur per se ignotae ac demonstrabiles, etiam illae, quae per axiomata significantur, per voces enim earum significatrices notificaremus res omnes, tanquàm ignotum ex noto per logica instrumenta probantes, hoc tamen nemo sanae mentis assereret.

15   Haec nostra sententia absque ullo dubio Aristotelis fuit, ut apertissimè in libro sequente demonstrabimus, in quo tractationem de methodis ab Aristotele in Posterioribus Analyticis factam declaraturi sumus. Nunc satis sit alium locum apud Aristotelem perpendere, qui est contextus quartus, et quintus primi libri de Anima, quaerit ibi Aristoteles an sit aliqua communis methodus investigandi quid est in omnibus substantiis, methodos etiam aliquas nominat, subiungit enim an sit demonstratio an divisio, an aliqua alia methodus. At verò si definitio, vel methodus definitiva esset proprium instrumentum et propria methodus, qua notificatur quid est, vana certè esset quaestio Aristotelis eo in loco, non enim opus erat dubitare et ad alias methodos confugere, cùm in promptu fuisset communis methodus investigandi omnes rerum quidditates, nempè definitio seu methodus definitiva. Igitur vidit Aristoteles definitionem non esse methodum vel instrumentum notificandi quid est, sed esse illud, quod per methodum notificatur, quando latet. Nam quaerere quid res aliqua sit, est definitionem eius quaerere. Proinde definitio quando est ignota, finis est methodorum, non methodus; quando autem nota est, notum est quid res sit, nec methodo eget, qua investigetur. Propterea ibi Aristoteles non ipsam definitionem, sed alias methodos nominat, nempè

when they are advanced orally, are words signifying something known *per se*. But they are, nevertheless, not said to be logical instruments making something known. And if we were to admit this, all things would be said to be unknown *per se* and demonstrable, even those that are signified by means of axioms. For we would make all things known by means of the words capable of signifying them, just like proving the unknown from the known by means of logical instruments. But no one of sound mind would assert this.

This position of ours was without any doubt Aristotle's, as we will demonstrate most plainly in the following book. In it we will make clear the treatment on methods made by Aristotle in the *Posterior Analytics*. For now it is enough to examine another passage in Aristotle, that is, texts no. 4 and no. 5 of the first book of *On the Soul*.[63] Aristotle there inquires whether there is some common method for investigating, in all substances, what [something] is. He even names some methods; for he adds, "whether it is demonstration or division or some other method." But of course if definition, or definitive method, were the proper instrument and proper method by which what [something] is [(its quiddity)] is made known, then certainly Aristotle's question in that passage would be vain. For he did not need to have a doubt and resort to other methods, since the common method for investigating all the quiddities of things, namely, definition or definitive method, would have been right at hand. Aristotle, therefore, saw that definition is not a method or an instrument for making known what [something] is, but is that which, when hidden, is made known by means of method. For to inquire what some thing is, is to inquire after its definition. Accordingly, when the definition is unknown, it is not a method but the end of methods; and when it is known, what the thing is, is known, and there is no need for a method by which it may be investigated. And so Aristotle there names not definition itself but other methods, namely, division according to

15

divisionem iuxta Platonis sententiam et demonstrationem secundùm propriam opinionem, Plato namque definitiones ignotas per divisionem venabatur, Aristoteles verò putavit demonstrationem esse methodum, qua omnes accidentium definitiones investigentur, substantiarum autem definitiones inquiri methodo resolutiva, quam significavit dum dixit, 'vel aliqua alia methodus,' sed de his postea accuratius disseremus.

16 Nunc animadversione dignum est, ne quis ea,quae diximus, perperam accipiat, definitionem investigari, quando est ignota et per logica instrumenta seu logicas methodos inveniri, ideò si eas methodos vellemus[36] à fine nuncupare definitivas, nil sequeretur absurdum, non enim ita definitivam methodum refellere volumus, ut methodum omnem, qua definitio ignota investigetur, de medio tollamus; sed solùm ut ostendamus, definitivam methodum ab aliis distinctam non dari. Cùm enim definitio ignota possit et per demonstrativam et per resolutivam methodum investigari, harum utramlibet si quis habita ratione finis definitivam appellare velit, non prohibemus, dicimus tamen eam, quam ille definitivam vocabit, esse revera demonstrativam, vel resolutivam, aliae namque praeter has duas methodi non dantur.

: XIV :

Quòd processus à partibus definitionis ad totam definitionem non sit methodus definitiva.

1 Considerandum superest an in progressu à partibus definitionis ad totam definitionem methodi definitivae natura sit constituta. Huius membri confutationem sumere possumus ex iis, quae

Plato's position and demonstration according to his own opinion. For Plato searched for unknown definitions by means of division, but Aristotle held that demonstration is the method by which all the definitions of accidents are investigated, and moreover that by resolutive method the definitions of substances are asked about; he indicated this when he said, "or some other method." But we will discuss this more precisely below.

Lest anyone accept what we said wrongly, it is now worth noting that a definition, when it is unknown, is investigated and discovered by means of logical instruments or logical methods. And so if we wanted to call these methods definitive, from [their] end, nothing absurd would follow. We do not want to refute definitive method so as to destroy every method by which an unknown definition is investigated, but only to show that there is no definitive method distinct from the others. For since an unknown definition can be investigated both by means of demonstrative and by means of resolutive method, if anyone wants to call either of them definitive, by taking account of [their] end, we do not object. Nevertheless, we say that what he will call definitive is in truth demonstrative or resolutive. For there are no other methods besides these two. 16

: XIV :

That proceeding from the parts of a definition to the whole definition is not definitive method.

It remains to be considered [(d)] whether the nature of definitive method is constituted in a progression from the parts of a definition to the whole definition. We can get a refutation of this branch [of the division] from the things that were conveyed by 1

traduntur ab Averroe in commentario primo[37] primi libri Posteriorum Analyticorum, ubi ab ipso multa doctè dicuntur ad definitionem et ad id, de quo in praesentia loquimur,[38] pertinentia. Ostendit enim nullum esse posse discursum à partibus definitionis ad totam definitionem tanquàm à noto ad ignotum, ita ut ex partibus praecognitis definitio per necessariam illationem colligatur et innotescat. Id enim si esset, definitiones naturaliter ignotae non egerent syllogismo, quo notificarentur, satis enim ex partibus suis tota definitio nota fieret. Quomodo autem definitio ignota per syllogismum colligatur, accuratè in sequentibus declarabimus.

2   Nunc satis est ad sententiam Averrois declarandam et ad id, quod proposuimus, demonstrandum, si dicamus tum partes definitionis tum etiam totam definitionem respectu rei definitae posse duobus modis considerari, aut enim cognoscuntur solùm ut praedicata illi inexistentia, aut cognoscuntur etiam ut essentialia et praedicata in eo quod quid est. Quatuor igitur membra consideranda nobis proponuntur, ut distinctè omnia declaremus, primum quidem sunt partes definitionis prout praedicantur de definito, idque vocemus A, secundum verò eaedem partes prout sunt partes definitionis et praedicata in eo quod quid est et vocetur B, tertio autem loco tota definitio, non ut definitio, sed ut praedicatum quoddam, quod appellemus C.

3   Demum ipsa definitio prout est eius rei definitio, et ipsius quidditatem significat et vocetur D.

4   Partes quidem non possunt esse cognitae dum definitio est ignota, quia totum non differt à suis partibus simul collectis, eo igitur modo cognita est definitio, vel incognita, quo sunt cognitae, vel incognitae partes, non est igitur cognitum A sine C neque est cognitum B sine D sed cognitio A est cognitio C et cognitio

Averroës in the first commentary to the first book of the *Posterior Analytics*,[64] where many things pertaining to definition, and to what we are at present speaking about, are said by him in a scholarly way. For he shows that there cannot be a discursive movement from the parts of a definition to the whole definition, as from known to unknown, such that the definition is gathered from the parts known beforehand and made known by means of a necessary inference. For if it were, then naturally unknown definitions would not need a syllogism by which they may become known. For the whole definition would become known enough from its parts. But now in what way an unknown definition is gathered by means of a syllogism, we will clarify precisely in the following.

For making Averroës' position clear and demonstrating what we   2
proposed, it is enough if we say that both the parts of a definition and also the whole definition can, with regard to the thing defined, be considered in two ways. For either they are known only as predicated things existing in it, or, instead, they are known as essential and as predicated by virtue of what it [i.e., the thing defined] is. There are set out, therefore, four branches [of a division] to be considered by us, so that we can make everything distinctly clear. First are the parts of the definition in that they are predicated of what is defined; let us call this A. Second then are the same parts in that they are parts of the definition and predicated by virtue of what it [i.e., the thing defined] is; let this be called B. In the third place is the whole definition, not as a definition, but as something predicated; let us call this C.

Lastly is the definition itself in that it is the definition of the   3
thing and signifies its quiddity; let this be called D.

The parts, of course, cannot be known when the definition is   4
unknown, because the whole does not differ from its parts gathered together. The definition, therefore, is known or unknown in the way in which the parts are known or unknown. A, therefore, is not known without C, nor is B known without D. But knowledge

B est cognitio D, qui enim cognoscit partes singulas de illa re praedicari, ille cognoscit totam definitionem de eadem re praedicari,
nondum tamen prout est eius definitio. Qui verò cognoscit partes
singulas de illa re in eo quod quid est praedicari, cognoscit etiam
totam definitionem esse illius rei definitionem. Quare nullus fit
discursus à partibus ad totam tanquàm à notis ad ignotam, partes
enim notas in unum colligere non est per necessariam illationem
procedere à noto ad ignotum, sed ab eodem ad idem et à noto ad
notum, erat enim actu cognita definitio in partibus suis vel ut definitio, si illae sint cognitae ut praedicata in eo quod quid est; vel
ut praedicatum quoddam, si illae sint notae solùm ut praedicata,
non ut praedicata in eo quod quid est.

5   Potest tamen A esse cognitum ante B quare et C ante D, prius
enim cognoscuntur haec praedicata ut illi rei competentia, postea
ut essentialia et eius definitionem constituentia. Qua igitur methodo ducimur à cognitione A ad cognitionem B seu (quod idem
est) à cognitione C ad cognitionem D? Certè nulla, ut Aristoteles
optimè ostendit in secundo libro Posteriorum Analyticorum, nullo
enim syllogismo colligere possumus hoc illi competere in eo quod
quid est, seu esse illius definitionem, nisi petatur illud idem, quod
ostendere volumus, ex eo enim quòd hoc illi inest, possumus quidem colligere aliquid alicui inesse, sed non possumus colligere
quòd insit in quid, nisi in propositionibus assumatur medium
minori inesse in quid, hoc autem est eiusdem rei definitionem ut
notam assumere, cuius quaerimus definitionem, quae est petitio
principii, cùm unius rei una tantùm sit definitio. Postquàm igitur
cognovimus haec de illa re praedicari, nullo syllogismo, proinde
nulla methodo probare possumus quòd praedicentur in quid et

of A is knowledge of C, and knowledge of B is knowledge of D. For whoever knows that each of the parts is predicated of that thing knows that the whole definition is predicated of the same thing, but not yet in that it is its definition. But whoever knows that all of the parts are predicated of that thing, by virtue of what it is, knows also that the whole definition is the definition of that thing. Therefore no discursive movement occurs from parts to whole as from knowns to unknown. For to gather known parts into one is not to proceed by means of a necessary inference from known to unknown, but from one thing to the selfsame thing and from known to known. For the definition was actually known in its parts, either as a definition, if they were known as predicated by virtue of what it [i.e., the thing defined] is; or as something predicated, if they were known only as predicated, not as predicated by virtue of what it [i.e., the thing defined] is.

Nevertheless A can be known before B, and therefore C before D. For these predicates are first known as coinciding with that thing, and afterward as essential and constituting its definition. By what method, therefore, are we led from knowledge of A to knowledge of B or (what is the same) from knowledge of C to knowledge of D? Certainly by none, as Aristotle shows very well in the second book of the *Posterior Analytics*.[65] For by no syllogism can we gather that this coincides with that by virtue of what it is or is its definition, unless the same thing we want to show is presupposed. For from the fact that this belongs to that, we can gather, of course, that something belongs to something else, but we cannot gather that it belongs by what it is, unless it is assumed that in the premises the middle [term] belongs to the minor by what it is. This however is to assume as known the definition of the thing whose definition we inquire after; this is a *petitio principii*, since there is only one definition of one thing. After, therefore, we have come to know that these are predicated of that thing, by no syllogism and accordingly by no method can we prove that they

quòd ex eis definitio illius rei constituatur. Restat itaque ut hoc
per seipsum nobis considerantibus innotescat, quando per aliud
ostendi non potest.

6   Neque res haec dubia videri debet, quandoquidem in sensu
quoque hoc idem evenire manifestum est, aliqua enim primo aspectu non videntur, quae postea, dum magis oculorum aciem in-
tendimus, intuemur. Pruna quoque cineribus abscondita non cer-
nitur, postea verò motione et tractatione° aliqua cinerum seu
ipsius prunae facta inspicitur et per se ipsa visum movet. Simile
quiddam menti nostrae contingere non est dubitandum, aliqua
enim prima ipsa mentis inspectione non apprehenduntur, quae
postea per diligentem ipsorummet considerationem nota fiunt,
proinde non per aliud, sed per seipsa notificantur. Quòd igitur
quae alicui rei competere novimus, sint ei essentialia et ei competant in eo quod quid est, non per aliud ostendimus, sed res ipsa
per se in lucem prodit, dum in eam oculos mentis efficaciter infigi-
mus. Talia namque attributa benè considerando et cum ipsa re
conferendo cognoscimus ei essentialia esse et ex eis definitionem
constituimus.

7   Hoc evenire solet in illa inductione, quam Averroes vocat demonstrativam, in qua non omnia particularia sumuntur, quia dum
aliqua pauca enumerare incipimus, statim apparet praedicatum illis essentiale esse, ideò dimissa reliquorum enumeratione colligimus universale, quod enim per se praedicatur, de omni praedicatur. Connexio igitur essentialis ipsa per se, non per aliud, menti
rem consideranti elucescit. Hinc fit ut per divisionem innotescere
partes definitionis videantur, quasi prius ignotae fuerint, non est

are predicated by what it is and that the definition of that thing is constituted from them. And so it remains that, by our considering [it], this becomes known by means of that thing itself—it cannot be shown by means of something else.

Nor ought this to appear doubtful, since it is manifest that this 6 same thing also happens in sense. For some things do not appear on first glance that we see afterward when we exert the eyes' acuity more. And hidden by ashes, a burning coal is not discerned, but then afterward, after there has been some moving and handling° of the ashes or the burning coal itself, it is beheld and moves the sight by means of that [coal] itself. It is not to be doubted that something similar happens with our mind. For some things that become known afterward by means of a careful consideration of them are not apprehended by the mind on the very first inspection. And so they are made known not by means of another, but by means of themselves. Therefore that those things that we knew appertain to some thing are essential to it and appertain to it by virtue of what it is — we do not show this by means of another; instead, the thing itself comes to light *per se* (by means of itself), when we effectually fix the eyes of the mind on it. For by considering such attributes well and comparing them with the thing itself, we come to know that they are essential to it and we constitute the definition from them.

This normally happens in the induction that Averroës calls de- 7 monstrative, in which not all the particulars are taken in, because when we start to enumerate a few, it is at once apparent that the predicate is essential to them. And so, an enumeration of the others being put aside, we gather the universal, for what is predicated *per se* is predicated of all (*de omni*). In a mind considering something, therefore, the essential connection shines forth by means of the thing itself, not by means of something else. And thus it happens that the parts of a definition appear to become known by means of division, as if they were first unknown. But it is not

tamen divisio, quae nobis eam cognitionem praestet, sed quia dividendo per praedicata essentialia transimus, proinde datur nobis occasio ea mente tractandi et considerandi, ideò facilè nobis innotescit ea essentialia esse, atque in definitione illius rei sumenda.[39] Nota quidem etiam ante divisionem erant, non tamen ut essentialia. Hoc autem sine discursu nobis manifestum fit, dum ea sive per divisionem sive etiam absque divisione intentè consideramus.

8   Non est igitur methodus definitiva processus ille ab A ad B sive à C ad D quia ubi res per se ipsa cognoscitur, ibi nulla illatio fieri dicitur. Sed est processus à cognitione eiusdem rei imperfecta ad perfectam per eiusmet diligentem inspectionem. Propterea rectè dicit Averroes tales definitiones modica egere declaratione sive per divisionem sive per inductionem sive aliquo alio modo quo in ipsarum consideratione mentis aciem intendamus, sic enim fit processus ab A ad B et à C ad D.

9   Hoc igitur modo fit notum B et D quando prius cognoscebatur A et C.

10   Sed quando A quoque ignoratur, tunc utique aliqua methodo id indagare oportet, quando enim partes definitionis aut omnes aut aliquas non modò quòd sint partes definitionis, sed etiam quòd omnino inexistant huic rei ignoramus, syllogismo opus est ad eas colligendas, in hoc enim tota syllogismi efficacitas posita est ut probet hoc huic inesse eò quòd illud illi inest, semperque τὸ ὅτι ostendat, sed non τὸ τί ἐστὶ. Ad has igitur definitiones extrahendas divisio est prorsus inutilis, quia non potest ostendere quòd hoc illi insit, ut antea demonstravimus. Sed ad resolutivam methodum confugere oportet, quod quomodo fiat, in sequenti libro declarabimus. Nunc satis sit ostendisse nullam dari methodum definitivam

division that really gets us this knowledge; instead [we get it] because in dividing we pass through the essential predicates, and so there is the occasion for us to treat and consider them in the mind. Therefore it easily becomes known to us that they are essential and have to be taken into the definition of that thing. They were known before the division, of course, but not as essential. But this becomes manifest to us without discursive movement when we consider them either by means of division or even intently without division.

This proceeding from A to B or from C to D, therefore, is not 8 definitive method, because where something is known by means of that thing itself, no inference is said to occur there. But it is a proceeding from imperfect to perfect knowledge of the same thing by means of a careful inspection of it. Averroës, therefore, correctly says that such definitions need a modicum of clarification, either by means of division or by means of induction or in some other way by which we may exert the mind's acuity in consideration of them. And thus proceeding occurs from A to B and from C to D.

In this way, therefore, B and D become known when A and C 9 were known first.

But when A is also unknown, then by all means it must be 10 tracked down using some method. For when we do not know either all or some parts of a definition, not only that they are parts of the definition but also that they exist in this thing, there is need of a syllogism to gather them. For the whole efficacy of the syllogism is located in this, that it proves that this belongs to this, because that belongs to that; and it always shows *to hoti* (the that-it-is-the-case), but not *to ti esti* (the what-it-is). For extracting these definitions, therefore, division is utterly useless, because it cannot show that this belongs to that, as we demonstrated earlier. Instead, one must resort to resolutive method. In the following book, we will make clear how this happens. For now let it be enough to

à partibus definitionis ad totam definitionem; quia si ignotum sit
A et C non possumus illud ostendere nisi methodo vel demonstrativa vel resolutiva, utraque enim vim habet syllogisticam et
potest ostendere hoc illi inesse. Si verò sit notum A et C nulla
potest esse illationis necessitas in progrediendo ad B et D, si enim
eò quòd inest aliquid alicui, ac de eo praedicatur, liceret colligere,
ergo praedicatur in eo quod quid est, eadem ratione dicere possemus, albedo inest homini, ergo inest in eo quod quid est. Id igitur
non per aliud, sed per seipsum innotescit, ut dictum est, quare
nulla datur methodus definitiva, quoniam ipsum quid est non potest ullo instrumento formaliter notificari, sed facta aliorum quaesitorum notificatione in lucem prodit per seipsum, ut dictum est.[40]

: XV :

*Solutio obiectionis ex primo capite
sexti libri Topicorum desumptae.*

1 Nonnulli eam sententiam tueri volentes quòd definitio sit methodus et sciendi instrumentum, existimarunt validum pro ea argumentum ex Aristotele sumi in primo capite sexti libri Topicorum,
ubi dicit omnem definitionem assignari rei cognoscendae gratia,
propterea ex prioribus et notioribus assignandam esse, quemadmodum in demonstrationibus et in omni doctrina ac disciplina
res sese habet. Fatetur itaque Aristoteles, definitionem esse
instrumentum cognoscendi, cùm per eam res cognoscatur; et

have shown that there is no definitive method from the parts of a definition to the whole definition, because if A and C are unknown, we cannot show that, except by demonstrative or resolutive method: each has syllogistic power and can show that this belongs to that. But if A and C are known, there can be no necessity of inference in progressing to B and D. For if because something belongs to something else and is predicated of it, if it is permitted to gather [the definition], it is, therefore, predicated by virtue of what it [i.e., the thing defined] is. By the same reasoning, we could say, "Whiteness belongs to man; therefore it belongs to him by virtue of what he is." It therefore becomes known by means of man himself, not by means of another, as was said. And so there is no definitive method, since what [something] is itself, cannot formally be made known with any instrument; it is instead brought to light by means of that thing itself, once there has been a making known of other things inquired after, as was said.

Solution to an objection drawn from the first
chapter of the sixth book of the Topics.

Some, wanting to uphold the position that definition is a method 1
and an instrument for knowing scientifically, judged that a valid argument in favor of this could be taken from Aristotle in the first chapter of the sixth book of the *Topics*,[66] where he says that every definition is assigned for the sake of knowing something; and so it has to be assigned from things prior and more known, just as it is in demonstrations and in every teaching and discipline. Thus [according to them] Aristotle acknowledges that definition is an instrument for knowing, since by means of it something is known,

procedere à notioribus ad ignotiora et talem esse methodum, qualis est demonstratio.

2   Qui argumentum hoc efficax esse putant, si id respicerent, quod maximè respiciendum est, utique cognoscerent, ipsum ita debile esse, ut nullam nobis difficultatem facere possit. Nos enim fortasse satis superque difficultatem hanc solveremus dicentes rationem hanc ex Topicis libris esse desumptam, quandoquidem Aristoteles aliter ut Dialecticus, aliter ut demonstrativus de definitione loqueretur. Nam Dialecticus definitionem sumit ut orationem quandam, quae de definito verè praedicatur et ei aequalis est, quare eam non considerat prout idem est, quod res ipsa definita, idem enim de seipso non rectè praedicaretur. At demonstrator, qui res potius, quàm voces respicit, sumit definitionem ut orationem significantem illam ipsam rem, quam etiam nomen significat, propterea definitionem nunquàm enunciat de re definita, dum eam sumit ut definitio est; et dicit, eam non significare quòd sit aliquid vel non sit, sed solùm quid sit.

3   Maximè igitur Dialectico convenit dicere definitionem ex causis rei esse, proinde ex prioribus et notioribus secundùm naturam, causa enim, quatenus est causa, non est idem quod effectus; neque idem est natura prius ac notius seipso. Ideò in secundo libro Posteriorum Analyticorum, cùm ostendisset Aristoteles in 37 contextu quòd definitio ut definitio est non potest de definito per demonstrationem concludi, quia oportet eiusdem rei definitionem in propositionibus assumi, cuius definitio quaeritur, quae est petitio principii; postea in 38 eiusmodi syllogismum admittit ut Dialecticum, quem ut demonstrativum refutaverat et vitiosum esse dixerat. Nam, ut ibi notat Averroes, apud Dialecticum non est inconveniens

and that it proceeds from [things] more known to [things] more unknown and is such a method as demonstration is.

If those who hold that this argument is effectual were to regard 2 what especially has to be regarded, then by all means they would come to know that this [argument] is so weak that it can give us no problem. Now perhaps we would do away with this problem more than enough, saying that this reasoning is drawn from the books of the *Topics,* seeing that Aristotle spoke about definition sometimes as a dialectician and sometimes as one being demonstrative. For the dialectician takes definition as some sort of speech that truly is predicated of what is defined and is equal in total extent to it; accordingly he does not consider it in that it is the same as the defined thing itself. For the selfsame thing is not correctly predicated of that thing itself. But a demonstrator, who gives regard to things rather than to words, takes definition as speech signifying that thing itself that the name also signifies. Therefore he never takes as a proposition a definition of something being defined when he takes it as it is a definition, and he says that it does not signify that it is or is not something, but only what it is.

It is, therefore, most appropriate for the dialectician to say that 3 definition is from the causes of the thing, and accordingly from things prior and more known according to nature. For a cause insofar as it is a cause is not the same as an effect, and the selfsame thing is not by nature prior to and more known than itself. And so in the second book of the *Posterior Analytics,* although Aristotle had shown in text no. 37[67] that the definition, as it is a definition, cannot by means of a demonstration be a conclusion about what is being defined—because then the definition of that same thing must be assumed in the premises whose definition is inquired after, which is a *petitio principii*—afterward, in no. 38,[68] as a dialectician, he admits a syllogism of the type that, as demonstrative, he had confuted and said was flawed. For, as Averroës notes there, for the dialectician, it is not inconsistent if two definitions of the same

quòd eiusdem rei duae ponantur definitiones et altera per alteram ostendatur sine petitione principii. Huius autem ratio ea est, quam diximus, quia Dialecticus considerat definitionem in enunciatione positam et praedicatam de re definita, quare non ut idem, quod illa, hoc cùm ita sit, quid mirum si Aristoteles in Topicis diceret in definitione processum quendam fieri à noto ad ignotum notificandum?

4   Veruntamen hoc praetereamus et propositi loci sententiam expendamus et videamus an nobis officiat necne.

5   In primis dicere debent adversarii quid putent se hoc argumento demonstrasse, aut enim arbitrantur Aristotelem constituere methodum ipsam definitivam in progressu à partibus definitionis ad totam definitionem, aut in progressu à definitione ad rem definitam declarandam. Illud quidem dici non potest, quia Aristoteles non dicit partes definitionis esse priores et notiores tota ipsa definitione, sed ipsa re definienda, hanc enim cognosci dicit per definitionem, sed non definitionem ipsam per cognitionem partium. Pro eodem igitur habet totam definitionem et eius partes et non minùs totam, quàm partes vult esse priores et notiores ipso definito, quod per definitionem videtur innotescere.

6   Hunc igitur sensum cùm Aristotelis verba non recipiant, an alterum habere possint consideremus, is enim fortasse nobis non adversatur neque absurdus est, si sano modo intelligatur. Certum est aliud esse rem aliquam per aliam rem notiorem notificari, aliud esse nomen per alterum nomen declarare. Res enim ignota ex alia re nota notificari non potest, nisi per methodum, proinde cum illationis necessitate, ut quando ex fumo notificatur ignis et ex

thing are posited and one is shown by means of the other without a *petitio principii*. And the reason for this is what we said, that the dialectician considers the definition posited in a proposition and predicated of the thing defined, and so not as the same as that [thing]. Since this is so, is it any wonder if in the *Topics* Aristotle says that in definition some sort of proceeding occurs from the known to making known something unknown?

Nevertheless, let us pass over this and weigh the position in the given passage and see whether it is detrimental to us or not.   4

In the first place, our opponents ought to say what they hold that they have demonstrated by this argument. For they think that Aristotle constitutes definitive method itself either to be in a progression from the parts of a definition to the whole definition or in a progression from the definition to clarifying the thing defined. This cannot be said, of course, because Aristotle does not say that the parts of the definition are prior and more known than the whole definition itself but than the very thing being defined. For he says that this is known by means of the definition, but not that the definition itself [is known] by means of knowledge of the parts. He has, therefore, the whole definition and its parts as the same, and he wants the whole no less than the parts to be prior and more known than the defined thing itself, which appears to become known by means of the definition.   5

Since, therefore, Aristotle's words are not receptive of this sense, let us consider whether they could have another. For perhaps it is not opposed to us and is not absurd, if it is understood in a sound way. It is certain that for some thing to be made known by means of some thing else more known is one thing; to clarify a name by means of another name is something else. For an unknown thing cannot be made known from another, known thing except by means of method and accordingly with the necessity of inference, as [for example] when fire becomes known from smoke and a   6

obiectione terrae eclipsis lunae. At nomen per aliud nomen vel per alia nomina declaratur et notificatur absque ullo discursu, absque illatione, absque methodo. Si quis enim petat, quid est merum? et alter respondeat, est vinum, notificat quidem significationem meri per significationem vini notiorem, sed nulla tamen utitur methodo. Methodus enim est via à re nota ad cognitionem rei ignotae, hic autem est processus à nomine ad nomen, non à re ad rem. Nomina autem ex arbitrio humano posita sunt, quare in hoc processu nulla fit probatio rei ignotae.

7   Idem penitus de notificatione definitionis dicendum est, quoniam enim idem est homo atque hominis quidditas, ideò nomen hoc, homo, non modò rem ipsam, sed et eius quidditatem significat, confusè tamen et obscurè. Ideò is, qui petit quid homo sit, petit sibi declarari significationem illius vocis per aliam vocem vel per alias voces notiores, quae clarius illam quidditatem significent. Definitio igitur constat ex partibus notioribus nomine ipso rei definitae, proinde notificat et declarat significationem illius nominis. Definitio quidem, quae nominalis dicitur, notificat significationem nominis confusam, quando ignoratur etiam quid eo nomine confusè significetur. Definitio autem essentialis declarat etiam distinctè significationem nominis, quae solùm confusè cognita erat.

8   Ad argumentum igitur respondendum est Aristotelem dicentem partes definitionis esse notiores et priores definito et eius cognitionem facere, aut intellexisse processum hunc tanquàm à re nota ad rem ignotam, quasi definitio significet quoddam distinctum à re definita; aut tanquàm ab eadem re ad eandem rem, sed à vocibus

lunar eclipse from obstruction of the earth. But a name is made clear by means of another name or by means of other names and is made known without any discursive movement, without inference, without method. For if someone asks, "What is *merum?*" and another responds, "It is *vinum* (wine)," he makes known the signification of *merum* by means of the signification of the more known [word] *vinum*, of course, but nevertheless he uses no method. For method is a way from something known to knowledge of something unknown. But this proceeding is from name to name, not from thing to thing. Names, however, are posited by human choice, and so in this procedure no proof of an unknown thing occurs.

What is completely the same has to be said about the making known of a definition. For since man and the quiddity of man are the same, this name, "man," signifies not only the thing itself but also its quiddity, though confusedly and obscurely. And so he who asks what man is asks that the signification of that word be made clear to him by means of another word or by means of other words more known that signify that quiddity more clearly. A definition, therefore, is composed out of parts more known than the name itself of the thing defined; accordingly, it makes the signification of that name known and clear. A definition, of course, that is said to be nominal makes known the confused signification of the name, when what is confusedly signified by the name is still unknown. But an essential definition also makes distinctly clear the signification of a name that had been known only confusedly.

To the argument, therefore, it has to be responded that Aristotle, saying that the parts of a definition are more known and prior to what is defined and [also] bring about knowledge of it, understood this proceeding either as from something known to something unknown, as if the definition signifies something distinct from the thing defined, or as from one thing to the selfsame

notioris significationis ad declarandam significationem ipsius nominis ignotiorem. Illud quidem si intellexit, Dialecticè locutus est
et eius dictum ut Dialectici admittendum est, alioqui falsum et
reiiciendum.

9   Si autem posteriorem sensum significare voluit, is omnino verus
est, sed nobis non obest, est enim processus à vocibus ad vocem,
qui solam notificat vocis significationem et nihil ignotum probat,
haec autem notificatio nil aliud est, quàm significatio, de qua superius loquebamur. Nam vera notificatio, quae non fit nisi per methodum, ea[41] est, quae rei ignotae per aliam rem notiorem fit.
Haec autem est veluti rei per seipsam notificatio, quae simplici
quadam expressione et sine ulla methodo fit, nam si partes definitionis ad ipsam quidditatem referamus, illam potius significant,
quàm notificent, sicut antea demonstravimus. Si verò ad ipsum
nomen rei definitae referantur, notificant quidem eius significationem, sed adhuc simpliciter et absque methodo, nisi methodus
dicatur etiam illa notificatio, qua nomen illud, merum, per vinum
declaratur. At certè à nomine ad nomen non est methodus, quia
nulla fit illatio huius ex illo et nominis declaratio pro probatione
nunquàm habetur, sed potius pro principio. Ideò etiam ille, qui
rem per suam definitionem declarat, in principiis adhuc versatur et
nondum aliquid ignotum ex principiis notis deducit.

10   Quomodo igitur definitio ex notioribus sive ex praecognitis sit,
manifestum est, ex notioribus enim est, ex quibus nulla fit illatio;
et per definitionem doctrina ac disciplina non fit per medium,

thing, but from words of more known signification to clarifying the more unknown signification of the name itself. If, of course, the former is what he understood, then he spoke as a dialectician and his dictum has to be admitted as that of a dialectician; otherwise, it is false and has to be rejected.

If, however, he wanted to signify the latter sense, then it is altogether true but is not an obstacle to us. For the proceeding is from words to a word; this makes known only the signification of a word and proves nothing unknown. But this making known is nothing other than signification, which we were speaking about above. For the true making known, which does not happen except by means of method, is that [i.e., the making known] which happens of something unknown by means of something else more known. This, however, is as it were the making known of something by means of that thing itself, which happens by some sort of simple expression and without any method. For if we refer the parts of a definition to the quiddity itself, then they signify it rather than make it known, as we demonstrated earlier. But if they are referred to the very name of the thing defined, they make known its signification, of course, but still absolutely and without method, unless even that making known by which the name *merum* is clarified by means of *vinum* is said to be a method. But certainly method is not from name to name, because no inference of the latter from the former occurs, and clarification of a name is never held as a proof but rather as a beginning-principle. And so even he, therefore, who makes something clear by means of its definition is still concerned with beginning-principles and he does not yet deduce something unknown from known beginning-principles.

It is manifest, therefore, in what way definition is from things more known or from things known beforehand, for it is from things more known, from which no inference occurs; and the teaching and learning of a discipline by means of definition does

sed res potius per se ipsam, quàm per aliud discitur. Quòd enim sine medio aliqua discamus, quae prius ignorabamus, testatur Aristoteles in primo capite primi libri Posteriorum Analyticorum. Imò etiam sine ipsius testimonio manifestum est, tum in iis, quae sensu, tum etiam in iis, quae mente, per seipsa discuntur.

11   Haec est Aristotelis sententia in illo primo capite sexti libri Topicorum, necnon in contextu 48 primi Metaphysicorum.

12   Quòd autem utroque in loco definitionem cum demonstratione comparet dicens definitionem ex notioribus esse, sicut etiam demonstratio, id nobis non officit, non enim vult Aristoteles eodem modo definitionem ac demonstrationem ex notioribus esse, nimirum cum illatione ignoti ex notis, id namque si ipse assereret, eius sententiam deserere ac refutare non vereremur.

13   Sed solùm in hoc communi vult earum similitudinem consistere, quòd utraque ex notioribus constat, alio tamen et alio modo. Demonstratio enim est ex notioribus, quoad esse, definitio verò est ex notioribus, non quoad esse, sed solum quoad significationem.[42] Quo fit ut altera quidem sit methodus et instrumentum sciendi, altera verò nequaquàm, eamque Aristotelis sententiam fuisse nos in libro sequente apertissimè demonstrabimus.

not happen by means of a middle [term]; instead, something is learned by means of that thing itself rather than by means of something else. For that we learn without a middle [term] some things that we earlier did not know, Aristotle attests in the first chapter of the first book of the *Posterior Analytics*.[69] But even without his testimony, it is manifest in those things that by sense and also those things that by the mind are learned by means of the things themselves.

This is Aristotle's position in that first chapter of the sixth book 11 of the *Topics*[70] and also in text no. 48 of the first [book] of the *Metaphysics*.[71]

Moreover, that in both passages he compares definition with 12 demonstration, saying that definition is from things more known, as demonstration is too, is not detrimental to us. For Aristotle does not want definition and demonstration to be from things more known in the same way, namely, with inference of unknown from knowns. And if he asserted this, we would not be afraid to abandon and confute his position.

But he wants their resemblance to consist only in what is com- 13 mon — that each is composed out of what is better known, but one in one way and one in another. For demonstration is from things more known, with respect to being, but definition is from things more known not with respect to being but only with respect to signification. Hence one is indeed a method and an instrument for knowing scientifically, and the other is not at all. And that this was Aristotle's position, we will demonstrate most plainly in the following book.

: XVI :

In quo aliorum sententia confutatur de methodo resolutiva.

1   Divisiva ac definitiva methodis refutatis duae relinquuntur, quas nos antea posuimus, demonstrativa et resolutiva. Methodum tamen resolutivam alii non rectè accepisse videntur, ideò non erit ab re si eorum de hac methodo sententiam breviter expendamus.

2   Duo resolutionis modi ab aliis statuuntur, unus ab Ammonio in suis commentariis in prooemium Porphyrii, alter ab Eustratio in praefatione sua in secundum librum Posteriorum Analyticorum. Posteriores verò utrumque modum recipientes in has duas species methodum resolutivam dividendam esse censuerunt.

3   Ammonius inquit methodum resolutivam esse quando hominem in caput, brachia, pedes et alia membra dissolvimus, haec rursus in partes homogeneas, carnem, ossa, nervos, deinde harum singulam in quatuor elementa et haec demum in materiam et formam. Posteriores verò hanc vocant resolutionem à notione finis; homine enim proposito et eius operationibus et officiis consideratis colligimus eorum gratia fuisse haec membra homini necessaria. Quare per notionem finis hominem in membra et eadem ratione haec in humores et homogeneas partes resolvimus et ita deinceps.

4   Eustratius verò nullam ponit aliam resolutionem, quàm illam, quae est ab individuis ad infimas species, deinde ad genera proxima, mox ad remotiora, donec tandem ad summum genus pervenerimus. Quam quidem resolutionis speciem constat esse directè contrariam divisioni; in eadem enim categoria à summo ad ima descendendo divisionem facimus, ab imis verò ad summum

: XVI :

In which others' position about resolutive method is confuted.

Divisive and definitive methods having been confuted, there re-　1
main the two that we posited earlier: demonstrative and resolutive.
Nevertheless, others appear to have accepted resolutive method
incorrectly. And so it will not be beside the point, if we briefly
weigh their position on this method.

Two types of resolution are set down by them: one from Am-　2
monius in his commentaries on the proem to Porphyry,[72] the
other from Eustratius in his preface to the second book of the
Posterior Analytics.[73] And those who came later, taking up both
types, deemed it that resolutive method has to be divided into
these two species.

Ammonius says that resolutive method is when we resolve man　3
into head, arms, feet, and other members, and these in turn into
homogeneous parts — flesh, bones, nerves — and then each of these
into the four elements, and lastly these into matter and form. And
those who came later call this resolution from a notion of the end.
For once [the nature of] man has been set out and his activities
and [bodily] functions considered, we gather that these members,
necessary to man, were for the sake of these [i.e., the activities and
functions]. Therefore by means of a notion of the end, we resolve
man into members and in the same way these [members] into the
humors and homogeneous parts, and so on.

Eustratius, on the other hand, posits no resolution other than　4
that which is from individuals to the lowest species, then to proximate genera, next to more remote, until finally we have arrived at
the highest genus. It is evident that this species of resolution, of
course, is directly contrary to division. For in the same category we
make a division by descending from the highest to the lowest and

ascendendo resolutionem. Utilem autem esse ait methodum hanc ad definitiones venandas, quoniam enim definitiones ex genere et differentiis constituuntur, per quae et dividendo et resolvendo transimus, ideò ad eas indagandas modò divisione, modò resolutione utimur.

5   Nonnulli verò finem omnis resolutionis eundem fermè statuunt, quem et divisionis, nempè numerum cognoscere, nam (dicunt) divisione numerum inferiorum, resolutione verò numerum superiorum ac priorum investigamus. Idcirco differentiam hanc in ipsius methodi resolutivae definitione expresserunt.

6   Alia quoque multa ab aliis de hac re dicuntur, quae consultò missa facimus.

7   Nos verò duas quidem resolutionis species his similes in sequentibus statuemus, tamen neutram ab aliis benè intellectam fuisse arbitramur. Cùm enim quatuor sint, quae hac in re in considerationem veniunt, terminus, à quo, terminus, ad quem, via ab illo ad hunc ac demum finis atque utilitas huius viae, ipsi in his duobus posterioribus maximè hallucinati sunt, praesertim in priore illa resolutionis specie, quae principem locum tenet. Quando enim resolutionem compositi faciunt in partes essentiales, petendum est ab eis an partes illae ante resolutionem sint notae an ignotae. Si notae, nulla opus est resolutione, siquidem finis huius resolutionis est partium seu principiorum inventio, ut mox ostendemus. Si verò sunt ignotae, via haec nihil habet efficacitatis ad eas notificandas, quod in eo ipso exemplo declarari potest, hominem enim in membra resolvimus et haec in carnem, ossa, nervos, quia hae omnes partes sensiles sunt et per se notae, quare non ex ipso homine concreto partium harum cognitionem consequimur tanquàm à noto ad ignoti notitiam° procedentes, sed sensu partes illae omnes

a resolution by ascending from the lowest to the highest. He says, moreover, that this method is useful in searching for definitions, since indeed definitions are constituted from genus and differentiae, through which we pass both by dividing and resolving, and so we use now division, now resolution for tracking them down.

But some think that the end of every resolution is nearly the 5 same as that of every division, namely, to know the number. For (they say) we investigate the number of the lower by division and the number of the higher and the prior by resolution. And so they expressed this difference in the definition of resolutive method itself.

Many other things are said on this matter by others; we inten- 6 tionally set these aside.

Now in what follows, we will ourselves lay down the two spe- 7 cies of resolution similar to these, but we think neither was well understood by others. Although there are four things that come into consideration in this issue—terminus *a quo* (from which), terminus *ad quem* (to which), the way from the former to the latter, and lastly the end and utility of this way—those [others] were most deceived in the latter two, especially in the first species of resolution; and it holds the foremost place. For when they perform resolution of a composite into essential parts, it has to be asked of them whether those parts are known or unknown before the resolution. If known, no resolution is needed, since indeed the end of this resolution is discovery of parts or beginning-principles, as we will soon show. And if they are unknown, this way has no efficacy for making them known; this can be made clear in the example itself. For we resolve man into members and these into flesh, bones, nerves, because all these parts are sensible and known *per se*. And so we do not gain knowledge of these parts from a concrete man himself, as proceeding from the known to knowledge° of the unknown; instead all those parts are known by sense. But

cognoscuntur, sensus autem illa, quae percipit, statim et sine medio percipit, ergo nulla ibi methodus est, cuius beneficio ignotum ex noto notificetur. Idque adhuc clarius est ubi partes per se ignotae et insensiles sunt, ut prima materia et forma, in quas omne corpus naturale resolvitur, ex ipso enim naturali corpore in materiae primae ac formae notitiam° non ducimur, nec dicere licet, ignis ibi est, ergo prima materia, dum ipsam materiam primam ignoramus. Ipsum igitur compositum non sufficit[43] ad partes suas essentiales notificandas, quando absconditae et insensiles sunt.

8   Huiusce autem rei ratio ex iis, quae antea diximus, colligitur, nulla enim via est, quae ex noto faciat rei ignotae cognitionem, nisi via syllogistica, ut Aristoteles clamat in calce secundi libri Priorum Analyticorum. Syllogismus autem omnis ex tribus terminis constat, ex duobus nullus fit syllogismus, quando igitur corpus naturale in partes essentiales resolvendum proponitur, duos tantùm terminos habemus, corpus ipsum notum et partes ignotas, ergo in his nulla potest illatio fieri, cùm tertius terminus desit, scilicet terminus medius, nam duo illi, quos habemus, solam conclusionem constituunt, quae est, omne corpus naturale ex materia et forma constat. Ideò qui dicunt ex composito noto nos duci per methodum resolutivam ad primae materiae ignotae cognitionem, dicunt conclusionem naturaliter ignotam posse seipsam notificare et solum subiectum propositionis posse notificare praedicatum quodcumque ipsi attribuatur, quae quidem falsa et absurda sunt. Medius igitur terminus praeter hos duos accipiendus est, nempè aliquod accidens in ipso corpore naturali causas illas consequens, ut generatio et interitus: hoc enim est medium idoneum ad partes illas essentiales notificandas, quo usus est Aristoteles ad primam materiam in omni corpore naturali demonstrandam.

now sense grasps those things that it grasps at once and without a middle [term]; there is, therefore, no method there thanks to which the unknown is made known from the known. And this is still clearer where the parts are unknown *per se* and insensible, as [are] first matter and form, into which every natural body is resolved. For we are not led from natural body itself into knowledge° of first matter and form, nor may it be said, "There is fire there, therefore there is first matter," when we are ignorant of first matter itself. The composite itself, therefore, is not sufficient to make its own essential parts known, when they are hidden and insensible.

The reason for this is gathered from the things that we said 8 earlier. For there is no way except the syllogistic way to bring about knowledge of something unknown from the known, as Aristotle claims at the end of the second book of the *Prior Analytics*.[74] Moreover, every syllogism is composed out of three terms; out of two, no syllogism occurs. When natural body, therefore, is set out to be resolved into essential parts, we have only two terms, the known body itself and the unknown parts; no inference, therefore, can occur in these, since a third term, that is, a middle term, is absent. For those two that we have establish only the conclusion, that is, that every natural body is composed out of matter and form. And so whoever says that we are led by means of resolutive method from a known composite to knowledge of unknown first matter is saying that a naturally unknown conclusion can make itself known and a premise's subject alone can make known whatever predicate is attributed to it; this, of course, is false and absurd. A middle term besides those two, therefore — namely, some accident in natural body itself ensuing from those causes, such as generation and passing away — has to be accepted. For this is a middle [term] fit for making known those essential parts; Aristotle used it for demonstrating first matter in every natural body.

9   Hoc modo sumenda est methodus resolutiva et sic eam intellexit Averroes interpretans tertium contextum primi libri Physicorum, cùm enim Aristoteles ibi dicat ad primorum principiorum
cognitionem progrediendum esse à confusis ut notioribus, confusa
intelligens ipsa naturalia corpora, quae composita nunc à philoso-
phis nostris appellantur, Averroes ea[44] verba declarans inquit, 'pos-
sibile est[45] ex rebus compositis, id est ex consequentibus earum
cognoscere causas,' deinde in commentario quinto clarius eundem
sensum referens dicit, 'species compositae apud nos sunt notiores
suis causis et ex istis speciebus procedimus ad cognitionem causarum mediantibus accidentibus existentibus in eis,' quae verba profert Averroes ad declaranda ea, quae dixerat Aristoteles in contextu
tertio. Ideò peccant ii, qui viam illam doctrinae resolutivam intelligunt à compositis ad principia absque accidentium consideratione.
Peccant etiam illi, qui eo in loco confusa intelligunt sola accidentia,
non corpora ipsa composita. Ipsa enim corpora propriè dicuntur
confusa, ut ex diversarum partium confusione constantia. Accidentia verò possunt quidem dici confusè cognita, quando ipsorum
causae ignorantur, at, cùm simplicia sint, non rectè confusa appellantur. Sed per se clara res est, ad invenienda principia opus esse
utraque praecognitione, scilicet, et subiecti et medii, Aristoteles
enim demonstrat in corpore naturali materiam primam inesse ex
generatione et interitu tanquàm ex medio nobis conspicuo. À
compositis igitur ad principia invenienda per resolutionem procedimus per accidentia media, quae in ipsis sunt. Quemadmodum
enim qui lignum aliquod in partes secare vult, eget gladio vel alio
ferreo instrumento, quo ipsum secet; ita humana mens volens

Resolutive method has to be taken in this way, and this is how   9
Averroës, commenting on text no. 3 of the first book of the *Physics*,[75] understood it. For although Aristotle there says that progressing has to be to knowledge of first beginning-principles from
the confused, as more known — understanding "confused" as the
natural bodies themselves, what are now called composite by our
philosophers — Averroës, clarifying these words, says, "It is possible to know causes from composite things, that is, from their
consequences." And then, recounting the same sense more clearly,
he says in the fifth commentary,[76] "In us, composite species are
more known than their causes, and we proceed from these species
to knowledge of the causes by the mediating accidents existing in
them." Averroës advances these words to clarify that which Aristotle had said in text no. 3. And so those who understand that the
way of teaching from composites to beginning-principles without
consideration of accidents is [the] resolutive [way] are mistaken.
Also mistaken are those who understand confused in that passage
as only the accidents and not the composite bodies themselves. For
the bodies themselves are properly said to be confused, as they are
composed from a confusion of different parts. Of course, accidents
can be said to be known confusedly when their causes are unknown, but since they are simple, they are not correctly called
confused. But it is *per se* clear that for discovering beginning-
principles, each [kind of] prior knowledge is needed, namely, both
of the subject and of the middle [term]. For Aristotle demonstrates from generation and passing away, as from a middle [term]
plainly apparent to us, that first matter belongs to natural body. By
means of resolution, therefore, we proceed from composites to
discovering beginning-principles by means of the accidents that
are in them [as] middle [terms]. For just as someone who wants to
split some wood into parts needs a sword or some other iron instrument by which he may cut it, so the human mind, wanting to

compositum in principia ipsum constituentia resolvere, eget accidentibus evidentioribus, per quae hanc resolutionem perficiat.

10   Haec si ita se habent, manifestum est methodum hanc resolutivam esse ipsam demonstrationem à signo sive ab effectu, ut in eo
prooemio ait Averroes, idque omnino, velint nolint, coguntur omnes confiteri.

11   Propterea non rectè faciunt illi, qui hanc demonstrationem sub
demonstrativa methodo collocant, aut enim una datur methodus
demonstrativa, si hanc quoque vocemus demonstrativam; aut, si
propriè demonstrationis nomen sumentes resolutivam methodum
ab ea distinguere velimus, non alia est methodus resolutiva à compositis ad simplicium inventionem progrediens, quàm demonstratio ab effectu.

12   Quod autem aliqui dicunt methodum hanc resolutivam esse à
notione finis, id fortasse in artibus concedi potest, ut mox conside-
rabimus; at in scientiis contemplativis nulla ratione admittendum
est, quando enim inquiunt ex officiis et operationibus singulorum
humani corporis membrorum ostendi quòd necessarium fuerit
hominem iis membris praeditum esse, haec non est methodus resolutiva, sed demonstrativa, est enim potissima demonstratio facta
per causam finalem, de qua in libro nostro, quem de medio demonstrationis scripsimus, copiosè disservimus. Extruitur autem
talis demonstratio quando cognoscimus tum illum effectum esse
tum à tali effectrice causa fuisse productum, ut talia membra homini à natura data esse et quaerimus cur illud efficiens effectum
illum produxerit, sic enim finalem causam quaerimus in iis, quae
nos cognoscere solùm, non producere possumus. Sic autem hominem in membra non resolvimus, nec invenire volumus quot illa
sint, sed cùm ea prius cognoscamus, rationem singulorum quaerimus, eamque ex operationibus et muneribus singulorum adduci-
mus tanquàm ex causa finali.

resolve a composite into the beginning-principles constituting it, needs accidents more evident, by means of which he may perfect this resolution.

However that may be, it is manifest that this resolutive method 10 is demonstration itself *a signo* or *ab effectu*, as Averroës says in the proem, and whether they want to or not, they are all forced to confess this completely.

And so those who include this demonstration under demon- 11 strative method act incorrectly. For either there is one demonstrative method—if indeed we may call this demonstrative—or, if, taking the name of demonstration properly, we want to distinguish resolutive method from it, then there is no resolutive method, [i.e., no] progressing from composites to discovery of simples, other than demonstration *ab effectu*.

But now some say that this resolutive method is from a notion 12 of the end. Perhaps this can be conceded in arts, as we will soon consider, but in no way is this to be admitted in contemplative sciences. For when they say that it is shown from the functions and activities of each of the members of the human body that it was necessary that man be endowed with these members, that is not resolutive method but demonstrative, for it is a demonstration *potissima* made by means of the final cause, which we discussed plentifully in the book that we wrote on the middle [term] of demonstration.[77] Such a demonstration is built up when we know both that there is this effect and that it was produced by such and such an efficient cause, as [for example when we know] that such members were given to man by nature, and we inquire why this efficient [cause] has produced this effect. Thus we inquire after the final cause in those things that we can only know, not produce. We do not, however, thus resolve man into members, nor do we want to discover how many there are, but since we know them first, we inquire after the reason for each, and we adduce it from the activities and jobs of each, as from a final cause.

13   Quòd si aliquod animal nobis antea incognitum offeratur, in quo an insit membrum aliquod, ut pulmo, ignoremus, illudque ex operatione, veluti ex ipsa respiratione inveniamus, ab effectu argumentari dicimur, non amplius à causa finali, quia membrum illud ignoramus et an sit quaerimus, est igitur demonstratio ab effectu et methodus resolutiva; cuius scopus est invenire et cognoscere aliquam rem esse, quae ignorabatur. At quando pulmonem inesse cognoscimus et ipsum ex respiratione demonstramus, non est demonstratio inventionis, sed demonstratio per causam finalem et methodus demonstrativa.

14   Methodi autem resolutivae nunc declaratae scopus quòd non sit cognoscere numerum partium constituentium, ut aliqui dicunt, sed solùm cognoscere quòd sint, tum manifestum est[46] ex iis, quae superius adversus methodum divisivam diximus, tum etiam ex iis, quae mox de hac methodo dicturi sumus, manifestius fiet.

15   Altera quoque resolutionis species, quae est ab inferioribus ad superiora in eadem categoria, mihi videtur non plenè ab aliis fuisse intellecta, siquidem quaenam in ipsa illatio fiat non declararunt. Imò secundùm eorum sententiam nulla illatio fieri videtur, mens enim nostra, quae patibilis dicitur, primo loco individua, quae sibi à sensibus offeruntur, intelligit, deinde externo lumine adiuta naturam universalem in illis intuetur. Ibi autem duo processus considerari possunt, unus est quando mens ab individuorum ad universalium intellectionem transit, alter quando plura universalia ordine quodam contemplatur. Prior quidem progressus nullam illationem habet, quia mens non dicit, est Socrates, ergo est homo, sed ordinem potius, quàm methodum servare videtur, dum prius hunc

Now if some previously unknown animal is presented to us, 13
and we do not know whether some member, as [for example] a
lung, belongs to it, and we discover it from an activity, such as
from respiration itself, we are said to argue from effect (*ab effectu*),
not any more from final cause, because we are ignorant of this
member and we inquire whether it exists. It is, therefore, demonstration *ab effectu* and resolutive method, whose goal is to discover
and know that some thing, of which we were ignorant, exists. But
when we know that the lung belongs [to this animal] and we demonstrate it from respiration, it is not a demonstration of discovery,
but a demonstration by means of final cause and a demonstrative
method.

Moreover, the goal of resolutive method, now clarified, is not to 14
know the number of constituent parts, as some say, but only to
know that they are. This is manifest from the things that we said
above against divisive method. And it will be even more manifest
from the things that we will soon say about this method.

Now it appears to me that the other species of resolution also, 15
which is from the lower to the higher in the same category, was
not fully understood by others, since they did not make clear what
sort of inference occurs in it. Indeed it appears that according to
their position no inference occurs. For our mind, which is said to
be passive,[78] understands in the first place the individuals that are
presented to it by the senses, and then, helped out by external
light, it sees in them a universal nature. Two procedures, however,
can be considered there. One is when the mind passes from
the understanding of individuals to that of universals, the other
when it contemplates many universals in some sort of order. The
first progression, of course, has no inference, because the mind
does not say, "There is Socrates, therefore there is a man"; instead it appears to maintain order rather than method, when it

hominem, deinde humanam naturam in hoc homine intelligit, hanc enim ex illo non colligit, sed hanc post illum contemplatur.

16   In altero autem progressu non modò idem dicendum est, sed etiam si concedamus eam esse methodum, est potius à superioribus ad inferiora, quàm è contrario, quod enim prius à mente nostra cognoscitur, est magis universale, à quo ad minùs universale transitum facit. Primo igitur loco naturam corporis in Socrate conspicatur, deinde viventis, postea animalis, tandem hominis, hic igitur processus neque est syllogisticus, neque ab inferioribus ad superiora, sed ordo quidam in contemplatione universalium transiens à superioribus ad inferiora.

17   Sed quaenam est huius resolutionis, qualem isti fingunt, utilitas? An ex inferioribus numerum superiorum invenire, ut aliqui dicunt? At verò mirabile dictu hoc est, quod equidem intelligere nequeo, ab inferioribus namque ad superiora ascendendo ad unitatem potius, quàm ad numerum ducimur, semper enim multa in unum colligimus et tandem ad summum genus pervenimus, quod ex necessitate unum est omnia inferiora complectens.

18   Qui verò methodum hanc ad venandam definitionem utilem esse dicunt, rectius sentiunt, imò rectissimè, si eius illationis ac venationis modum declarassent. Nam à particularibus ad universalia progressio fit non ut ipsa universalia genera vel species cognoscantur, sed potius ut aliquid inesse alicui universali colligatur eò quòd illud idem omnibus particularibus inest, ut si omnem hominem bipedem esse ostendamus propterea quòd singuli homines bipedes esse inspiciuntur. Haec igitur resolutio nil aliud est,quàm inductio, quam Aristoteles in secundo libro Posteriorum Analyticorum compositionem vocat, quia ascendendo multa componimus

understands first this man and then human nature in this man. It does not gather the latter from the former, but it contemplates the latter after the former.

Now in the other progression, not only does the same thing have to be said, but even if we concede that it is a method, it is rather from the higher to the lower than vice versa. For what is first known by our mind is the more universal, from which it makes a passage to the less universal. In the first place, therefore, it sees in Socrates the nature of body, then of living, afterward of animal, finally of man. This proceeding, therefore, is not syllogistic, nor from the lower to the higher; it is instead a sort of order passing from the higher to the lower in the contemplation of universals.

But what is the utility of this resolution, the sort they fancy? That from the lower the number of the higher is discovered, as some say? But surely this is a wonderfully strange thing to say, something I cannot, for my part, understand. For by ascending from the lower to the higher, we are led to a unity rather than to a number. For we always gather many into one, and we finally arrive at the highest genus, which is, out of necessity, one, encompassing all of the lower.

And those who say this method is useful for the search for a definition sense things correctly—indeed very correctly, had they made clear the way of the inference and the search. For progression occurs from particulars to universals, not so that the universal genera or species themselves may be known, but rather so that it may be gathered that something belongs to some universal, because that same thing belongs to all the particulars—as if we were to show that every man is a biped, and on that account that all men are beheld to be bipeds. This resolution, therefore, is nothing other than induction, what Aristotle calls composition in the second book of the *Posterior Analytics*,[79] because by ascending, we

16

17

18

et in unum colligimus. De hac nos inferius loquemur. Nunc pauca haec contra aliorum sententiam de methodo resolutiva dicere voluimus, ut sano modo haec methodus intelligatur.

: XVII :

In quo ostenditur duas methodos ad res omnes cognoscendas sufficere.

1 Quòd autem ad res omnes cognoscendas duae methodi sufficiant, demonstrativa et resolutiva, facilè ostendi potest, nam omne, quod cognoscendum proponitur, aut est substantia aut accidens, substantia quidem tunc plenè cognoscitur, quando perfecta ipsius definitio habetur, haec si nota sit, nulla eget methodo ut investigetur. Si verò ignota, per aliquam methodum venanda est, per demonstrationem quidem venari eam non possumus, ut ait Aristoteles in contextu 42 secundi libri Posteriorum Analyticorum, ea enim sola à priori et per causam notificari possunt, quorum essentia pendet ab aliqua externa causa, at essentia substantiae à nulla externa causa pendet, nulla igitur causa datur, per quam definitio substantiae, si ignota fuerit, demonstrari possit. Relinquitur eam non posse nisi à rebus posterioribus et ab effectu aliquo notiore declarari, quae est methodus resolutiva.

2 Accidens autem aliud proprium est, aliud commune, commune quidem sub scientiam non cadit, proprium verò semper habet certam aliquam externam causam, à qua pendet, proinde per eam potest demonstrari. Externam dico non quòd loco et subiecto

bring many together and gather [them] into one. We will speak about this below. For now, we wanted to say these few things about resolutive method, against the position of others, so that this method may be understood in a sound way.

: XVII :

In which it is shown that two methods are
sufficient for knowing all things.

Moreover, that two methods — demonstrative and resolutive — are 1
sufficient for knowing all things can easily be shown. For everything set out to be known is either substance or accident. A substance, of course, is fully known whenever its perfect definition is had. If this is known, then no method is needed for it to be investigated. And if it is unknown, it has to be searched for by means of some method. We cannot, of course, search for it by means of demonstration,[80] as Aristotle says in text no. 42 of the second book of the *Posterior Analytics*.[81] For only those things whose essence depends on some external cause can be made known *a priori* and by means of the cause. But the essence of a substance depends on no external cause. There is no cause, therefore, by means of which the definition of a substance, if it was unknown, could be demonstrated. It remains that it cannot be made clear unless from posterior things and from some effect. This is resolutive method.

But now accidents: some are proper, others common. The com- 2
mon, of course, does not fall under science, but the proper always has some definite external cause on which it depends and by means of which it can accordingly be demonstrated. I say external, not because it [i.e., the cause] is always separated in place and

semper seiuncta sit, sed ratione essentiae ipsius accidentis, extra quam necessarium est eam esse. At quicquid talem causam habet, nulla ratione sciri potest, nisi per illam demonstretur, omne igitur accidens demonstrabile est, et nulla via est, quae ad ipsius perfectam scientiam ducere possit, nisi methodus demonstrativa.

3   Cùm itaque substantias omnes definitione cognoscamus, earum autem definitiones ignotas per solam resolutionem investigemus; accidentia verò per solam demonstrationem innotescant, hae duae methodi ad rerum omnium cognitionem comparandam sufficiunt, nec alio ullo logico instrumento indigemus ad probandum et colligendum ex noto ignotum. Egemus quidem ad disponendum, quia methodus non disponit, sed probat et notificat singillatim ea, quae convenienti ordine disposita sint. Ideò neque solus ordo neque sola methodus perfectam rerum scientiam praebere potest, sed ambo requiruntur, ordo enim solus nihil docet, methodus verò sine ordine docet quidem, sed perfectam scientiam non parit vel cum summo discentis labore et summa cum difficultate. Nam si philosophus naturalis propria animalium accidentia per proprias causas antè demonstrare niteretur, quàm de iis, quae corpori naturali communiter competunt et postea de elementis et de mistis generaliter disseruisset, vanum certè opus aggrederetur, rebus enim universalioribus ignoratis nunquàm animalium perfectam cognitionem assequeremur.

4   At si conveniente ordine res omnes tractandae dispositae fuerint, methodus in singularum declaratione operam suam egregiè praestat perfectamque scientiam parit, in qua discentium animus conquiescit. Propterea divisione quoque aliquando methodus eget,

subject [from the accident], but by reason of the essence of the accident itself; it is necessary that it [i.e., the cause] is outside this [i.e., outside the essence of the accident]. But whatever has such a cause, can in no way be known scientifically unless it is demonstrated by means of that [cause]. Every accident, therefore, is demonstrable and there is no way except demonstrative method that could lead to perfect scientific knowledge of it.

And so, since we know all substances by a definition and we investigate their unknown definitions by means of resolution alone, and accidents become known by means of demonstration alone, these two methods are sufficient for procuring knowledge of all things. Nor are we in need of any other logical instrument for proving and gathering the unknown from the known. Of course, we need [such] for disposing, because method does not dispose, but it proves and makes known one by one those things that have been disposed using appropriate order. And so neither order alone nor method alone can provide perfect scientific knowledge of things; instead both are required. For order alone teaches nothing, and method without order teaches, of course, but does not bring forth scientific knowledge or [does so] with the greatest effort in learning and with the greatest difficulty. For if the natural philosopher endeavored to demonstrate the proper accidents of animals by means of proper causes before he had discussed what appertains to natural body in general and afterward the elements and mixed [bodies] generally, he certainly would have undertaken the work in vain. For the more universal being unknown, we would never secure perfect knowledge of animals.

But if every thing being treated has been disposed using appropriate order, method really does its work very well in making each of the things clear and brings forth perfect scientific knowledge in which the soul of learners rests. Accordingly, method sometimes

quia divisio ordinatio quaedam est, quae rectam rerum probandarum seriem methodis subministrat.

5   Ex ipsa igitur rerum cognoscendarum natura clarum est duas illas methodos ad tradendam rerum omnium cognitionem sufficere.

6   Idem ex ipso methodi progressu ostenditur, quemadmodum antea dicebamus, omnis enim à noto ad ignotum scientificus progressus vel à causa est ad effectum, vel ab effectu ad causam, illa quidem est methodus demonstrativa, haec autem resolutiva. Alius processus, qui certam rei notitiam° pariat, non datur. Nam si ab aliquo ad aliquod progrediamur, quorum neutrum alterius causa sit, non potest inter illa esse connexus essentialis ac necessarius, quare nulla certa cognitio illum progressum consequi potest. Patet igitur nullam dari scientificam methodum praeter demonstrativam et resolutivam.

: XVIII :

In quo utriusque methodi definitio ponitur.

1   Harum duarum methodorum definitiones colligere ex iis, quae diximus, non difficile erit, si prius de utriusque fine atque utilitate quaedam pauca dixerimus, natura enim et essentia cuiusque instrumenti in ipso fine et usu potissimùm constituta est.

2   Dictum est antea, communem utriusque methodi finem esse rei ignotae cognitionem, res autem ignota, quam in scientiis quaeri contingat, duplex est, aut enim est causa aut effectus, praeter haec

also needs division, because division is a sort of ordering that furnishes to methods the correct sequence of the things that are to be
proved.

From the very nature of the things to be known, therefore, it is 5
clear that those two methods are sufficient for conveying knowledge of all things.

The same is shown from the very progression of method, as we 6
said earlier. For every scientific progression from known to unknown is either from cause to effect or from effect to cause. The
former, of course, is demonstrative method, the latter resolutive.
There is no other procedure that brings forth certain knowledge°
of something. For if we progress from something to something, of
which neither is the cause of the other, there cannot be an essential and necessary connection between them, and so no certain
knowledge can ensue from that progression. It is patent, therefore,
that there is no scientific method besides demonstrative and resolutive.

: XVIII :

In which the definition of each method is posited.

It will not be difficult to gather definitions of these two methods 1
from the things that we said, if we first say a few things about the
end and utility of each. For the nature and the essence of any instrument is constituted chiefly in [its] very end and use.

It was said earlier that the common end of each method is 2
knowledge of something unknown. What is unknown, however, what happens to be inquired after in the sciences, is twofold.
For it is either cause or effect. Besides these two, I see nothing

duo nil video in scientiis quaeri, imò ne considerari quidem. Aristoteles quoque in contextu 73 primi libri Posteriorum Analyticorum, in duo tantùm ea omnia secat, quae in scientiis tractantur, nempè in principia et ea quae ex[47] principiis pendent, hoc modo trimembrem à se prius factam divisionem ad bimembrem redigens. Nam antea dixerat tria esse, quae in scientia considerantur, principia, subiectum et affectiones, postea principiorum nomine subiectum quoque comprehendens duo esse dicit, principia et affectiones illa consequentes.

3   Ad haec duo, si latuerint, cognoscenda duae nobis methodi inserviunt, ut dictum est, nam affectiones demonstratione cognoscimus per propria principia, ipsa verò principia, si ignota fuerint, per resolutionem venamur. Hinc fit ut methodus resolutiva sit serva demonstrativae et ad eam dirigatur, non enim finem talem resolutio habet, quo invento quiescamus, sed à quo invento exordium compositionis sumamus. Principia enim ideò per resolutionem indagamus, ut per ea cognita effectus consequentes demonstremus, ultimus enim finis et scopus omnium, qui in scientiis speculativis versantur, est per methodum demonstrativam duci à principiorum cognitione ad scientiam perfectam effectuum, qui ab illis principiis prodeunt. Et certum est, si nos ad aliquam scientiam accedentes principia omnia nota haberemus, supervacuam ibi resolutionem fore, quia statim methodo demonstrativa à principiis notis ad effectus, qui semper secundùm naturam sunt ignoti, absque resolutionis usu progrederemur, quod in mathematicis evenit, proinde in ipsis locum non habet methodus resolutiva. Nam

inquired after in the sciences, indeed not even considered. Aristotle too, in text no. 73 of the first book of the *Posterior Analytics*,[82] splits all things that are treated in the sciences into just two, namely into beginning-principles and those things that depend on beginning-principles, in this way reducing to a two-branch division the three-branch division made by him earlier. For he had earlier[83] said that there are three things that are considered in a science — beginning-principles, subject, and affections. Afterward, comprehending also the subject under the name of beginning-principles, he says that there are two — beginning-principles and affections ensuing from them.

For knowing these two, if they should be hidden, two methods 3 serve us, as was said. For we know affections using demonstration by means of proper beginning-principles. And beginning-principles themselves, if they were unknown, we search for by means of resolution. And thus it happens that resolutive method is a servant to demonstrative and is directed toward it. Resolution does not have the sort of end that, once it is discovered, we rest. Instead, once it is discovered, we take from it the beginning of composition. We therefore track down beginning-principles by means of resolution, so that by means of them, once they are known, we may demonstrate the ensuing effects. For the ultimate end and goal of everyone who is concerned with speculative sciences is to be led by means of demonstrative method from knowledge of beginning-principles to perfect scientific knowledge of the effects that issue from those beginning-principles. And it is certain that if we, reaching for some scientific knowledge, had all the beginning-principles as known, resolution there would be superfluous, because we would progress at once, without use of resolution and by demonstrative method from known beginning-principles to effects that are always unknown according to nature. This happens in mathematics, and so resolutive method has no place in it.[84] For

resolutio illa mathematica, qua post factas omnes demonstrationes retrocedimus, et posteriora theoremata in priora et haec denique in prima principia resolvimus, est potius quaedam eruditorum exercitatio, quàm methodus resolutiva, de qua in praesentia loquimur, est enim processus ab ignotioribus ad notiora, qui cuilibet rudi ad eam scientiam capessendam accedenti esset prorsus inutilis, quia nullam cognitionem pararet, nos autem de illa resolutiva methodo sermonem facimus, quae rerum ignotarum ex notioribus cognitionem parit et in aliis scientiis locum habet, praesertim in scientia naturali. Cùm enim propter ingenii nostri viriumque nostrarum imbecillitatem ignota nobis occurrant principia, ex quibus demonstrandum est, ab ignotis autem progredi non possimus, ideò necessitate coacti ad secundariam quandam viam confugimus, quae est methodus resolutiva ad principiorum inventionem ducens, ut ex eis inventis postea effectus naturales demonstremus.

4 Quare methodus resolutiva secundaria est et ministra demonstrativae. Quam sententiam apud Aristotelem legere possumus in prooemio primi libri Physicorum, in ipso enim eius libri initio methodum servandam proponit demonstrativam, qua praecipuè uti vult, dicens res naturales ex principiorum suorum cognitione esse cognoscendas; deinde videns non esse nobis nota illa principia, subiungit utendum esse alia secundaria methodo à notioribus nobis ad principia notiora natura, quae est methodus resolutiva, ad quam certè non confugisset, si principia rerum naturalium statim nobis nota occurrissent.

5 Ex his colligere possumus finem methodi demonstrativae esse perfectam scientiam, quae est rei cognitio per suam causam;

that mathematical resolution by which we go back after all demonstrations have been made and resolve the later theorems into prior ones and those, lastly, into the first beginning-principles, is an exercise for the learned rather than the resolutive method about which we are presently speaking. For it is a proceeding from [things] less known to [things] more known; this would be utterly useless to anyone uneducated who was just starting to get hold of scientific knowledge, because it would bring forth no knowledge. We are here discoursing on that resolutive method that brings forth knowledge of things unknown from [those] more known and that has a place in other sciences, especially in natural science. For since, on account of the feebleness of our wit and our powers, the beginning-principles, from which the demonstration is to be, come to us as unknown, and since, moreover, we cannot progress from unknowns, therefore, forced by necessity, we resort to some sort of secondary way, which is resolutive method, leading to discovery of the beginning-principles, with the result that we may afterward demonstrate natural effects from these discovered [beginning-principles].

Resolutive method, therefore, is secondary and a minister to the 4 demonstrative. We can read this position in Aristotle in the proem of the first book of the *Physics*.[85] For at the very start of the book, he sets out to maintain demonstrative method, which he principally wants to use, saying that natural things have to be known from knowledge of their beginning-principles, and then seeing that those beginning-principles are not known to us, he adds that he has to use another, secondary method, from [things] more known to us to beginning-principles more known by nature; this is resolutive method. He certainly would not have resorted to it, if the beginning-principles of natural things came to us known straightaway.

From all this we can gather that the end of demonstrative 5 method is perfect scientific knowledge, which is knowledge of

methodi autem resolutivae finem esse inventionem potius, quàm
scientiam. Quoniam enim resolutione causas inquirimus ex effectis, ut postea ex causis effecta cognoscamus, non ut in ipsarum
causarum cognitione quiescamus, ideò inventionem causarum dicimus finem esse methodi resolutivae, idque plurium bonorum interpretum testimonio comprobari potest.

6   Averroes in Epitome logica in capite de demonstratione loquens
de demonstratione à signo eam vocat demonstrationem inventionis.

7   Themistius de eadem loquens in calce primi contextus secundi
libri Posteriorum dicit signa posteriora esse quidem principia inventionis cùm per ea causas inveniamus, tamen non esse veras
causas rei, quae demonstratur.

8   Idem in contextu 93 primi libri Posteriorum Analyticorum dicit, resolvere conclusionem in sua principia est conclusione vera
proposita invenire principia, ex quibus colligatur, quare finem resolutionis in inventione constituit.

9   Eustratius quoque in sua praefatione in secundum librum dicit
demonstrationem ideò vocari resolutionem, quia eius principia per
resolutionem inveniuntur, postea exemplo hanc resolutionem declarans facit demonstrationem ab effectu, per quam dicit inveniri
causam ignotam.

10   Haec de fine utriusque methodi in praesentia sufficiant, nam in
libro sequente declarantes tractationem Aristotelis de methodis
fusius de eodem loquemur, nunc ad colligendas harum methodorum definitiones satis sint[48] ea, quae modò diximus.

something by means of its cause, and that the end of resolutive method is discovery rather than scientific knowledge. Now since in resolution we ask after causes from effects, so that afterward we may know effects from causes, and not so that we may rest in knowledge of the causes themselves, we therefore say that discovery of causes is the end of resolutive method. And this can be confirmed by the testimony of many good commentators.

Averroës in the *Epitome of Logic* in the chapter on demonstra- 6 tion,[86] speaking about demonstration *a signo*, calls it the demonstration of discovery.

Themistius speaking about the same thing at the end of text 7 no. 1 of the second book of the *Posterior [Analytics]*,[87] says posterior signs are, of course, the beginning-principles of discovery, since we may discover causes by means of them, although they are nevertheless not the true causes (*verae causae*) of the thing that is demonstrated.

He says the same in text no. 93 of the first book of the *Posterior* 8 *Analytics*,[88] that to resolve a conclusion into its beginning-principles is to discover, once a true conclusion has been set out, the beginning-principles from which it is gathered. He thereby establishes the end of resolution in discovery.

And so Eustratius too, in his preface to the second book,[89] says 9 demonstration is called resolution, because its beginning-principles are discovered by means of resolution. Afterward, clarifying this resolution by an example, he makes a demonstration *ab effectu*, by means of which he says the unknown cause is discovered.

All this about the end of each method is at present sufficient. 10 For in the following book we will speak at greater length about the same thing, clarifying Aristotle's treatment on methods. For now let what we have just said be enough for gathering the definitions of these methods.

11   Methodus quidem demonstrativa, in qua declaranda praecipuè versatur Aristoteles in Posterioribus Analyticis, ab ipso etiam ibidem definitur, nam proposita in primis tali demonstrationis definitione per causam finalem tradita, demonstratio est syllogismus scientiam pariens, ex ea investigat materiam demonstrationis, scilicet principiorum conditiones, quibus inventis haec assignari potest perfecta et omnibus numeris absoluta definitio, Methodus demonstrativa est syllogismus scientiam pariens ex propositionibus necessariis, medio carentibus, notioribus et causis conclusionis. Quae quidem definitio obscura esse non potest iis, qui Posteriores Analyticos Aristotelis intellexerint, ideò ipsius declaratione in praesentia supersedebimus; praesertim cùm in aliis libris logicis, quos de rebus ad demonstrationem attinentibus conscripsimus, has principiorum conditiones satis abundè declaraverimus.

12   Methodus autem resolutiva est syllogismus ex propositionibus necessariis constans, qui à rebus posterioribus, et effectis notioribus ad priorum et causarum inventionem ducit. Esset quidem declarandum quomodo haec methodus ex propositionibus necessariis constet, sed tum hac de re tum de aliis multis ad hanc methodum attinentibus satis superque dictum est à nobis in libro nostro de speciebus demonstrationis necnon in libro de propositionibus necessariis, ideò inde omnia petenda sunt. Nunc an hae duae methodi aliquam divisionem recipiant consideremus.

Demonstrative method, of course, which Aristotle is principally 11
concerned with clarifying in the *Posterior Analytics*, is also defined
by him there. Having, in the first place, set out such a definition
of demonstration conveyed by means of a final cause—"Demonstration is a syllogism bringing forth scientific knowledge"[90]—he
then investigates the matter of demonstration, that is, the characteristics of beginning-principles. Once these have been discovered,
this definition, perfect and on all counts complete, can be assigned: "Demonstrative method is a syllogism bringing forth scientific knowledge from premises that are necessary, lacking a middle,
more known, and the causes of the conclusion."[91] This definition,
of course, cannot be obscure to those who have understood Aristotle's *Posterior Analytics*. And so we will refrain at present from a
clarification of it, especially since, in other books on logic that we
have written about things pertinent to definition, we clarified
abundantly enough these characteristics of beginning-principles.

Resolutive method, moreover, is a syllogism composed out of 12
necessary premises, which leads from things posterior and from
effects more known to discovery of causes and things prior. Of
course, it would have to be made clear in what way this method is
composed out of necessary premises, but both about this issue and
about many others pertinent to this method, more than enough
was said by us in our book *On the Species of Demonstrations* and also
in the book *On Necessary Premises*.[92] So everything will have to be
sought there. Let us now consider whether these two methods are
receptive to some division.

: XIX :

De speciebus methodi resolutivae, et earum differentiis.

1 Natura utriusque methodi declarata considerandum manet an hae in species dividantur. Sed methodus quidem demonstrativa, cùm sit sola illa demonstratio, quae potissima dicitur, nullam divisionem admittit, nisi fortè illam, quae ex divisione generum causarum derivatur, haec autem ad praesentem considerationem parum pertinere videtur, sed de ea dictum à nobis est in libro de medio demonstrationis.

2 Methodus autem resolutiva in duas species dividitur efficacitate inter se plurimum discrepantes, altera est demonstratio ab effectu, quae in sui muneris functione est efficacissima et ea utimur ad eorum, quae valdè obscura, et abscondita sunt, inventionem. Altera est inductio, quae est multò debilior resolutio et ad eorum tantummodò inventionem usitata, quae non penitus ignota sunt et levi egent declaratione.

3 Utramque esse resolutivam methodum et sub tradita à nobis definitione contineri manifestum est, utraque enim est via syllogistica et à posterioribus ad priora et ad principiorum inventionem progreditur tanquàm ad proprium finem. De demonstratione à signo nemo dubitare potest, est enim syllogismus procedens ab effectu ad inventionem causae, quare proprium eius munus est inventio causae ignotae.

4 Inductionem verò esse syllogismum docet Aristoteles in secundo libro Priorum Analyticorum in capite de inductione, sicuti nos suprà declaravimus, non enim necessitatem colligendi haberet, nisi in bonum syllogismum verti posset. Quòd autem inductio sit instrumentum cognoscendi principia Aristoteles multis in locis testatur, in calce secundi libri Priorum Analyticorum in capite de

: XIX :

On the species of resolutive method and their differentiae.

The nature of each method having been made clear, it remains to 1
be considered whether they are divided into species. But demonstrative method, of course, since it alone is that demonstration
said to be *potissima*, admits no division except perhaps that which
is derived from division of the kinds of causes. This, however, appears to pertain little to the present consideration, and we spoke
about this in the book *On the Middle [Term] of a Demonstration.*[93]

Resolutive method, however, is divided into two species, very 2
different among themselves in efficacy. One is demonstration *ab
effectu*, which is most effectual in the performance of its job and is
used for discovery of things that are highly obscure and hidden.
The other is induction, which is a much weaker resolution, and
normally used only for discovery of those things that are not completely unknown and need [only] a light clarification.

And it is manifest that each is resolutive method and is con- 3
tained under the definition conveyed by us. For each is a syllogistic
way and progresses from posteriors to priors and toward discovery
of beginning-principles as toward its proper end. No one can have
doubts about demonstration *a signo*. For it is a syllogism proceeding from effect (*ab effectu*) to discovery of a cause. Its proper job,
accordingly, is discovery of an unknown cause.

But in the second book of the *Prior Analytics*, in the chapter on 4
induction,[94] Aristotle teaches that induction is a syllogism, as
we made clear above; for it would not have the necessity of gathering, unless it could be turned into a good syllogism. That induction is, moreover, an instrument for knowing beginning-principles,
Aristotle testifies in many passages — at the end of the second
book of the *Prior Analytics* in the chapter on induction, in text

inductione in contextu 134 primi Posteriorum et in ultimo capite secundi et capite tertio sexti libri de moribus. In his enim omnibus locis asserit Aristoteles, proprium inductionis officium esse, ut per eam principia confirmentur.

5   Est autem inductio processus à posterioribus ad priora, quia universale est natura prius particularibus et habet rationem° causae, ideò à particularibus ad universale progredi est à posterioribus ad priora procedere, idque dicit clarè Aristoteles in capite de inductione in secundo libro Priorum Analyticorum.

6   Est autem inter has duas resolutiones magnum discrimen, quia inductione non inveniuntur nisi illa principia, quae sunt nota secundùm naturam et levi egent comprobatione. At demonstratio à signo est multò efficacior, per eam enim illa principia inveniuntur, quae secundùm naturam sunt ignota, ad quorum inventionem inductio est prorsus inutilis. Haec autem differentia clara erit si intelligatur quidnam sit notum, vel ignotum secundùm naturam. De hac quidem re superius aliqua diximus, aliqua etiam hîc dicenda sunt. Notum secundùm naturam illud dicitur, quod sensile est, eiusmodi autem sunt non ea solùm, quae singularia sunt, sed ea quoque universalia, quorum singularia sensu percipi possunt, hominem enim rem sensilem esse dicimus, non quòd hominem universalem sensus cognoscat, sed quia singuli individui homines sensiles sunt, propterea haec propositio, homo est bipes, dicitur nota secundùm naturam, quia quocumque individuo homine oblato statim cognoscit sensus eum esse bipedem. Haec autem iure vocantur nota secundùm naturam, quia proprio lumine cognoscuntur, neque egent alia re notiore, per quam mediam demonstrentur.[49]

no. 134 of the first [book] of the *Posterior* [*Analytics*],[95] and in the last chapter of the second[96] and the third chapter of the sixth book of the *Ethics*.[97] For in all these passages, Aristotle asserts that it is the proper function of induction that by means of it beginning-principles are confirmed.

Induction, moreover, is a proceeding from posteriors to priors, because the universal is by nature prior to particulars and has the relationship° of a cause. And so to progress from particulars to universal is to proceed from posteriors to priors. And Aristotle says this clearly in the chapter on induction in the second book of the *Prior Analytics*.[98]

Now there is a great discriminating difference between these two resolutions, because nothing is discovered by induction except those beginning-principles that are known according to nature and need [only] a light confirmation. Demonstration *a signo*, however, is much more effectual, for by means of it those beginning-principles are discovered that are unknown according to nature; induction is utterly useless for their discovery. This difference, moreover, will be clear if what is known or unknown according to nature is understood.[99] Regarding this issue, of course, we said some things above and others have to be said here also. What is sensible is said to be known according to nature, and of this type are not only those that are singular, but also those universals whose singulars can be grasped by sense. For we say that man is a sensible thing, not because sense knows universal man, but because each of the individual men are sensible. Therefore this premise, "Man is biped," is said to be known according to nature, because when any individual man whatsoever is presented, sense at once knows that he is biped. These are justly called "known according to nature," because they are known by their own light and need nothing else more known [as] a middle [term] by means of which they are demonstrated.

7   Contrà verò ignotum secundùm naturam illud dicitur, quod in suis singularibus sensile non est, ideò eget alio medio notiore, per quod demonstretur, et cùm ipsum proprio lumine non cognoscatur, per alterius lumen innotescit, veluti prima materia, quae cùm sensum penitus lateat, per se nunquàm cognosceretur, nisi per generationem notificaretur. Ita haec propositio, triangulum habet tres angulos duobus rectis aequales, dicitur ignota secundùm naturam, quia eius praedicatum sensu discerni non potest, sed innotescit per aliud, ex longa enim trianguli inspectione nunquàm cognosceremus tres illos angulos esse duobus rectis aequales, sed ratio id nobis demonstrat. Itaque notum secundùm naturam idem significat ac per se notum. Ignotum autem secundùm naturam illud est, quod per se ignotum dicitur et cognoscitur per aliud, quemadmodum declarat Averroes in 25 et 29 commentariis secundi libri Posteriorum Analyticorum et Aristoteles in secundo Priorum in capite de petitione principii.

8   His igitur differentiis invicem dissident inductio et demonstratio ab effectu, utraque enim est methodus resolutiva à rebus posterioribus ad principia progrediens; sed duo principiorum genera nobis offeruntur, alia quidem naturaliter nota sunt, ideò nullo egent instrumento logico, nisi inductione, qua sola notificantur. Omnis enim nostra cognitio à sensu originem ducit,° nec potest aliquid à nobis mente cognosci, quin prius sensu cognitum fuerit, proinde inductione omnia eiusmodi principia nobis innotescunt, nec propterea demonstrari seu probari dicuntur, ea namque propriè dicuntur probari, quae demonstrantur per aliud. Inductio autem non probat rem per aliam rem, sed modo quodam

On the contrary, something is said to be unknown according to   7
nature that is not sensible in its singulars [i.e. single instances],
and therefore needs something else, a more known middle [term]
by means of which it is demonstrated. And since it is not known
by its own light, it becomes known by means of the light of another, just as first matter, which, because it is completely hidden
from sense, would never be known *per se*, unless it were made
known by means of generation. Thus this premise, "A triangle has
three angles equal in total magnitude to two right [angles]," is said
to be unknown according to nature, because its predicate cannot
be discerned by sense, but becomes known by means of another.
For from long inspection of a triangle, we would never come to
know that those three angles are equal in total magnitude to two
right [angles]; instead, reason demonstrates this to us. And so
"known according to nature" and "known *per se*" signify the same
thing. Moreover, "unknown according to nature" is that which is
said to be unknown *per se*, and is known by means of another, as
Averroës makes clear in commentaries 25 and 29 to the second
book of the *Posterior Analytics*,[100] and Aristotle [does] in the second
[book] of the *Prior* [*Analytics*] in the chapter on *petitio principii*.[101]

By these differentiae, therefore, induction and demonstration *ab*   8
effectu are each distinguished. For each is a resolutive method progressing from posterior things to beginning-principles, but two
kinds of beginning-principles are presented to us. The first, of
course, are known naturally. They therefore need no logical instrument except induction; by it alone they become known. For all our
knowledge draws° [its] origin from sense. Nor can anything be
known by us with the mind that was not first known with sense.
And so all beginning-principles of this type become known to us
by induction and accordingly are said not to be demonstrated or
proved. For those things are properly said to be proved that are
demonstrated by means of something else. Induction, however,
does not prove something by means of something else but in some

eam per se ipsam declarat, universale enim à singulari reipsa non distinguitur, sed ratione solùm. Et quia res notior est ut singularis, quàm ut universalis, quoniam sensilis dicitur ut singularis, non ut universalis, ideò inductio est processus ab eodem ad idem, ab eodem ea ratione, qua evidentius est, ad idem cognoscendum ea ratione, qua obscurius est atque latentius. Propterea non modò principia rei, sed etiam principia scientiae seu principia cognoscendi, quae dicuntur indemonstrabilia, inductione cognoscuntur, ut videre possumus apud Aristotelem in contextu undecimo primi libri Physicorum et in omnibus antea memoratis locis, praecipuè verò in ultimo capite secundi libri Posteriorum Analyticorum.

9   Alia verò principia sunt naturaliter ignota, quia insensilia, ideò ad eorum inventionem inductio nihil penitus efficacitatis habet, sed egent demonstratione à signo, qua[50] per effectum notiorem inveniantur ac notificentur, cùm ipsa per se nequeant innotescere, ut prima materia, quae est principium methodi demonstrativae, per eam enim tanquàm per causam demonstrat Aristoteles plures effectus naturales, ut in primo libro de Ortu et interitu generationis possibilitatem et eiusdem aeternitatem. Sed quia materia naturaliter ignota proponebatur, non potuit Aristoteles ea ut principio ad aliquid demonstrandum uti, nisi prius ex effectu notiore ipsam demonstrasset. Ita primus motor aeternus est causa, per quam aeternus motus demonstratur in octavo libro Physicorum, sed quia naturaliter ignotus ipse primus motor nobis offerebatur, prius ipsum invenit Aristoteles per demonstrationem ab effectu.

way makes it clear by means of that thing itself. For the universal is not in reality, but only in reasoning, distinguished from the singular. And because a thing is more known as singular than as universal, since it is said to be sensible as singular, not as universal, induction is, therefore, a proceeding from one thing to the selfsame thing, from the one thing in the way in which it is more evident, to knowing the selfsame thing in the way in which it is more obscure and hidden. Therefore, not only are the beginning-principles of something known by induction, but so also the beginning-principles of scientific knowledge or the beginning-principles of knowing that are said to be indemonstrable,[102] as we can see in Aristotle in text no. 11 of the first book of the *Physics*[103] and in all the passages referred to earlier, but principally in the last chapter of the second book of the *Posterior Analytics*.[104]

The other beginning-principles are unknown naturally, because [they are] insensible, and so for their discovery induction has no efficacy at all. They need, instead, demonstration *a signo*, by which they are discovered and made known by means of an effect more known, since they themselves cannot become known *per se* — as [(a)] first matter, which is a beginning-principle of demonstrative method, for by means of it, as by means of a cause, Aristotle demonstrates many natural effects, as [for example], the possibility of generation and its eternity, in the first book of *On Coming to Be and Passing Away*. But because matter was set out as naturally unknown, Aristotle could not use it as the beginning-principle for demonstrating something else, unless he had first demonstrated it from an effect more known. And [(b)] thus the eternal first mover is the cause by means of which eternal motion is demonstrated in the eighth book of the *Physics*,[105] but because the first mover was presented to us as naturally unknown, Aristotle first discovered it by means of demonstration *ab effectu*.

9

10   Principia igitur methodi demonstrativae resolutiva methodo inveniuntur, alia quidem sola inductione alia verò demonstratione à signo.

11   Hanc differentiam etiam in illa definitione, quae scientiae principium est, notare possumus, eius enim partes aliae sola inductione cognoscuntur, aliae verò, quae non possunt nisi per aliud nota fieri, egent demonstratione ab effectu, quod significavit Aristoteles in contextu undecimo primi libri de Anima, dicens accidentia plurimum conferre ad cognoscendas substantiarum definitiones.

12   Sed de definitionibus omnibus quomodo per methodos innotescant et ad eas methodi dirigantur, docuit egregiè Aristoteles in Posterioribus Analyticis, idque nos in libro sequente declarare statuimus.

: XX :

Solutio difficultatis de ordine resolutivo.

1   Posteaquam differentias omnes, ac naturas methodorum declaravimus, non erit ab re si dubium quoddam solvamus tactum à nobis antea cùm de ordine resolutivo loqueremur. Dicebamus enim cum illo ordine methodum quandam esse coniunctam, ut alia ratione dicatur ordo, alia ratione methodus, quatenus enim prius de fine agitur, postea de iis, quae sunt ante finem, ordo servari dicitur resolutivus. Quatenus autem ex fine praecognito colliguntur per necessariam et syllogisticam illationem ea, quae ad finem ducunt, eatenus methodus quaedam est, haec enim non est dispositio, sed

The beginning-principles of demonstrative method are, there- 10
fore, discovered by resolutive method, some by induction alone
and some by demonstration *a signo*.

We can note this differentia also in that definition that is the 11
beginning-principle of scientific knowledge. For some parts of it
are known only through induction; but others, which cannot become known except by means of another, need demonstration *ab
effectu*. Aristotle indicated this in text no. 11 of the first book of *On
the Soul*,[106] saying that accidents contribute much to knowing the
definitions of substances.

But Aristotle taught very well in the *Posterior Analytics* the way 12
in which all definitions become known by means of methods, and
in what way methods are directed toward them [i.e., the defini-
tions]; we have decided to make this clear in the following book.

⁚ XX ⁚

Solution to a problem regarding resolutive order.

After we have made clear the natures of methods and all their dif- 1
ferentiae, it will not be beside the point if we do away with a
doubt touched upon by us earlier when we were speaking about
resolutive order. For we said some sort of method is conjoined
with that order, so that in one way it is said to be order and in
another way method. Insofar as an end is dealt with first, and afterward those things that are before the end, resolutive order is
said to be maintained. And insofar as those things that lead to an
end are gathered from the end known beforehand, by means of
necessary and syllogistic inference, to that extent there is some sort
of method. This is not a disposition, but making clear and known

ignoti ex noto declaratio et notificatio. Quaerere igitur non absque ratione quispiam posset qualisnam ea methodus sit, an demonstrativa, an resolutiva, nam pro utraque parte° argumenta non desunt.

2 Primùm quidem finis respectu eorum, quae sunt ante finem, causae locum habet, dicitur enim causa finalis, quae causa causarum nuncupari solet. Quoniam igitur finis in qualibet arte notus proponitur quatenus est finis et ab eo proceditur tanquàm à fine, methodus videtur esse demonstrativa à causa finali ad effectum, nam ex causa finali potissimam demonstrationem fieri posse docuit Aristoteles in secundo libro Posteriorum Analyticorum. Videtur etiam huic sententiae favere Aristoteles ipse in capite octavo septimi libri de Moribus, quando dicit finem in actionibus esse principium, sicut in mathematicis suppositiones.

3 Ex altera verò parte° videtur esse methodus resolutiva, quoniam à fine proceditur ad invenienda principia ad ipsius generationem et productionem idonea, ergo est processus à fine potius ut ab effectu, quam ut à causa; et ad principia quatenus principia sunt, ex quibus producatur, non quatenus habent locum effectus, quare est methodus resolutiva ab effectu ad causas.

4 Pro huius dubii solutione sciendum° est quòd quando de methodis loquimur, scientias contemplativas respicimus, quarum finis est scire, methodi namque scientiam rerum ignotarum pariunt, quare scientiarum speculativarum instrumenta sunt, in his enim scientia propriè dicta locum habet, non in artibus et aliis omnibus[51] operatricibus disciplinis, quae in rebus contingentibus versantur et actionem, vel effectionem, non scientiam quaerunt. In his igitur propriè dicta methodus non datur, sicuti neque propriè

the unknown from the known. Not without reason, therefore, someone could inquire what sort of method it is, whether demonstrative or resolutive. And arguments are not lacking on either side.°

First, of course, the end, with regard to those things that are 2 before the end, has the place of a cause. For it is said to be the final cause, what is normally called the cause of causes. Since, therefore, in any art a known end is set out insofar as it is an end, and there is a proceeding from it as from an end, the method appears to be demonstrative, from final cause to effect. For in the second book of the *Posterior Analytics*, Aristotle taught that a demonstration *potissima* can occur from a final cause.[107] It appears Aristotle is himself even favorable to this position in the eighth chapter of the seventh book of the *Ethics*,[108] when he says that in actions, as in the suppositions of mathematicians, the end is the beginning.

But, on the other side,° the method appears to be resolutive, 3 since there is a proceeding from the end to discovering beginning principles fit for its generation and production. Therefore, the proceeding is from the end, as from an effect rather than as from a cause, and to beginning-principles, not insofar as they have the place of an effect, but insofar as they are the beginning-principles from which it [i.e., the effect] is produced. And so the method is resolutive, from effect to causes.

For a solution to this doubt, it has to be understood° that when 4 we speak about methods, we give regard to the contemplative sciences, whose end is to know scientifically. For methods bring forth scientific knowledge of things unknown and so are the instruments of speculative sciences. Scientific knowledge, said properly, has a place in them but does not in the arts and all the other practical disciplines, which are concerned with contingent things and seek after action or bringing about an effect, not scientific knowledge. In these, therefore, there is, properly speaking, no

dicta scientia neque vera necessitas. Si quam enim necessitatem habent, eam tantùm habent, quae dicitur ex suppositione finis, ut si hic homo sanandus sit, necessarium est talibus remediis uti, simpliciter autem iis uti non est necessarium. Scientia igitur propriè dicta ibi locum non habet, sed cognitio quaedam rei non necessariae, nisi ex suppositione finis et ipsa non per se quaesita, sed propter operationem.

5   Quoniam igitur methodi sunt instrumenta acquirendi certam scientiam, ideò huiusmodi disciplinis propriè non competunt, aptari tamen illis aliquo modo possunt per similitudinem pro subiectae materiae conditione, quemadmodum etiam scientiae nomen impropriè acceptum ipsis quandoque tribuitur. Nam, ut ait Alexander in principio commentariorum suorum in primum librum Priorum Analyticorum, logica philosophiae instrumentum est et propter eam tradita, non propter alias disciplinas. Attamen postquàm scripta fuit inventa est aliis quoque disciplinis prodesse posse, ideò aliae non prohibentur logica uti eo modo, quo possunt.

6   Dicimus itaque in disciplinis utentibus ordine resolutivo nullam dari demonstrationem neque à causa neque ab effectu: nam utraque scientiam parit immutabilem rei simpliciter necessariae. Dantur tamen in illis syllogismi ignotum ex noto colligentes, qui cùm propriè loquendo non sint demonstrationes, in illis tamen facultatibus locum habent demonstrationum, siquidem efficaciores demonstrationes ibi non dantur. Non est igitur mirum si huiusmodi syllogismi aliquando et demonstrationis potissimae et demonstrationis à signo conditiones prae se ferunt, cùm solùm similitudine et proportione quadam, non simpliciter dicantur demonstrationes.

method, just as, properly speaking, no scientific knowledge and no true necessity. If they have any necessity, they have only that which is said to be from supposition of the end,[109] as if this man is to be healed, it is necessary to use such and such remedies. But absolutely, it is not necessary to use them. Scientific knowledge, therefore, properly speaking, does not have a place there; [what does have a place is] some sort of knowledge of something not necessary (unless from supposition of an end) and not itself inquired after for its own sake but for the sake of practical activity.

Therefore since methods are instruments for acquiring certain 5 scientific knowledge, they do not properly apply to disciplines of this type. They can, nevertheless, be applied to them in some way by means of a resemblance to a characteristic of the subject matter, just as the name "scientific knowledge (scientia)," accepted improperly, is sometimes ascribed to them. For as Alexander says in the beginning of his commentaries on the first book of the Prior Analytics,[110] logic is an instrument of philosophy and conveyed for the sake of it, not for the sake of other disciplines. But nevertheless, after it had been written down, it was discovered that it could be of profit to other disciplines also. And so [these] others are not prohibited from using logic in whatever way they can.

We say, therefore, that in disciplines using resolutive order, 6 there is no demonstration, either from cause or from effect (ab effectu), for each brings forth immutable scientific knowledge of something absolutely necessary. In those [disciplines], however, there are syllogisms gathering the unknown from the known; although, properly speaking, they are not demonstrations, they nevertheless have the place of demonstrations in those branches of learning, since indeed there are not demonstrations more effectual there. It is, therefore, not a wonder, if syllogisms of this type sometimes exhibit characteristics of both demonstration potissima and demonstration a signo, although they are said to be demonstrations only by resemblance and in some sort of comparative

Idque causa[52] fuit praesentis dubitationis, nam processus à fine artis ad colligenda principia partim videtur posse appellari demonstratio potissima, partim demonstratio ab effectu, neutra tamen propriè, sed utraque per similitudinem. Mihi tamen videtur eam potius esse vocandam methodum resolutivam cum ordine resolutivo coniunctam, quàm demonstrativam.

7   Quam quidem rem melius intelligemus, si hanc demonstrationem cum aliqua philosophi naturalis demonstratione ex causa finali facta contulerimus et ipsarum discrimen consideraverimus, data est animalibus vis coquendi alimentum propter nutritionem tanquàm coctionis causam finalem et haec omnia sunt simpliciter necessaria, animalia nutriri et habere vim coctricem alimenti, neque in nostro arbitrio est constitutum facere ut animalia nutriantur, vel non nutriantur et ut vim coctricem habeant, vel non habeant, ergo si per nutritionem demonstremus animalia vim habere coctricem alimenti, erit potissima demonstratio per causam finalem facta et propositiones erunt simpliciter necessariae, quia[53] in ipsa re necessitas inest, non ex nostra tantùm constitutione. Ideò nutritio ut causa tantùm sumitur, non ut effectus, quia nos efficere ipsam non possumus, ideò proposito effectu noto, ipsa vi coctrice à natura animalibus tradita, demonstratio illa non habet pro fine inventionem facultatis coctricis, haec enim iam cognita proponitur, neque ad eius effectionem, sive ad effectionem nutritionis dirigitur cognitio illa, quam per eam demonstrationem quaerimus, quandoquidem haec facere nos minimè possumus, sed ad solam facultatis coctricis cognitionem per suam causam finalem, quare non est alia methodus, quàm demonstrativa.

relation, not absolutely. And this was the cause of the present doubt. For it appears that a proceeding from the end of an art to gathering the beginning-principles could [partly] be called demonstration *potissima* and partly demonstration *ab effectu* — neither, however, properly, but each by means of resemblance. Nevertheless it appears to me that this has to be called resolutive method conjoined with resolutive order, rather than demonstrative [method].

We will, of course, understand this issue better if we compare   7 this demonstration with some demonstration of the natural philosopher made from final cause and consider their discriminating difference. In animals there is a power to digest nourishment for the sake of nutrition as the final cause of digestion. And all these are necessary absolutely: that animals be nourished and [that they] have the power to digest nourishment. It is not our choice to make the determination that animals are nourished or are not nourished and that they have or do not have digestive power. Therefore if we demonstrate by means of nutrition that animals have the power to digest nourishment, it will be a demonstration *potissima* made by means of a final cause, and the premises will be necessary absolutely, because the necessity belongs to the thing itself, not just from our determination. And so nutrition is taken only as cause, not as effect, because we cannot effect it. And so the known effect having been set out — [i.e. that] digestive power itself has been conveyed to animals by nature — that demonstration does not have as its end the discovery of the digestive faculty. For this is set out as already known. Nor is the knowledge that we inquire after by means of the demonstration directed toward bringing it into effect or effecting nutrition since we cannot at all make it, but only toward knowledge of the digestive faculty by means of its final cause. It is, therefore, no method other than demonstrative.

8   At quando in arte medica demonstramus, si talem febrem cu-
rare et sanitatem recuperare velimus, talibus remediis utendum
esse et ita ex febre remedia colligimus, non proponitur nobis sanitas ut existens, sed potius ut non existens et ut à nobis recupe-
randa et efficienda, quare potius ut effectus, quàm ut causa nobis
proponitur. Cùm enim duo ibi consideranda sint, cognitio remediorum ex illius finis praecognitione, deinde illius finis productio
per illa remedia, certè hoc posterius longè praecipuum est, nam
remediorum cognitionem propter effectionem sanitatis quaerimus.
Hanc igitur effectionem ab initio concipientes ad remedia cognoscenda progredimur potius ut ab effectu, quàm ut à causa.

9   Quòd si medicam artem solùm ut cognoscentem, non ut ope-
rantem spectemus, fortasse sanitas locum habet causae finalis, per
quam cognoscimus illa remedia. Attamen cùm tota illa cognitio ad
operationem dirigatur, sanitas videtur non ut causa, sed ut effectus
considerari et progressus ille potius dicendus est methodus resolutiva, quàm demonstrativa. Adde quòd demonstratio per causam
finalem facta non demonstrat ex fine causam effectricem, sed effectum à causa effectrice productum, ut alibi declarabimus, quare non
videtur ille processus posse vocari demonstratio à causa finali, sed
potius ab effectu.

10   Praeterea ordo resolutivus ideò appellatur resolutivus, quia est à
posteriori ad prius, ergo si qua methodus cum eo coniuncta est, ea
est resolutiva, proinde ab effectu, non à causa, nam omnis causa
quatenus causa est, prior est effectu, finis igitur quatenus est causa
finalis non est posterior iis, quae sunt ante finem, sed prior.

11   Solemus etiam dicere in arte medica remedia inveniri ex praecognitione sanitatis et morbi, utimur igitur nomine inventionis et

But when we demonstrate in medical art, [for example,] "if we   8
want to cure a fever and restore health, such and such remedies
have to be used," and thus gather the remedies from the fever,
health is not set out to us as existing, but instead as not existing,
as [something] to be restored and effected by us. It is, therefore,
set out to us as effect rather than as cause. For although two
things have to be considered there — knowledge of remedies from
the prior knowledge of that end and then production of that end
by means of those remedies — certainly the latter is by far the most
important. For we inquire after knowledge of remedies for the
sake of effecting health. Conceiving this effecting from the start,
therefore, we progress to knowing remedies as from effect rather
than as from cause.

Now if we look at medical art only as knowing, not as practic-   9
ing, perhaps health has the place of the final cause by means of
which we know those remedies. But nevertheless, since that whole
knowledge is directed toward practical activity, it appears health is
considered not as a cause but as an effect, and that progression is
said to be resolutive method rather than demonstrative. Add in
that demonstration made by means of final cause does not demonstrate efficient cause from the end but the effect produced by the
efficient cause, as we will make clear elsewhere. Therefore it appears that this procedure cannot be called demonstration from final cause, but rather [demonstration] from effect (*ab effectu*).

Moreover, resolutive order is accordingly called resolutive be-   10
cause it is from posterior to prior. Therefore, if some method is
conjoined with it, it is resolutive, and so from effect, not from
cause. For every cause, insofar as it is a cause, is prior to the effect.
The end, therefore, insofar as it is the final cause, is not posterior,
but prior to those things that are before the end.

For we usually say that in medical art remedies are discovered   11
from prior knowledge of health and disease. We use the name

fatemur illam esse demonstrationem, quae est methodus resolutiva, nam methodum demonstrativam non admodum solemus demonstrationem inventionis nominare; etenim inventionis nomen solam respicit cognitionem an sit, methodus autem demonstrativa consistit in declaratione quamobrem sit. Illa igitur methodus ad resolutivam potius, quàm ad demonstrativam videtur esse redigenda.

12   Ad Aristotelem autem in septimo libro de Moribus dicentem finem esse principium dicimus ipsum intelligere principium cognitionis, non principium rei. Vel, si admittamus etiam principium rei significari, id non propriè dicitur, sed solùm per quandam similitudinem, nam ille processus speciem° quandam habet methodi demonstrativae, sed multò tamen magis resolutivae, ut dictum est.

"discovery," therefore, and we acknowledge that that is a demonstration, which is resolutive method. For we do not normally call demonstrative method demonstration of discovery; indeed the name of discovery regards only knowledge of whether [something] is, but demonstrative method consists in clarification of why it is. It appears, therefore, that this method has to be reduced to resolutive rather than to demonstrative.

And about Aristotle saying in the seventh book of the *Ethics*[111] that the end is the beginning-principle, we say that he understands the beginning-principle of knowledge, not the beginning-principle of the thing. Or even if we should admit that the beginning-principle of the thing is being signified, it is not said properly, but only by means of some sort of resemblance. For that procedure has some sort of appearance° of demonstrative method, but nevertheless much more of resolutive, as was said

12

LIBER QUARTUS

: I :

Aliorum sententiae de consilio Aristotelis in Posterioribus Analyticis.

1   Posteaquam ratione duce de methodis earumque numero ac differentiis verba fecimus, non erit ab re si eam methodorum traditionem, quam Aristoteles in Posterioribus Analyticis conscripsit, aliquantum pro occasione declaraverimus. Cùm enim nullum aliud
ibi consilium eius fuisse arbitremur, quàm de methodis agere,
idque optimè et artificiosissimè ipsum praestitisse non dubitemus,
nostram de methodis sententiam testimonio tanti viri mirificè
comprobatam iri speramus. Praeterea verò in consideranda Aristotelis de methodis tractatione occasio nobis dabitur de singularum
methodorum utilitate ac fine multa dicendi, quae maximè digna
cognitu sunt et in praecedentibus dicta non fuere, ea namque facilioris doctrinae gratia ad sequentem contemplationem remittenda
esse censuimus.

2   Quoniam autem verum collatione° falsi melius conspici firmius-
que cognosci solet, primo loco sententias aliorum de Posterioribus
Analyticis referemus eorumque errores detegemus. Deinde quaenam revera sit in iis libris intentio Aristotelis declarabimus to-
tumque eius artificium in tractatione de methodis explanabimus.

3   Duae circumferuntur hac de re interpretum opiniones, una est
Latinorum omnium excepto Linconiense et Graecorum praeter
Themistium, quam his temporibus pauci secuntur. Altera, quam
omnes ferè posteriores tuentur et Averroi ac Themistio attribuunt.

BOOK FOUR

: I :

The positions of others regarding Aristotle's intent in the Posterior Analytics.[1]

Now that, led by reason, we have offered up some words on meth- 1
ods and their number and differentiae, it will not be beside the
point if we take the opportunity to clarify a little the conveying of
methods as Aristotle wrote it in the *Posterior Analytics*. Since we
think his intent there was nothing other than to deal with methods—and do not doubt that he really did so optimally and most
skillfully—we trust our position on methods will be confirmed
wonderfully well by the testimony of such a man. And moreover,
in considering Aristotle's treatment on methods, we will take the
opportunity to say a good deal about the utility and end of each of
the methods. These are especially worth knowing, but they were
not talked about in the preceding, for we deemed it that, for the
sake of easier teaching, they should be left for the following contemplation.

Now since normally the true is better seen and more firmly 2
known in collation° with the false, we will in the first place recount
others' positions regarding the *Posterior Analytics* and expose their
errors. We will then make clear what, in truth, Aristotle's intention is in those books, and explain the whole of his skill in the
treatment on methods.

On this issue two opinions of commentators circulate. One is 3
that of all the Latins except Grosseteste[2] and [all] the Greeks besides Themistius. These days, few follow this. The other [is that]
which nearly all the later [commentators] uphold and attribute to
Averroës and Themistius.

4   Latini dicunt intentionem Aristotelis in iis libris hanc unam esse, agere de demonstratione et in primo quidem libro tractare[1] de principiis eius complexis, de propositionibus earumque conditionibus; in secundo autem de principio simplici, nempè de medio ac de facili eius inventione.

5   Posteriores verò putant, in primo quidem libro agi de demonstratione, in secundo autem non amplius, sed de definitione tanquàm de altero instrumento sciendi reipsa distincto à demonstratione, proinde duo instrumenta in iis libris tradi, demonstrationem, qua accidentia et definitionem, qua substantiae cognoscantur. Quibus autem argumentis utraque secta nitatur, considerandum est.

: II :

In quo prioris sectae argumenta exponuntur.

1 Latini et Graeci sententiam suam probant multis argumentis, quorum aliqua praecipua in medium afferemus.

2   Primum quidem dicunt, demonstrationem esse instrumentum sciendi traditum propter usum, propterea eius tractationem has duas partes postulasse, unam, in qua de eius constructione sermo fieret; alteram, in qua modus facilè construendi traderetur, non enim satis est, docere ex quibusnam propositionibus demonstratio extruenda sit, sed modus quoque declarandus est, quo eiusmodi propositiones facilè invenire possimus, quando demonstrare aliquid voluerimus. Talis fuit tractatio Aristotelis de syllogismo in primo libro Priorum Analyticorum, prius enim egit de ipsius generatione, postea de facili eiusdem generatione, ut ipse testatur

The Latins say that Aristotle's intention in these books is this   4
one: to deal with demonstration and to treat in the first book its
compound beginning-principles, [that is,] premises and their characteristics, and in the second, the simple beginning-principle, that
is, the middle [term] and of the easy discovery of it.

Those who came later hold that, in the first book, demonstra-   5
tion is dealt with, and in the second, not that any more, but instead definition as another instrument for knowing scientifically,
distinct in reality from demonstration; and that accordingly there
are two instruments conveyed in these books — demonstration, by
which accidents are known, and definition, by which substances
are. On what arguments each school relies [now] has to be considered.

:   II   :

In which the arguments of the earlier school are laid out.

The Latins and Greeks prove their position by multiple argu-   1
ments; of these, we will bring up the principal ones.

First, of course, they say that demonstration is an instrument   2
for knowing scientifically, conveyed for the sake of use. And so a
treatment of it demanded these two parts: one, in which there was
discourse about its construction, the other, in which the way of
easily constructing [it] was conveyed. For it is not enough to teach
from what premises a demonstration has to be built up; the way
by which we can easily discover premises of this type, whenever we
want to demonstrate something, also has to be made clear. Such
was Aristotle's treatment on syllogism in the first book of the *Prior
Analytics*. For he dealt first with its [i.e., syllogism's] generation
and afterward with the easy generation of the same, as he himself

in principio secundae ac tertiae sectionis eiusdem libri. Facilis autem constructio syllogismi consistit in facili inventione medii, ex quo proposita conclusio per necessariam illationem colligatur.

3   Quoniam igitur demonstrationis quoque tractatio talis esse debuit, et Aristoteles in primo Posteriorum libro priorem tantùm partem prosecutus est, egit enim ibi de generatione et constructione demonstrationis, dum conditiones omnes docuit illarum propositionum, ex quibus demonstratio construenda est, ergo sequebatur ut de facili eiusdem generatione loqueretur, quod quidem facit in secundo libro docens facilem inventionem medii. Hoc enim si negemus, sequitur Aristotelem in tractatione de demonstratione mancum et diminutum fuisse.

4   Secundo argumento idem confirmant, medium in demonstratione pars praecipua est et principem locum tenet, ergo in libris de demonstratione fuit omnino agendum de medio, imò praecipuè de medio, ergo Aristoteles vel de medio egit in primo libro vel in secundo; non in primo, quia in eo egit de propositionibus, quae sunt principia complexa, medium verò est principium simplex, quod non est propositio, ergo in secundo, quare secundus liber est de medio demonstrationis.

5   Tertio loco ita argumentantur, certum est Aristotelem in secundo Posteriorum libro agere de definitione, at definitio duobus modis considerari potest, vel prout est definitio vel prout est medium demonstrationis. Prout est definitio non à logico, sed à primo philosopho consideratur, ut videre possumus in septimo et octavo Metaphysicorum, ergo in secundo Posteriorum non

attests in the beginning of the second and the third sections of the same book.[3] And moreover, easy construction of a syllogism consists in easy discovery of the middle [term] from which a given conclusion is gathered by means of necessary inference.

Since, therefore, a treatment of demonstration ought to be of 3 this sort also and in the first book of the *Posterior* [*Analytics*] Aristotle pursued only the first part—for he dealt there with generation and construction of a demonstration when he taught all the characteristics of those premises from which the demonstration has to be constructed—it will, therefore, follow that he spoke on the easy generation of the same, which, of course, he does in the second book, teaching easy discovery of the middle [term]. And if we deny this, it follows that Aristotle was deficient and wanting in [his] treatment on demonstration.

They confirm the same thing with a second argument: In a 4 demonstration, the middle [term] is the principal part and holds the foremost place. Therefore in the books on demonstration, what had to be dealt with altogether was the middle [term]—indeed, principally the middle [term]. Therefore Aristotle dealt with the middle [term] either in the first book or in the second; but not in the first, because in that [book] he dealt with premises, which are compound beginning-principles, but the middle [term] is a simple beginning-principle, which is not a premise; therefore, in the second. And so the second book is about the middle [term] of a demonstration.

In the third place they make the following argument. It is certain 5 that Aristotle deals with definition in the second [book] of the *Posterior* [*Analytics*]. But definition can be considered in two ways, either in that it is a definition or in that it is the middle [term] of a demonstration. In that it is a definition, it is considered not by the logician but by the philosopher of metaphysics, as we can see in the seventh and eighth [books] of the *Metaphysics*;[4] therefore in the second [book] of the *Posterior* [*Analytics*], it is not

consideratur nisi ut est medium demonstrationis, quare ille secundus liber est de medio.

6   Quarto argumento utuntur aliqui, nunquàm Aristoteles in Posterioribus Analyticis de definitione loquitur, quin eam referat ad demonstrationem, dicit enim omnem definitionem aut esse demonstrationis principium aut conclusionem aut demonstrationem positione differentem, ergo non vult definitionem in iis libris aliquo modo considerare nisi cum relatione ad demonstrationem et propter demonstrationem, proinde tota utriusque libri tractatio est de demonstratione, quia omnia propter demonstrationem tractantur. Igitur quando in secundo libro agit Aristoteles de definitione, eam considerat prout ad demonstrationem dirigitur, ergo prout est medium. Propterea videmus Aristotelem in secundo libro de definitione loquentem semper demonstrationis mentionem facere, quia omnia tractat propter demonstrationem.

7   Quintum argumentum addere volumus quo utitur contra Alexandrum Eustratius in principio libri secundi Posteriorum, dicebat enim Alexander illum secundum librum de definitione esse, adversus quam sententiam hunc in modum ab inscriptione argumentatur Eustratius, libri illi à re considerata inscribuntur resolutorii, quia sunt de demonstratione, quae est resolutio. At si secundum librum de definitione esse dicamus, illi ea inscriptio non aptabitur, quia definitio non potest vocari resolutio, sed potius ille liber definitivus, quàm resolutivus appellandus erit. Attamen Aristoteles utrumque vocavit resolutorium, ergo secundus quoque est de demonstratione, non de definitione. Non potest autem alia ratione dici de demonstratione, nisi quia est de medio demonstrationis et de definitione quatenus est medium, igitur non est de definitione, sed de medio.

considered except as the middle [term] of a demonstration, and accordingly, that second book is about the middle [term].

Some use a fourth argument: In the *Posterior Analytics*, Aristotle never speaks about definition without referring it to demonstration. For he says[5] that every definition is either the beginning-principle or the conclusion of a demonstration or is a demonstration differing in position. Therefore, in these books he does not want to consider definition in any way other than in relation to demonstration and for the sake of demonstration. Accordingly, the whole treatment in each book is on demonstration, because everything is treated for the sake of demonstration. When in the second book, therefore, Aristotle deals with definition, he considers it in that it is directed toward demonstration, therefore, in that it is a middle [term]. And so in the second book we see Aristotle speaking about definition and always making mention of demonstration, because he treats everything for the sake of demonstration.

We want to add a fifth argument, which Eustratius uses against Alexander in [his commentary on] the beginning of the second book of the *Posterior* [*Analytics*].[6] Alexander said that the second book is on definition. Eustratius argues against this position by way of the title. From the issue under consideration, these books are titled "analytics" (*resolutorii*) because they are about demonstration, which is [a] resolution. But if we say that the second book is about definition, this title would not apply to it, because definition cannot be called resolution. The book would instead have to be called definitive rather than resolutive. But Aristotle nevertheless called each "analytic" (*resolutorius*).[7] Therefore the second too is about demonstration, not about definition. For no other reason, moreover, can it be said to be about demonstration except because it is about the middle [term] of a demonstration and about definition insofar as it is a middle [term]. It is, therefore, not about definition but about the middle [term].

8   Alia quoque argumenta adducere possemus, quae consultò omittimus, quia ex horum, quae attulimus, solutione omnium aliorum solutio facilè desumetur. Videntur autem argumenta omnia ad secundum librum dirigi, siquidem de primo nulla fuit apud alios altercatio, licèt etiam quid in primo libro agatur non omnino intellexerint.

∶ III ∶

Prima dictae sententiae impugnatio.

1   Sententiam hanc falsam esse et ab Aristotele alienam possumus duplici ratione demonstrare, quando enim de intentione authoris in aliquo libro seu de eius libri materia controversamur, è duobus fontibus argumenta haurire possumus ad veritatem declarandam, aut enim eum librum intentè legendo consideramus quid revera author ille tractet, aut ratione duce quid agere debeat, indagamus. Primum igitur ex iis, quae ab Aristotele in primo libro Posteriorum dicuntur, argumentum sumemus; deinde ex iis, quae in secundo. Demum ratione ex natura rei sumpta demonstrabimus quid in his libris Aristoteles agere debuerit.

2   In primis quando adversarii dicunt Aristotelem in secundo libro agere de medio demonstrationis, et ipsum medium esse definitionem, petendum est ab eis an intelligant medium ibi tractari ut pariens scientiam propter quid est, an ut pariens scientiam quid est. Ipse enim in principio secundi libri asserit ab eodem medio

We could also adduce other arguments; we intentionally omit 8
them because a solution to all the others will easily be drawn from
the solution to those that we brought forward. Moreover, all the
arguments appear directed toward the second book, since indeed
in other [authors] there was no wrangling regarding the first—
even granting that they did not entirely understand what is dealt
with in the first book, either.

: III :

First attack on the said position.

We can demonstrate that this position is false and alien to Aris 1
totle in two ways. For when we dispute an author's intention in
some book or the material of the book, we can draw up arguments
for making the truth clear from two sources. Either we consider
the book by intently reading what in fact the author treats, or, led
by reason, we track down what he ought to deal with. We will,
therefore, first take an argument from the things that are said by
Aristotle in the first book of the *Posterior* [*Analytics*] and then from
those that [were said] in the second. Lastly, we will demonstrate
by reasoning, taken from the nature of the issue, what Aristotle
ought to have dealt with in these books.

In the first place, when our opponents say that in the second 2
book Aristotle deals with the middle [term] of demonstration and
that this middle [term] is a definition, it has to be asked of them
whether they understand the middle [term] to be treated there
as bringing forth scientific knowledge *propter quid* [i.e., scientific
knowledge of what something is on account of] or as bringing
forth scientific knowledge *quid est* [i.e., scientific knowledge of
what something is].[8] For in the beginning of the second book[9] he

utramque nobis praeberi cognitionem, ut per terrae interpositionem cognoscimus et cur lunae insit eclipsis et quid sit ipsa eclipsis,
quam sententiam postea fusius declarat in eodem libro à contextu
36 usque ad 47.

3   Si dicant medium tractari ut pariens scientiam quid est, sequitur tractari medium prout est definitio, quod ipsi negant, dicunt
enim definitionem potius tractari ut medium, quàm medium ut
definitionem. Cùm enim nominatio sumenda sit à modo considerandi, sequeretur tractationem non de medio, sed de definitione
appellandam esse. Consequentia verò clara est, quoniam tradere
cognitionem quid est, est proprium definitionis officium, igitur
tractari ut faciens cognitionem quid est, nil aliud significat quàm
tractari ut definitionem.

4   Si verò dicant tractari medium in secundo libro, ut pariens scientiam propter quid est, contrà ita argumentor, de medio Aristoteles hac ratione in primo libro satis superque locutus est, ut nihil
amplius de eo dicendum maneat, ergo supervacuum est de eodem
eadem ratione in secundo libro tractare. Antecedens ita probatur,
ibi agitur de medio ut pariente scientiam propter quid est, ubi
declarantur conditiones omnes, quae in ipso ad talem scientiam
pariendam requiruntur, at hae omnes in primo libro declarantur,
ergo in primo agitur de medio ut pariente scientiam propter quid
est. Maiorem propositionem nemo sanae mentis negaret, minor
probatur, tres conditiones in medio requiruntur ut talem scientiam praestet, prima ut sit notius, secunda ut sit causa illius rei,

himself asserts that each [type of] knowledge is provided to us by the same middle [term], as [for example] by means of interposition of the earth, we know both why an eclipse belongs to the moon and what an eclipse itself is. He makes this position clear at greater length afterward in the same book, from text no. 36 up to no. 47.[10]

If they say that the middle [term] is treated as bringing forth 3 scientific knowledge *quid est*, it follows that the middle [term] is treated in that it is a definition. But they deny this, for they say that the definition is treated as a middle [term] rather than the middle [term] as a definition. And since naming has to be taken from the way in which [something] is considered, it would follow that it has to be called a treatment not on the middle [term] but on definition. And the consequence is clear, since to convey knowledge *quid est* is definition's proper function; therefore, to be treated as bringing about knowledge *quid est* indicates nothing other than to be treated as a definition.

But if they say that in the second book the middle [term] is 4 treated as bringing forth scientific *propter quid*, I make the following argument to the contrary. Aristotle spoke more than enough about the middle [term] in this latter way in the first book, so nothing more would remain to be said about it. Therefore it is superfluous to treat the same thing in the same way in the second book. The antecedent is proved as follows: The middle [term], as bringing forth scientific knowledge *propter quid*, is dealt with wherever all the characteristics that are required in it [i.e., the middle term] for it to bring forth such scientific knowledge are made clear; but these are all made clear in the first book; therefore in the first, the middle [term] is dealt with as bringing forth scientific knowledge *propter quid*. No one of sound mind would deny the major premise. The minor is proved thus: Three characteristics are required in a middle [term] if it is to ensure such scientific knowledge — first, that it be more known; second, that it be the cause of

quam quaerimus, tertia ut sit causa prima, id est proxima et ae-
quata, certum est nullam aliam conditionem postulari, quia si hae
tres adsint, scientiam perfectam acquirimus propter quid res sit, ut
considerantibus manifestum est et ut Aristoteles testatur in contextu 95 primi libri Posteriorum, quando dicit, 'scientia propter
quid est fit per primam causam,' nam cui dubium, si medium sit
causa proxima rei quaesitae et praeterea etiam notior, quòd scientiam tradet praestantissimam, quae dicitur propter quid est? Hae
autem omnes conditiones tractantur ab Aristotele diligenter in
primo libro, ut videre possumus in secundo capite et in memorato
contextu 95 et aliis sequentibus.

5   At respondebunt adversarii tractari quidem in primo libro eas
conditiones, non tamen ut conditiones medii, sed ut principiorum
complexorum. Ego[2] verò ostendo eas tractari ut conditiones medii,
nam in principio illius secundi capitis definiens ipsum scire Aristoteles dixit, 'scire est causam cognoscere, propter quam res est,'
nomine autem causae omnes intelligunt medium, quare in ea definitione exprimitur medium ut causa, propter quam res est, id est
ut faciens scientiam propter quid est. Deinde ex ea definitione
colligit principiorum conditiones, inter quas nominavit duas illas
praecipuas, principia esse prima seu immediata et esse causas, id
est esse causas immediatas rei, quae demonstratur. Quas conditiones certum est medio primùm competere, deinde propositionibus propter medium, imò ne intelligi quidem in propositionibus

the thing we are inquiring after; third, that it be the first cause, that is, proximate and coextensive. It is certain that no other characteristic is demanded, because if these three are present we acquire perfect scientific knowledge *propter quid*, as is manifest to those considering [the issue] and as Aristotle attests in text no. 95 of the first book of the *Posterior* [*Analytics*], when he says, "scientific knowledge *propter quid* occurs by means of the first cause."[11] For who doubts, if the middle [term] is the proximate cause of the thing inquired after, and is, moreover, also more known, that this will convey the most excellent scientific knowledge, what is said to be [scientific knowledge] *propter quid?* And all these characteristics are treated by Aristotle carefully in the first book, as we can see in the second chapter, and in the text no. 95 referred to, and the others that follow.

But our opponents will respond that of course these characteristics are treated in the first book, but not as characteristics of the middle [term] and instead as characteristics of compound beginning-principles. But I show that they are treated as characteristics of the middle [term]. For in the beginning of that second chapter, defining that "to know scientifically," Aristotle said, "to know scientifically is to know the cause on account of which something is."[12] And by the name of cause everyone understands the middle [term]. And so in the definition is expressed the middle [term] as the cause on account of which something is, that is, as bringing about scientific knowledge *propter quid*. Then, from the definition, he gathers the characteristics of beginning-principles, among which he named these two as principal: beginning-principles are first or immediate, and they are causes, that is, they are immediate causes of something that is demonstrated. It is certain that these characteristics appertain first to the middle [term], and then to the premises on account of the middle [term]. Indeed they cannot even be understood in the premises unless they are

5

possunt, nisi prius in ipso medio intelligantur. Ideò Aristoteles medium praecipuè respexit.

6   Etenim nomen causae potest ibi multa significare, quia propositiones possunt esse causae inferendi, possunt esse causae cognoscendi; possunt denique esse causae essendi. Quonam igitur modo intellexit Aristoteles principia esse causas? Nonne his omnibus simul, potissimùm verò essendi? Hoc certè omnes fatentur. Atqui causae, quibus res sunt, non sunt propositiones, sed sunt res simplices, ut Deus, coelum, materia, forma, hae namque res aliarum rerum causae sunt. Nomen igitur causae non competit nisi medio termino, propositionibus autem tribuitur ratione medii, quoniam ex eo utraque propositio conflatur, ideò rectè ait in commentario undecimo eius libri Averroes, 'principia sunt causae, quia medium est causa maioris extremi,' quod ipse quoque Aristoteles ibi significavit, declarans enim eam conditionem subiunxit, 'causas quidem esse oportet, quia tunc scimus, cùm causam cognoscimus,' singulari namque numero medium significavit, et per medium, quod est causa, declaravit principia esse causas.

7   In contextu autem 95 discrimen statuere volens inter scientiam propter quid et scientiam quòd est, ex alterius conditionis defectu, eas nominat in singulari numero, quia in medio ipsas considerat, nam in principio illius capitis dicit fieri tunc scientiam solùm quòd est, sed non propter quid est, quando non sumitur prima causa, deinde addit, 'scientia enim propter quid est fit per primam causam,' per hanc autem conditionem ibi separat demonstrationem propter quid ab illa demonstratione quòd, cuius medium est causa

first understood in the middle [term] itself. And so Aristotle regarded the middle [term] as principal.

Now indeed the name of cause can signify many things there, 6 because the premises can be causes of inferring, they can be causes of knowing, and lastly they can be causes of being. In which way, therefore, did Aristotle understand beginning-principles to be causes? Is it not in all these at once, but chiefly [as causes] of being? Certainly everyone acknowledges this. But yet the causes by which things are, are not premises, but are simple things, such as God, heaven, matter, form, and these things are causes of other things. The name *cause* therefore, does not appertain to anything except the middle term, and is ascribed to the premises by reason of the middle [term], since either premise can be assembled from it. Because of this, in the eleventh commentary to the book, Averroës correctly says, "Beginning-principles are causes, because the middle is the cause of the major extreme."[13] Aristotle himself indicated this there also, for, clarifying this characteristic, he added, "There must, of course, be causes, because we know scientifically whenever we know the cause."[14] He indicated the middle [term] in the singular, and by means of the middle [term], which is a cause, he made clear that beginning-principles are causes.

Moreover, in text no. 95,[15] wanting to fix the discriminating dif- 7 ference between scientific knowledge *propter quid* and scientific knowledge *quòd* [i.e., scientific knowledge that something is the case[16]], from the absence of another characteristic, he names them [i.e., the two forms of scientific knowledge] in the singular, because he is considering them in the middle [term]. For in the beginning of that chapter he says that when the first cause is not given, there is then only scientific knowledge *quòd* and not *propter quid*. He then adds, "For scientific knowledge *propter quid* occurs by means of the first cause."[17] So by means of this characteristic, he there separates demonstration *propter quid* from that former, demonstration *quòd*, whose middle [term] is the remote cause. He

remota, quod facere non posset, nisi prius in demonstratione propter quid eam conditionem declarasset. At nullibi id fecit nisi in secundo capite, quare ibi tradens conditiones principiorum tradit conditiones medii, per quas postea distinguit demonstrationem propter quid à demonstratione quòd.

8   In contextu etiam 99 declarans Aristoteles eam demonstrationem, quae fit per causam remotam, in singulari numero nominat causam primam et causam non primam, quam inquit esse medium distans et remotum à maiore extremo. Quare dubitare minimè debemus Aristotelem conditiones omnes medii termini in primo libro declarasse; quod ex eo quoque manifestum esse potest, quòd dè illis ne verbum quidem in secundo libro facit, attamen eas esse conditiones in^3 medio requisitas, inficiari nemo potest, quòd enim principia complexa debeant esse immediata et causae concedimus, at nonne etiam medium debet esse immediata causa eius, quod quaeritur? Igitur si agere de principiis non fuit agere de medio, debuit Aristoteles non solùm de principiis, sed etiam seorsum de medio eas conditiones declarare. Ergo si secundus liber de medio est, ubinam in secundo libro docet Aristoteles medium esse causam immediatam? Certè nullibi. Secundus igitur liber non est dicendus de medio, si conditiones, quae in medio sunt penitus necessariae, in secundo libro non declarantur.

9   Verùm erroris istorum causa^4 fuit quòd principia complexa à medio distinguere voluerunt, quae Aristoteles in Posterioribus Analyticis nunquàm distinxit, in iis enim materia demonstrationis, non forma consideratur, materiam autem sive ipsum medium esse

could not have done this unless he had first clarified that characteristic in a demonstration *propter quid*. But he did this nowhere
except in the second chapter. Therefore conveying the characteristics of beginning-principles there, he conveys the characteristics of
the middle [term], by means of which [characteristics] he afterward distinguishes demonstration *propter quid* from demonstration *quòd*.

Also, in text no. 99,[18] Aristotle, clarifying the demonstration 8
that occurs by means of remote cause, names in the singular the
first cause and the cause that is not first, what he says is a middle
[term] distant and remote from the major extreme. And so we
ought not to doubt at all that Aristotle made all the characteristics
of the middle term clear in the first book; this can also be manifest
from the fact that he offers not even a word about all this in the
second book. Nevertheless, no one can deny that in the middle
[term], these are required characteristics. For we concede that
compound beginning-principles ought to be immediate and ought
to be causes. But ought not the middle [term] to be the immediate
cause of that which is inquired after? If, therefore, to deal with
beginning-principles was not to deal with the middle [term], Aristotle ought not to have clarified only those characteristics regarding beginning-principles, but also, separately, those regarding the
middle [term]. Therefore if the second book is about the middle
[term], where at all in the second book does Aristotle teach that
the middle [term] is the immediate cause? Certainly nowhere. The
second book, therefore, should not be said to be about the middle
[term], if the characteristics that are completely necessary in the
middle [term] are not made clear in the second book.

But the cause of their error was surely that they wanted to dis- 9
tinguish compound beginning-principles from the middle [term];
Aristotle never distinguished this in the *Posterior Analytics*, for in
it, the matter, not the form, of demonstration is considered.
But whether we say the matter is the middle [term] itself or the

dicamus sive propositiones, idem dicimus, ex medio enim constat utraque propositio et praeter medium nihil habet nisi terminos conclusionis. Discrimen autem inter medium et propositiones est solum penes formam logicam, quia medium est vox simplex vel saltem conceptum significat simplicem, principia verò sunt enunciationes, verum vel falsum[5] significantes. Quae quidem forma in libro de Interpretatione et in Prioribus Analyticis consideratur, sed non in Posterioribus, ubi sola materia demonstrationis respicitur. Quo fit, quemadmodum diximus, ut de principiis loquens Aristoteles de medio loquatur et de medio loquens loquatur de principiis, haec enim absque dubio in iis libris confundit, ut videre est in 48 et 49 contextibus primi libri, cùm enim proposuisset ostendendum principia demonstrationis esse necessaria, probat medium esse necessarium. Idem legere possumus in contextu 51, sic etiam in 54 ostendit principia demonstrationis esse per se et universalia, deinde id declarans et concludens in 56 dicit medium tertio et primum medio per se inesse, significans nil aliud esse principia ponere per se, quàm dicere medium terminum per se cum utroque extremo connexum esse. Nam conditiones medii quatenus est medium sunt conditiones principiorum, quia vox ista, medium, relationem habet ad extrema, medium enim aliquorum est medium et nisi extrema sint, medium esse non potest, sicuti medio existente necesse est extrema quoque actu esse. Praesentibus autem extremis et medio, iam propositiones actu adsunt.°

premises, we say the same thing. For each premise is composed out of the middle [term] and has nothing besides the middle [term] except the terms of the conclusion. Moreover, the discriminating difference between the middle [term] and the premises is only in logical form, because the middle [term] is a simple word, or at least signifies a simple concept, but beginning-principles are propositions signifying true or false. This form, of course, is considered in the book *On Interpretation* and in the *Prior Analytics*, but not in the *Posterior* [*Analytics*], where only the matter of demonstration is considered. Hence, just as we said, when speaking about beginning-principles, Aristotle speaks about the middle [term], and when speaking about the middle [term], he speaks about beginning-principles. For without doubt, in those books he conflates these, as appears in texts no. 48 and no. 49 in the first book.[19] For although he had proposed to show that the beginning-principles of demonstration are necessary, he proves that the middle [term] is necessary. We can read the same in text no. 51.[20] And in no. 54,[21] too, he shows that the beginning-principles of demonstration are *per se* and universal. Then, clarifying and concluding this in no. 56,[22] he says that the middle [term] belongs *per se* to the third, and the first to the middle [term], indicating that to posit beginning-principles as *per se* is nothing other than to say that the middle term is connected *per se* with each extreme. For the characteristics of the middle [term], insofar as it is a middle [term], are the characteristics of beginning-principles, because that word, "middle [term]," is relative to the extremes. For a middle [term] is a middle between some things, and unless there are extremes, there can be no middle, just as, with the middle existing, it is necessary that the extremes actually are also, and with the extremes and middle present, the premises then actually exist.°

10   Quoniam igitur conditiones principiorum complexorum consistunt in relatione terminorum ad se invicem et conditiones medii consistunt in relatione ad extrema, necesse est ut conditiones principiorum sint conditiones medii et conditiones medii sint conditiones principiorum. Ideò Aristoteles in primo libro conditiones principiorum considerans modò in plurali numero loquitur, modò in singulari, falsum est igitur id, quod Latini dicunt, Aristotelem distinguere tractationem de principiis complexis à tractatione de medio et in primo libro ita de principiis loqui, ut non de medio et in secundo ita de medio, ut non de principiis, haec enim, si rectè considerentur, pugnantia sunt, quae simul consistere nequeunt.

11   Notare etiam possumus, Aristotelem in contextibus 73, 74, 75 primi libri de praecognitis loquentem et dicentem principia esse praecognita, nomine principiorum non modò propositiones, sed etiam subiectum comprehendisse, tamen subiectum non est principium complexum. Non est igitur verum quòd ita separaverit tractationem de principiis complexis à tractatione de simplicibus, ut in primo libro de complexis tantùm, in secundo de simplicibus tantùm agere constituerit.

Since, therefore, the characteristics of the compound beginning- 10
principles consist in the relation of terms to each other in turn,
and the characteristics of the middle [term] consist in the relation
to the extremes, it is necessary that the characteristics of the
beginning-principles are characteristics of the middle [term], and
characteristics of the middle [term] are characteristics of the
beginning-principles. Because of this, Aristotle, considering the
characteristics of beginning-principles in the first book, speaks
now in the plural, now in the singular. What the Latins say, therefore, is false, that Aristotle distinguishes a treatment on compound
beginning-principles from a treatment on the middle [term], and
speaks in the first book about beginning-principles and so not
about the middle [term], and in the second about the middle
[term] and so not about beginning-principles. For these [state-
ments], if they are considered correctly, are in conflict; they cannot
hold at the same time.

We can also note that Aristotle, in texts no. 73, no. 74, and no. 11
75 of the first book,[23] speaking about what is known beforehand
and saying that beginning-principles are known beforehand, comprehended under the name, "beginning-principles," not only the
premises, but also the subject. Nevertheless a subject is not a compound beginning-principle. It is not true, therefore, that he separated a treatment on compound beginning-principles from a treatment on simple [beginning-principles] and so decided to deal only
with compounds in the first book and only with simples in the
second.

: IV :

Eiusdem sententiae impugnatio secunda.

1   Possumus etiam illa spectando, quae in secundo libro Posteriorum tractantur, eius sententiae falsitatem ostendere. Aristoteles in primo capite secundi libri Posteriorum Analyticorum declarat quomodo una et eadem causa, unum et idem medium satisfaciat omnibus quaestionibus, quae de eadem re fieri possunt, sive an sit sive quid sit sive an insit sive cur insit, quo fit, ut omnis quaestio sit unius medii quaestio, ut ipse ibi colligit. Postea de definitione tractationem aggreditur, in qua usque ad contextum 47 eius libri versatur.

2   Videamus igitur an in illa parte definitionem consideret quatenus est medium, an potius medium quatenus est definitio, hoc autem facilè dignoscemus,° si intelligamus quidnam sit medium considerari ut medium et medium ut definitionem et definitionem ut medium et definitionem ut definitionem. Officium definitionis est significare quid est, officium autem medii est alterius in altero inhaerentiam declarare, medium enim extremorum est medium; quare per ipsum, quatenus est medium, non ostenditur nisi extremorum connexio et alterius in altero inhaerentia. Quando autem medium est vera causa illius inhaerentiae, declarat etiam cur alterum alteri inhaereat, ut evenit in demonstratione, quae dicitur propter quid. Itaque medii quatenus est medium nulla est alia vis et natura, quàm ut declaret quaestiones complexas, quòd est et

: IV :

Second attack on the same position.

We can also show the falsity of this position by looking at those 1
things that are treated in the second book of the *Posterior* [*Analytics*]. In the first chapter of the second book of the *Posterior Analytics,* Aristotle makes clear how one and the same cause, one and
the same middle [term], may answer all questions that there can
be regarding some one thing—whether it is, what it is, whether it
belongs, why it belongs.[24] Hence every question is a question of
one middle [term], as he himself there gathers. Afterward he undertakes a treatment on definition, with which he is concerned up
to text no. 47[25] of the book.

Let us see, therefore, whether in that part he considers defini- 2
tion insofar as it is a middle [term], or, instead, the middle [term]
insofar as it is a definition. We will easily discern° this if we understand what it is for a middle [term] to be considered as a middle, and a middle [term] as a definition, and a definition as a
middle [term], and a definition as a definition. The function of a
definition is to signify what [something] is (*quid est*), and the function of the middle [term] is to make clear the inherence of one
thing in another. For the middle [term] is a middle between extremes, and therefore, insofar as it is a middle, there is nothing
shown by means of it except a connection of extremes and the inherence of something in something else. But now when the middle
[term] is the true cause of that inherence, it also clarifies why
something inheres in something else, as happens in a demonstration that is said to be *propter quid.* And so, insofar as it is a middle,
there is no other force and nature of a middle [term] than to
clarify the compound questions that [something] is the case (*quòd*)
and what [something] is on account of (*propter quid*). For it is like

propter quid est, est enim veluti mediator quidam inter duos, qui eos componit atque coniungit.

3   Quòd si medium alterius extremi essentiam et quidditatem declaret, id non praestat quatenus medium, sed quatenus est definitio. Officium enim definitionis, ut omnes norunt, est declarare quid res sit et quaestioni simplici satisfacere, quare definitio, ut est definitio, non est media inter duo; quemadmodum medium ut est medium,[6] respicit duo extrema et quaestionibus tantùm complexis satisfacere potest. Definitionem igitur sumere ut definitionem est eam considerare ut declarantem quid res sit. Eandem autem sumere ut medium, est eam considerare ut notificantem propter quid inest hoc in illo. Similiter sumere medium ut medium est ipsum[7] considerare ut faciens scientiam propter quid est. At sumere ipsum ut definitionem est ipsum considerare ut declarans quid est.

4   Quòd si Aristoteles in primo capite secundi libri Posteriorum dicit quaestionem quid est esse quaestionem medii, non propterea vult medium, quatenus est medium, praebere cognitionem quid est, de medio enim demonstrationis loquitur et ostendere vult quòd unum et idem medium largitur cognitionem propter quid, et quid, at non eadem ratione, praestat enim scientiam propter quid, quatenus est medium demonstrationis, declarat autem quid est quatenus est definitio vel pars definitionis, non quatenus est medium seu pars demonstrationis, quandoquidem eadem oratio, ut inquit Aristoteles, alio modo dicitur demonstratio, alio modo definitio, ut nos infrà declarabimus. Igitur non ad modum considerandi ipsum medium, sed ad medium et ad rem ipsam sunt

a sort of mediator between two things — it brings them together and conjoins them.

Now if the middle [term] clarifies the essence and quiddity of another extreme, it does not really do this insofar as it is a middle but insofar as it is a definition. For the function of a definition, as everyone knows, is to make clear what a thing is (*quid est*) and to answer a simple question, and so a definition, as it is a definition, is not a middle between two things, just as a middle [term], as it is a middle, is with regard to two extremes and can only answer compound questions. To take a definition, therefore, as a definition, is to consider it as making clear what something is. And to take the same thing as a middle [term] is to consider it as making known on account of what (*propter quid*) this belongs to that. Similarly, to take the middle [term] as a middle is to consider it as bringing about scientific knowledge *propter quid*. But to take it as a definition is to consider it as making clear what [something] is (*quid est*).

Now if Aristotle says, in the first chapter of the second book of the *Posterior* [*Analytics*],[26] that the question of what [something] is (*quid est*) is the question of a middle [term], that does not mean he wants the middle [term], insofar as it is a middle, to provide knowledge *quid est*. For he is speaking about the middle [term] of a demonstration and wants to show that one and the same middle [term] yields knowledge *propter quid* and [knowledge] *quid*, but not in the same way. For insofar as it is the middle [term] of a demonstration, it ensures scientific knowledge *propter quid*, but clarifies what [something] is (*quid est*) insofar as it is a definition or part of a definition — not insofar as it is a middle [term] or part of a demonstration — since, as Aristotle says,[27] the same speech is said in one way to be demonstration and in another way to be definition, as we will make clear below. Those words and Aristotle's position, therefore, have to be referred, not to the way of considering the middle [term] itself, but to the middle [term] and to the thing

3

4

referenda verba illa et sententia Aristotelis, idem enim medium
dat nobis cognitionem propter quid et quid, alia tamen et alia ratione.

5   His omnibus declaratis quae ex ipsorum vocabulorum significatione manifesta sunt, falsitas illius sententiae facillimè demonstratur, nam Aristoteles postquàm declaravit omnes quaestiones
esse unius medii quaestiones proponit in fine illius primi capitis
investigandum quomodo veniamus in cognitionem quid est, an per
demonstrationem an per aliquam aliam viam et in hoc declarando
versatur usque ad 47 contextum.

6   Ut igitur concedamus, Aristotelem in tota illa parte agere de
medio, hoc enim nos quoque asseveramus, tamen clarum est de
medio ibi agi non prout est medium, sed prout est definitio, quare
si à modo considerandi tractationi nomen imponere convenit, liber
ille de definitione potius vocandus est, quàm de medio, quia de
medio agitur ut ducente ad cognitionem quid est, proinde quatenus est definitio, non quatenus est medium.

7   Praeterea triplicem esse definitionem docet ibi Aristoteles, alia
namque est principium demonstrationis, alia est demonstrationis
conclusio, alia demum est ipsamet demonstratio positione differens à demonstratione. Dicant igitur adversarii, si in ea parte de
definitione ageretur quatenus est medium, de quanam definitione
ageretur? Certè respondere coguntur, de illa, quae est principium
demonstrationis, idem enim est principium dicere demonstrationis
et medium demonstrationis, nec alia ratione definitio principium
demonstrationis esse dicitur, nisi quatenus est ipsius medium. Id
tamen est manifestè falsum et Aristoteli adversatur, qui in illa
parte in contextu 42 asserit se non loqui de illa definitione, quae

itself. For the same middle [term] gives us knowledge *propter quid* and [knowledge] *quid est*, one in one way, the other in another.

Once these things (manifest from the signification of the words 5 themselves) have been made clear, the falsity of that position is most easily demonstrated. For Aristotle, after he made clear that all questions are questions of one middle, sets out, at the end of that first chapter,[28] to investigate in what way we come into knowledge of what [something] is (*quid est*), whether by means of demonstration or by means of some other way, and concerns himself with clarifying this up to text no. 47.[29]

As we concede, therefore, that Aristotle deals in that whole part 6 with the middle [term] — for we too have averred this — it is nevertheless clear that he deals there with the middle [term], not in that it is a middle, but in that it is a definition. Therefore, if it is appropriate to impose a name on a treatment from [its] way of considering something, that book has to be called "on definition" rather than "on the middle [term]," because it deals with the middle [term] as leading to knowledge *quid est*, and so insofar as it is a definition, not insofar as it is a middle [term].

Moreover, Aristotle there[30] teaches that definition is threefold: 7 it is the beginning-principle of a demonstration; it is the conclusion of a demonstration; and lastly, it is the very demonstration itself, differing in position from a demonstration.[31] Let our opponents say, therefore, if in that part [of the *Analytics*] definition was being dealt with insofar as it is a middle [term], what definition was being dealt with? Certainly, they are forced to respond: that which is the beginning-principle of a demonstration, for it is the same thing to say "the beginning-principle of a demonstration" and [to say] "the middle [term] of a demonstration," and there is no other reason a definition is said to be the beginning-principle of a demonstration except insofar as it is its middle [term]. This, however, is manifestly false and conflicts with Aristotle, who in that part, in text no. 42,[32] asserts that he is not speaking about that

est immediata et demonstrationis principium. Et totam illam partem legentibus manifestum est Aristotelem considerare definitionem et ipsum quid est, prout est ignotum quoddam, quod sub quaestionem cadit. Illa verò definitio, quae est medium, est principium notum et indemonstrabile.

8   Sed utcumque res haec sese habeat, haec enim omnia in sequentibus diligentius expendentur, illud certum esse debet, Aristotelem in tota illa parte loqui de ea definitione, quae est demonstratio positione differens, non de illa, quae est demonstrationis principium. Et haec quidem est principium ante demonstrationem cognitum, illa verò est naturaliter ignota et quaeritur et ex facta demonstratione extrahitur et innotescit, quare non ad demonstrationem dirigitur, sed potius demonstratio ad ipsam. Qui igitur dicunt agi in ea parte de definitione prout est medium et prout dirigitur ad demonstrationem, decipiuntur et illam Aristotelis tractationem non intelligunt.

: V :

Eiusdem sententiae impugnatio tertia.

1   Verùm non modò quid egerit Aristoteles, sed etiam quid agere debuerit considerando possumus adversus eam sententiam argumentari. Ars logica, ut perfecta dici possit, tradere nobis debet instrumenta sive methodos, quibus ad omnium eorum, quae quaeri ac sciri possunt, cognitionem pervenire possimus.[8] Quaeruntur autem saepe rerum definitiones ignotae, imò inquit Aristoteles

definition that is immediate and the beginning-principle of a demonstration. And to those reading the whole part it is manifest that Aristotle considers definition and what something itself is (*quid est*), in that it is something unknown, as something falling under the question. But that definition, which is the middle [term], is a known and indemonstrable beginning-principle.

But however this may be—and all of it will be weighed very 8 carefully in what follows—this ought to be certain: in that whole part, Aristotle speaks about the definition that is a demonstration differing by position, not about that which is the beginning-principle of a demonstration. And the latter, of course, is a beginning-principle known before the demonstration. And the former is naturally unknown, and is inquired after, and is extracted from the demonstration performed, and is made known. It is thereby not directed toward demonstration, but instead demonstration [is directed] toward it. Those, therefore, who say that definition is dealt with in that part, in that it is a middle [term] and in that it is directed toward demonstration, are deceived and do not understand Aristotle's treatment.

: V :

Third attack on the same position.

We can make an argument against this position not only by con- 1 sidering what Aristotle dealt with, but also what he ought to have dealt with. So that it can be said to be perfect, logical art ought to convey to us instruments or methods by which we can come through to knowledge of everything that can be inquired after and known scientifically. And often the unknown definitions of things are inquired after, and indeed Aristotle says that many

multas definitiones usque adeò esse ignotas, ut demonstratione indigeant, qua innotescant, propterea etiam in principio secundi libri Posteriorum Analyticorum inter illa, quae in scientiis ignorari et quaeri solent, nominavit quaestionem quid est. Ergo nisi in libris logicis doceat Aristoteles methodum aliquam, quae ducat in cognitionem ipsius quid est, quando ignoratur, absque dubio logicam artem mancam atque imperfectam tradidit.

2   Quod quidem absurdum sententiam eorum sequitur, qui dicunt Aristotelem in secundo Posteriorum libro non agere de definitione nisi prout est medium demonstrationis, nam medium quatenus est medium non praestat nobis aliam scientiam, quàm quòd hoc insit in illo et quamobrem insit, sed non quid sit. Hoc enim si praestet, id facit quatenus est definitio, non quatenus est medium. Ergo si de definitione agat Aristoteles ea tantùm ratione, qua est medium demonstrationis, non considerat eam nisi ut nos ducit in cognitionem quaesitorum complexorum. Igitur instrumentum, quo ducamur ad cognitionem quid est, desideratur et ab Aristotele fuit praetermissum. Quod absurdum ausus est confiteri Albertus in suis commentariis in Porphyrium tractatu primo capite quinto, dicens definitionem à primo quidem philosopho considerari, sed tamen alia ratione etiam à logico esse consideradam, nempè quatenus tradit cognitionem quid est, quam tractationem neque ab Aristotele neque ab ullo alio antiquorum factam reperiri, proinde desiderari tractationem logicam de modo definiendi; unde fit, ut Aristoteles logicam artem imperfectam et mancam tradiderit et partem eius valdè praestantem et praecipuam praetermiserit.

3   Cognovit itaque Albertus tractandum fuisse in logica de definitione etiam quatenus est definitio, sed deceptus est dum putavit,

definitions are so unknown that they are in need of a demonstration by which they may become known. And accordingly, in the beginning of the second book of the *Posterior Analytics*,[33] he named the question of what [something] is (*quid est*) among those things that, in sciences, are normally unknown and inquired after. Therefore, unless in the logic books Aristotle teaches some method that leads into knowledge *quid est* when it is unknown, without doubt he conveyed a deficient and imperfect logical art.

This absurdity, of course, follows from the position of those who say that Aristotle, in the second book of the *Posterior* [*Analytics*], does not deal with definition except in that it is the middle [term] of a demonstration. For the middle [term], insofar as it is a middle, does not ensure for us other scientific knowledge than that this belongs to that and why it belongs, but not what it is (*quid est*). And if it does do the latter, it does so insofar as it is a definition, not insofar as it is a middle [term]. Therefore, if Aristotle deals with definition only in the way in which it is the middle [term] of a demonstration, he does not consider it except as it leads us into knowledge of compound inquiries. The instrument, therefore, by which we are led to knowledge *quid est* is wanting and was overlooked by Aristotle. Albert dared to confess this absurdity in his commentary on Porphyry's first treatise, in the fifth chapter,[34] saying that definition is, of course, considered by the philosopher of metaphysics, but it also has to be considered in another way by the logician, that is, insofar as it conveys knowledge *quid est*; and that such a treatment is found to have been made neither by Aristotle nor by any other of the ancients; and accordingly that a logical treatment on the way of defining is wanting; and so it happens that Aristotle conveyed an imperfect and deficient logical art, and overlooked its principal and highly excellent part.

Albert thus knew that indeed, in logic, definition had to be treated insofar as it is a definition, but he was deceived when he

tractationem hanc ab Aristotele praetermissam fuisse, ut nos postea demonstrabimus.

4   Videntur etiam Latini solam accidentium cognitionem respexisse, dum solam demonstrationem in Posterioribus Analyticis tractari asseruerunt, demonstratione enim sola accidentia cognoscimus, non substantias. Tamen substantias quoque ignorari et quaeri contingit, quare ipsae[9] quoque instrumento logico egent, quo innotescant.

5   Ratio igitur nos fateri compellit tractanda esse in logica instrumenta declarantia non modò quaestiones complexas, sed etiam quaestionem quid est; et non solùm ad accidentium, verùm etiam ad substantiarum cognitionem conferentia. Attamen si eorum sententiam admittamus, non tradidit Aristoteles instrumentum ducens ad cognoscendum quid est, neque instrumentum idoneum ad substantias notificandas, sed solùm ad accidentia et ad solas quaestiones complexas.

∴ VI ∴

In quo posterioris sectae argumenta proponuntur.

1   Alteram sententiam plurimi posteriores secuntur, quam Averrois fuisse profitentur et eam videtur sequi Linconiensis in principio secundi libri Posteriorum Analyticorum. Hi dicunt, duo esse instrumenta sciendi, quae tractantur ab Aristotele in duobus libris Posterioribus Analyticis, demonstrationem, quae in primo et definitionem, quae in secundo libro consideratur, eaque esse instrumenta re distincta, nam demonstratione accidentia, definitione verò substantias cognoscimus. Et quia demonstrationem in primo

held that this treatment had been overlooked by Aristotle, as we will demonstrate below.

The Latins too appear to have given regard only to knowledge 4 of accidents when they asserted that only demonstration is treated in the *Posterior Analytics*. For by demonstration we know only accidents and not substances. Nevertheless, it happens that substances too are unknown and inquired after, and so they too need a logical instrument by which they may become known.

Reason, therefore, compels us to acknowledge that in logic 5 there has to be treatment of instruments for making clear not only compound questions but also the question of what [something] is (*quid est*) and for contributing to knowledge not only of accidents but also of substances. But if we admit their position, Aristotle did not convey an instrument leading to knowing what [something] is (*quid est*), nor an instrument fit for making substances known, but only [an instrument for making known] accidents and only compound questions.

: VI :

In which the arguments of the later sect are set out.

Many of those who came later follow another position, which they 1 claim was Averroës' and which Grosseteste appears to follow in the beginning of the second book of the *Posterior Analytics*.[35] They say that there are two instruments for knowing scientifically and they are treated by Aristotle in the two *Posterior Analytics* books — demonstration, which is considered in the first book, and definition, which is considered in the second — and that each instrument is distinct in reality, for we know accidents by demonstration and substances by definition. And because no one ever denied that

libro tractari nemo unquàm negavit, sed tota difficultas manet in secundo an in eo tractetur definitio ut instrumentum substantiae cognoscendae re distinctum à demonstratione, ideò hanc partem pluribus argumentis comprobare nituntur.

2   Primùm dicunt, in unaquaque re duo sunt ad summum, quae nobis cognoscenda proponuntur, substantia ipsius rei et accidentia propria, quae in re insunt, ut significat Aristoteles in tertio contextu primi libri de Anima, ergo si ars logica debet esse perfecta, debet instrumenta tradere, quibus utrumque cognoscamus, at certum est substantiam demonstratione non innotescere, sed accidentia solùm; substantiam autem nullo alio instrumento cognosci, quàm definitione. Ergo de definitione ut de instrumento substantiae cognoscendae debuit omnino Aristoteles agere in logica. At nullus est alius liber, in quo id agatur, nisi secundus Posteriorum et id negando dicimus Aristotelem in logica fuisse diminutum.

3   Praeterea logica est facultas instrumentalis, quia instrumenta sciendi tradit, ergo quicquid à logico tractatur, instrumentum sciendi est et tractatur ut instrumentum, atqui tractatur in secundo libro Posteriorum definitio et longo quidem sermone, igitur tractatur ut instrumentum sciendi.

4   Praeterea definitio potest duobus modis considerari, vel ut est essentia rei vel ut est instrumentum tradens cognitionem quid est, primo quidem modo consideratur à divino philosopho; secundo autem modo vel ab eodem consideratur vel à logico, alius enim praeter hos duos qui id facere possit, non manet. Sed à primo philosopho ea ratione considerari non potest, quia sequeretur

demonstration is treated in the first book—the whole problem lies in the second—[the issue is] whether definition is treated in it as an instrument for knowing substance, distinct in reality from demonstration; they try to confirm this latter part with many arguments.

First they say that in any thing there are at most two things 2 that are set out for us to know: the substance of the thing and the proper accidents that belong to it, as Aristotle indicates in text no. 3 of the first book of *On the Soul*.[36] Therefore, if logical art is to be perfect and complete, it ought to convey instruments by which we may know each. But it is certain that by demonstration, not substance but only accidents become known, and substance is known by no instrument other than definition. Therefore, in [his writings on] logic, Aristotle ought to have dealt altogether with definition as an instrument for knowing substance. But there is no other book in which this is dealt with except the second of the *Posterior* [*Analytics*], and by denying this, we say that Aristotle was wanting in logic.

Moreover, logic is an instrumental faculty because it conveys in- 3 struments for knowing scientifically. Therefore, whatever is treated by the logician is an instrument for knowing scientifically and is treated as an instrument. Yet definition is treated in the second book of the *Posterior* [*Analytics*] and in quite a long discourse too. Therefore it is treated as an instrument for knowing scientifically.

Furthermore, definition can be considered in two ways, either 4 as the essence of something or as an instrument conveying knowledge *quid est*. In the first way, of course, it is considered by the philosopher of divinity, and in the second way, it is considered either by the same or by the logician. There is no other besides these two who could do this. But for the same reason, it cannot be considered by the philosopher of metaphysics, because it would follow that the science of divinity is an instrumental faculty. Therefore [definition is considered] by the logician, who is

divinam scientiam esse facultatem instrumentalem, ergo à logico, qui est artifex instrumentalis, proinde in secundo libro Posteriorum.

5   Confirmant hanc sententiam authoritate Averrois multis in locis in suis magnis commentariis in Posteriores Analyticos, praesertim in sua magna praefatione et in principio secundi libri, ubi asserit primum librum esse de demonstratione, secundum de definitione. Sed videtur apertissimè hanc sententiam protulisse Averroes in commentario 42 septimi libri Metaphysicorum, ubi dicit definitionem considerari à logico quatenus est instrumentum dirigens ad cognoscendum quid est, à metaphysico autem quatenus est essentia et quidditas rei.

6   Alia quoque pro hac opinione comprobanda possent argumenta adduci, quae cùm et apud alios legi et cognita veritate facilè solvi queant, missa facere voluimus.

: VII :

In quo ea sententia et ratione et Aristotelis
authoritate refellitur.

1   Huius sententiae falsitas ex iis, quae antè de methodis diximus, manifesta est, ostendimus enim, definitionem logicum instrumentum non esse. Nunc autem demonstrabimus eam secundùm Aristotelis et Averrois placita falsam esse et omnino reprobandam.

2   In primis quando dicunt definitionem esse instrumentum substantiae cognoscendae, videndum est, quid nomine substantiae intelligant, substantia namque duobus modis accipi solet, quandoque

the practitioner of instruments, and accordingly in the second book of the *Posterior [Analytics]*.

They confirm this position by the authority of Averroës in 5 many passages in his large commentary on the *Posterior Analytics*, especially in his large preface[37] and in the beginning of the second book,[38] where he asserts that the first book is about demonstration and the second about definition. And Averroës appears to have advanced this position most plainly in commentary 42 to the seventh book of the *Metaphysics*,[39] where he says that definition is considered by the logician insofar as it is an instrument directing [us] toward knowing what [something] is (*quid est*), and by the metaphysician insofar as it is the essence and quiddity of something.

Other arguments too could be adduced for confirming this 6 opinion; we want to set them aside since they can also be read in others and, once the truth is known, can easily be done away with.

: VII :

In which this position is refuted both by reason and by the authority of Aristotle.

The falsity of this position is manifest from the things that we said 1 earlier about methods. For we showed that definition is not a logical instrument. And so, we will now demonstrate that, according to the preferences of Aristotle and Averroës, it [i.e., the position of the later sects] is false and has to be altogether condemned.

In the first place, when they say that definition is an instrument 2 for knowing substance, what they understand by the name of substance has to be considered. For substance is normally accepted in two ways. Sometimes it signifies the first category, distinct from

significat primam categoriam ab aliis novem distinctam, quandoque latius pro omni essentia sumitur tam substantiae quàm accidentis, quemadmodum usus est hac voce Aristoteles in declaratione primi modi dicendi per se, quando dixit substantiam trianguli ex linea constare et substantiam lineae ex puncto. Nomen igitur substantiae si latè pro essentia sumant, non separant, ut ipsi separare volunt, scopum definitionis à scopo demonstrationis, quia si in eo sensu definitio est instrumentum cognoscendi substantiam, id est essentiam tam substantiae quàm accidentis, eorum ratio nihil roboris habet, quando ex eo quòd in rebus duo sunt cognoscenda, substantia et accidens, colligunt[10] duobus instrumentis re distinctis ad eorum notitiam° acquirendam opus esse. Dicam enim, duo quidem esse cognoscenda, utraque tamen unico instrumento cognosci definitione, quare demonstratio supervacua erit.

3   Si verò substantiam strictius accipiant pro substantia, ut vocant, praedicamentali distincta à novem generibus accidentium, ut definitio dicatur instrumentum, quo substantiam cognoscimus, demonstratio verò instrumentum, quo accidentia primùm quidem in hoc dogmate haec absurditas notari potest quòd ita substantiam definitione cognosci aiunt, ut inficiari videantur accidentia definitione cognosci posse, quod omnino falsum est, traduntur enim in scientiis non minùs accidentium, quàm substantiarum definitiones, neque definitio his limitibus arctanda et includenda est, ut solis substantiis tribuatur, nam si libros Aristotelis naturales scrutemur, videbimus plurima accidentia definitione declarari, inquirit enim quid sit motus, quid locus, quid tempus, quid generatio, quid mistio, quid putredo, quid coctio et alia complura accidentia naturalia, quae videntur ab Aristotele non demonstratione, sed

the other nine.[40] Sometimes it is taken more broadly for every essence, as much of substance as of accident. Aristotle used the
word in this way in clarification of the first way of saying *per se*,
when he said that the substance of a triangle is composed out of
line and the substance of line out of point.[41] If, therefore, they take
the name of substance in the broad sense for essence, they do not
separate, as they want to separate, the goal of definition from the
goal of demonstration. Because if definition is an instrument for
knowing substance in this sense, that is, as much [for knowing]
the essence of substance as [for knowing the essence] of accident,
their reasoning has no weight, when, from the fact that in things,
two things — substance and accident — have to be known, they
gather that two instruments distinct in reality are needed for acquiring knowledge° of them [i.e., of the two]. Now let me say that
two have to be known, of course, but both are known by only one
instrument, definition. Demonstration, therefore, will be superfluous.

But if they accept substance more strictly for predicamental   3
substance, as they call it,[42] distinct from the nine kinds of accidents, then definition is said to be the instrument by which we
know substance, and demonstration is the instrument by which
[we know] accidents. First, of course, in this doctrine can be noted
this absurdity: they say that substance is thus known by definition, and so they appear to deny that accidents can be known by
definition; this is altogether false. For in the sciences, definitions
of accidents are conveyed no less than are [those] of substances;
definition does not have to be narrowed down and enclosed in
these limits such that it is ascribed only to substances. For if we
scrutinize Aristotle's natural books, we will see many accidents
made clear by definition. For he asks about what motion, place,
time, generation, mixture, putrefaction, digestion, and many other
natural accidents are, and they appear to be made clear by Aristotle not by demonstration but by definition. These, however, are

definitione declarari, sunt tamen naturaliter ignota et aliqua etiam cognitu difficillima, ut locus, tempus, putredo et alia multa, quae si adversarii benè considerarent,[11] certè inextricabili nodo se implicitos esse animadverterent.

4   Propterea ego distinctionem illam vitiosam esse semper existimavi, qua dicitur, definitione substantiam cognosci, demonstratione accidentia, dum sumitur substantia praedicamentalis appellata, quae contra novem accidentium genera distinguitur. Quemadmodum id verissimè dicitur, si substantia latissimè pro cuiusque rei essentia accipiatur, sic enim sensus est quòd dum rei essentia secundùm se consideratur, ea per definitionem declaratur, sive de substantia sive de accidente loquamur. Demonstratione verò accidentia cognoscimus non prout secundùm se considerantur, sed prout accidentia sunt, id est prout alteri ut subiecto inhaerent, quid enim sit generatio definitione significamus, utrum autem in corporibus naturalibus insit vel cur insit, demonstratione, haec etenim accidentium scientiam tradit prout alteri accidunt, definitio verò eorum essentiam declarat prout simpliciter proponuntur, ideò rectè dicitur definitionem quaestioni simplici satisfacere, demonstrationem verò complexae.

5   Praeterea si in secundo Posteriorum libro de definitione ageretur tanquàm de altero instrumento re distincto à demonstratione, liber Posteriorum non esset unus liber, quia duas haberet subiectas materias seu duos diversos scopos, quorum neuter ad alterum dirigeretur. Nam demonstratio ad definitionem substantiae nihil confert. Definitio verò substantiae quatenus ad substantiae cognitionem refertur, nil confert ad demonstrationem, confert quidem dum ad accidentia refertur tanquàm ipsorum causa, sed dum substantiam cognoscendam respicit, eo uno munere fungitur

naturally unknown, and some even, such as place, time, putrefaction, and many others, are most difficult to know. If our opponents considered the latter well, they would certainly have noticed that they had wrapped themselves up in an inextricable knot.

I, therefore, always judged to be flawed the distinction by which   4
it is said that substance is known by definition, accidents by demonstration, when substance is taken as what is called predicamental and distinguished from the nine kinds of accidents. It is said in the most correct way, if substance is accepted in the broadest sense for the essence of anything; the sense is such that when the essence of something is considered in and of itself, it is made clear by means of definition, whether we speak of substance or accident. But we know accidents by demonstration, not in that they are considered in and of themselves, but in that they are accidents, that is, in that they inhere in something else as in a subject. For what generation is, we indicate by definition, but whether it belongs to natural bodies or why it belongs, [we indicate] by demonstration. For indeed this [demonstration] conveys scientific knowledge of accidents in that they are accidents of something else. But definition makes their essence clear in that they are set out absolutely. And so, it is correctly said that definition answers a simple question, and demonstration a compound question.

Moreover, if in the second book of the *Posterior* [*Analytics*], defi-   5
nition were dealt with as another instrument, distinct in reality from demonstration, then [this] book of the *Posterior* [*Analytics*] would not be one book, because it would have two subject matters or two different goals, neither of which would be directed toward the other. For demonstration does not contribute to definition of a substance. And definition of a substance, insofar as it refers to knowledge of a substance, does not contribute to demonstration. It does contribute, of course, when it is referred to accidents as their cause, but when it regards the substance to be known, it is performing the one job of clarifying what the defined substance

declarandi quid sit substantia ipsa definita, in cuius muneris functione nullum habet cum demonstratione commercium.

6   Ad hanc obiectionem aliqui respondere volentes, ut libri unitatem tuerentur, dixerunt demonstrationem et definitionem reipsa non differre, sed solo terminorum situ et ordine, ut Aristoteles ipse asserit, quare ita duo instrumenta sunt, ut sit etiam unum.

7   Sed qui hoc dicunt, proprium dogma evertunt et pugnantia dicunt. Prius enim dixerant demonstrationem ac definitionem esse duo instrumenta re distincta, postea verò dicunt esse unum re.

8   Praeterea non sunt boni Averroistae, ut se esse hac in re profitentur, neque ea, quae ab Aristotele in secundo libro Posteriorum dicuntur, intelligunt, Averroes enim in primo commentario sexti libri Metaphysicorum clara voce testatur eam definitionem, quae est demonstratio positione differens, non esse substantiae definitionem, sed accidentis. Et haec est Aristotelis sententia in secundo Posteriorum manifestissima. Quare de definitione substantiae haudquaquam verum est, imò ne excogitabile quidem quòd sit demonstratio positione differens et idem re quod demonstratio, cùm penitus distinctae sint, siquidem substantiae non est demonstratio. Differunt autem et ratione finis et ratione formae et ratione materiae. Ratione finis, quia illius definitionis finis est cognitio substantiae, finis autem demonstrationis est cognitio accidentis. Ratione formae, quia forma demonstrationis est syllogismus, quae definitioni non convenit, quia definitio ne enunciativa quidem est, tantum abest ut sit ratiocinativa, ratione autem materiae, quia demonstrationis materia est terminus medius, ut Themistius, et Aristoteles asserunt in contextu 48 secundi libri Posteriorum,

itself is. In the performance of this job it has no commerce with demonstration.

Those wanting to respond to this objection, so as to maintain   6
the unity of the book, said that demonstration and definition do not differ in reality, but only in arrangement and order of the terms, as Aristotle himself asserts. Therefore the two instruments are such that they are also one.

But those who say this overturn their own doctrine and say   7
things that are in conflict. For first they said that demonstration and definition are two instruments distinct in reality, but then afterward they say they are in reality one.

Moreover, they are not good Averroists, as they profess them-   8
selves to be in this issue, nor do they understand what is said by Aristotle in the second book of the *Posterior* [*Analytics*]. For Averroës, in the first commentary to the sixth book of the *Metaphysics*,[43] attests in a clear voice that a definition that is a demonstration differing by position is not the definition of a substance but of an accident. And this is most manifestly Aristotle's position in the second [book] of the *Posterior* [*Analytics*]. Therefore, it is not at all true about the definition of a substance, indeed not even imaginable, that it is a demonstration differing in position and the same in reality as demonstration, since they are completely distinct, since indeed there is no demonstration of a substance. Moreover, they differ by reason of end, by reason of form, and by reason of matter. By reason of end, because the end of that definition is knowledge of a substance, whereas the end of demonstration is knowledge of an accident. By reason of form, because the form of a demonstration is the syllogism, which is not appropriate for a definition, because a definition is not even propositional, so far is it from being ratiocinative. By reason of matter, because the matter of a demonstration is the middle term, as Themistius and Aristotle assert in text no. 48 of the second book of the *Posterior*

ex eo enim ambae propositiones constant. Materia verò definitionis sunt partes ipsae, quae in definitione sumuntur, at non per eandem causam mediam accidentia demonstrantur, per quam substantia definitur, vult enim Averroes medium demonstrationis rarò esse definitionem subiecti et ut plurimum esse alterum accidens, quod alterius accidentis causa est. Ex eadem igitur materia non conflatur definitio substantiae, ex qua fit demonstratio accidentis. In omnibus igitur differunt, quare eorum solutio vana est.

9 Alii verò Averroistae alio pacto respondere conati sunt, dixerunt enim duo quidem instrumenta esse, demonstrationem accidentis et definitionem substantiae, ea tamen fieri unum quatenus sub uno communi genere, quod analogum vocant, continentur, est enim commune utriusque genus instrumentum sciendi, sic autem una est intentio Aristotelis docere instrumentum perfectae scientiae acquirendae, quod postea in duas species dividitur, in demonstrationem et definitionem.

10 Haec quoque responsio vana et commentitia est, quoniam Aristoteles nullum consideravit demonstrationis genus, nisi syllogismum, ut patet legentibus initium primi libri Priorum Analyticorum et initium capitis de prima figura in eodem libro. In secundo etiam capite primi libri Posteriorum in definitione demonstrationis non accepit instrumentum loco generis, sed syllogismum. Et in secundo libro definiens definitionem sumpsit pro genere orationem, non instrumentum. Erat tamen accipiendum instrumentum in utriusque definitione, si utraque ut species instrumenti consideratur. Sed revera commune omnium logicorum instrumentorum genus apud Aristotelem est syllogismus; ut antea demonstravimus et ut ex definitione demonstrationis manifestè colligimus, cùm igitur sub eo genere definitio non contineatur,

[*Analytics*],[44] for both premises are composed out of it. And the matter of a definition is the parts themselves that are taken in the definition. But accidents are not demonstrated by means of the same middle cause by means of which a substance is defined. For Averroës wants the middle [term] of a demonstration rarely to be the definition of the subject and mostly to be some accident that is the cause of another accident. The definition of a substance, therefore, is not assembled from the same matter out of which the demonstration of an accident is made. They differ, therefore, in everything. And so their solution is vain.

Other Averroists tried to respond another way. For they said 9 that of course there are two instruments, demonstration of an accident and definition of a substance, but they become one insofar as they are contained under one common genus, which they call analogous, for that is the common genus of both instruments for knowing scientifically. And so it is Aristotle's one intention to teach the instrument for acquiring perfect scientific knowledge, and it is afterward divided into two species, into demonstration and definition.

This response too is vain and counterfeit, since Aristotle con- 10 sidered no genus of demonstration except the syllogism, as is patent to those reading the start of the first book of the *Prior Analytics* and the start of the chapter on the first figure in the same book.[45] Also, in the second chapter of the first book of the *Posterior [Analytics*], in the definition of demonstration,[46] he did not accept instrument in the place of the genus, but syllogism. And in the second book, defining definition,[47] he took speech, not instrument, to be the genus. But instrument would have to have been accepted in the definition of each, if each were considered as a species of instrument. But in truth, in Aristotle, the common genus of every logical instrument is the syllogism, as we demonstrated earlier and as we manifestly gather from the definition of demonstration. Since, therefore, definition is not contained under that genus, it is

non est instrumentum logicum. Ideò Aristoteles in contextu centesimo secundi libri Posteriorum epilogo colligens omnia, quae in
libris Analyticis docuerat, nullam fecit definitionis mentionem, sed
solum syllogismum et demonstrationem nominavit, tanquàm genus et speciem. Quòd si in eo secundo libro de definitione tractas-
set tanquàm de specie instrumenti distincta à demonstratione,
eam quoque exprimere in illo epilogo debuisset, imò et ipsum
commune utriusque genus instrumentum, de quo etiam aliqua
prius tractanda erant, quàm ad species accederet, quemadmodum
etiam prius de syllogismo egit, quàm de demonstratione.

11   Sed certè demonstratio et definitio nullum habent commune
genus, nisi remotissimum, orationem, ut significant verba Aristotelis in contextu 45 secundi libri Posteriorum, de qua etiam aliqua
dixit Aristoteles in principio libri de Interpretatione. Quare si instrumentum sciendi putasset esse earundem proximum genus,
multò magis de ipso, quàm de oratione loqui debuisset, genus
enim proximum principalius est, quàm remotum. Et oratio non
solùm à logico, sed etiam à Grammatico tractatur, at instrumentum sciendi à logico solo considerari potest. Propterea mihi certissimum atque compertissimum est nullum esse apud Aristotelem
logicorum instrumentorum genus, nisi syllogismum, de quo ipse
diligentissimè scripsit in Prioribus Analyticis. Ratio igitur nostra
non parum roboris habet, cùm enim definitio species syllogismi
non sit, certè, si ut instrumentum sciendi consideraretur, eius tractatio tractationi de demonstratione non benè cohaereret, quare
librum unum constituere non possent.

12   Praeterea reprehensione dignus esset Aristoteles, qui in secundo
illo libro sibi proponens de definitione agendum tanquàm de altero
instrumento penitus distincto à demonstratione, saepius in eo libro
ad tractationem demonstrationis revertitur et has tractationes

not a logical instrument. So Aristotle in the summary in text no. 100 of the second book of the *Posterior* [*Analytics*],[48] gathering everything he had taught in the *Analytics* books, made no mention of definition; he instead named only syllogism and demonstration, as genus and species. Now if in the second book he had treated definition as a species of instrument distinct from demonstration, he ought to have expressed this in that summary also—indeed also their common genus itself, instrument. And regarding this [instrument], some things would have to have been treated first, before he went on to the species, just as he dealt with syllogism first, before demonstration.

But certainly demonstration and definition have no common   11
genus—except a very remote one, speech—as Aristotle's words in text no. 45 of the second book of the *Posterior* [*Analytics*][49] indicate; Aristotle also spoke about this in the beginning of the book *On Interpretation*.[50] And so if he had held that the instrument for knowing scientifically was their proximate genus, he ought to have spoken much more about that than about speech. For the proximate genus is more important than the remote. And speech is treated not only by the logician, but also by the grammarian, while the instrument for knowing scientifically can be considered only by the logician. And so it is to me most certain and assuredly known that in Aristotle there is no genus of logical instruments except the syllogism, which he wrote about most carefully in the *Prior Analytics*. Our reasoning, therefore, has not a little weight, for since definition is not a species of syllogism, then certainly, if it were not considered as an instrument for knowing scientifically, his treatment would not cohere well with a treatment on demonstration. Therefore they could not constitute one book.

And Aristotle would deserve censure: in that second book [of   12
the *Posterior Analytics*], intending to deal with definition as another instrument completely distinct from demonstration, he often returns in the book to a treatment of demonstration and conflates

confundit, postquàm enim longo sermone egit in secundo libro de definitione, ad demonstrationem redit et docet quomodo ipsius medium possit esse quodlibet ex quatuor generibus causarum; deinde iterum ad definitionem revertitur, postea iterum ad demonstrationem. Debuisset enim in primo libro totum de demonstratione sermonem absolvere, deinde in secundo seorsum de definitione loqui, si duo distincta instrumenta sunt, quemadmodum illi dicunt.[12] Quam difficultatem non effugiunt nonnulli, qui dicunt definitionem tractari in secundo libro ut instrumentum re distinctum à demonstratione, non tamen ut instrumentum substantiae tantùm cognoscendae, sed simpliciter ut instrumentum cognoscendi quid est sive substantiarum, sive accidentium. Hi namque multò minùs peccant, sed tamen ipsi quoque decipiuntur, quia definitio ut instrumentum cognoscendi quid est in eo libro non consideratur, ut argumentum modò factum et alia prius adducta demonstrant.

13   Sed huius sententiae falsitas non potest clarius et efficacius ostendi, quàm totum illum secundum librum percurrendo et singulas partes breviter expendendo, nam si nullum in eo libro locum invenerimus, in quo Aristoteles de definitione agat ut de instrumento ducente ad cognitionem quid est, hoc uno argumento ea sententia satis superque confutata erit. Aristoteles ab initio eius libri usque ad contextum 47 diligentem tractationem facit de definitione et de via ducente ad cognitionem quid est, considerandum igitur est, quaenam sit illa via et de quanam definitione loquatur in ea parte Aristoteles. Certè vel caecus inspicere posset totam illam tractationem esse de definitione accidentium, non de definitione substantiarum, viam autem ducentem ad eiusmodi definitionem non esse aliam, quam demonstrationem potissimam, haec enim in definitionem ipsius accidentis demonstrati convertitur et ita nos ducit ad cognoscendum quid est accidens unumquodque. Qua de

these treatments. For after he deals with definition in a long discourse in the second book, he goes back to demonstration and teaches in what way its middle [term] can be any one of the four kinds of causes;[51] he then returns again to definition, and afterward again to demonstration. He ought to have completed the whole discourse on demonstration in the first book, and then spoken separately on definition in the second, if they are two distinct instruments, as they say. This difficulty is not escaped by the several who say that definition is treated in the second book as an instrument distinct in reality from demonstration, but not as an instrument only for knowing substance and instead absolutely as an instrument for knowing what [something] is (*quid est*), either of substances or of accidents. Now these men go astray much less, but nevertheless they too are deceived, because definition as an instrument of knowing what [something] is (*quid est*) is not considered in that book, as the argument just made and others adduced earlier demonstrate.

But the falsity of this position cannot be shown more clearly or more effectually than by running through that whole second book and briefly weighing each of its parts. For if we discover in the book no passage in which Aristotle deals with definition as an instrument leading to knowledge *quid est*, then by this one argument, the position will be confuted more than enough. Aristotle, from the start of the book up to text no. 47,[52] makes a careful treatment on definition and the way leading to knowledge *quid est*. It has to be considered, therefore, what that way is and what definition Aristotle is speaking about in that part. Certainly even a blind man could observe that the whole treatment is on definition of accidents, not about definition of substances, and the way leading to a definition of this type is none other than demonstration *potissima*, for this is converted into the definition of the demonstrated accident itself and thus leads us to knowing what any accident whatever is. We will speak at greater length about this issue below.

re postea fusius loquemur. Ergo in ea parte Aristoteles non loquitur de definitione substantiae, imò eius considerationem reiicit in
contextu 42 et in 47 dicens eam esse principium indemonstrabile,
quare, si ignoretur, non potest per demonstrationem indagari, sed
aliqua alia via, quam ipse ibi non considerat, quia in ea parte loquitur solùm de illa definitione, quae per demonstrationem innotescit, haec autem est definitio affectionis, ut in ea parte mani-
festissimum est et ut asserit etiam Averroes in primo commentario
sexti libri Metaphysicorum.

14   Similiter clarum est viam ducentem ad cognoscendum quid est
in accidentibus non esse definitionem, sed demonstrationem, in
fine enim primi capitis proponit Aristoteles declarandum quomodo per demonstrationem declaretur quid est et in contextu 47
epilogum faciens colligit se declarasse quomodo sit demonstratio
ipsius quid est et quomodo quod quid est monstretur. Sed totam
illam partem legentibus manifestum est Aristotelem nil aliud docere, quàm quomodo demonstratio ducat ad cognitionem ipsius
quid est. Non est igitur definitio instrumentum ducens ad cognoscendum quid est, sed demonstratio.

15   Deinde Aristoteles in contextu 48 incipit tractationem de generibus causarum et docet eorum quodlibet posse in demonstratione
medium esse; et in ea causarum consideratione versatur usque ad
contextum 68 quae tota tractatio absque ullo dubio de demonstratione est, quia nihil ibi dicitur de definitione.

16   Postea in contextu 69 incipit declarare viam venandi praedicata
essentialia, quae praedicantur in eo quod quid est et constituunt
illam definitionem, quae est principium indemonstrabile. Nec magis est definitio substantiae, quàm accidentis, praedicata enim,
quae vocantur quidditativa, non possunt demonstrari de illo, de

Therefore Aristotle does not speak in this part about the definition of a substance. In fact, in texts no. 42[53] and no. 47,[54] he rejects consideration of it, saying that it is an indemonstrable beginning-principle. And so if it is unknown, it can be tracked down not by means of demonstration but in some other way that he does not there consider, because in that part he is speaking only about that definition that becomes known by means of demonstration. This, however, is the definition of an affection, as is most manifest in that part and as Averroës too asserts in the first commentary to the sixth book of the *Metaphysics*.[55]

It is similarly clear that the way leading to knowing, in accidents, what [something] is (*quid est*), is not definition but demonstration. For at the end of the first chapter Aristotle sets out to clarify in what way what [something] is (*quid est*) is made clear by means of demonstration and in text no. 47,[56] making the summary, gathers that he has clarified in what way demonstration may be of what something is (*quid est*) and in what way what something is (*quid est*) is demonstrated. But to those reading that whole part it is manifest that Aristotle teaches nothing other than in what way demonstration leads to knowledge *quid est*. Definition, therefore, is not an instrument leading to knowing what something is (*quid est*), but demonstration is.

Then, in text no. 48,[57] Aristotle starts a treatment on the kinds of causes, and teaches that any of these can be the middle [term] in a demonstration; and he is concerned with consideration of causes up to text no. 68.[58] This whole treatment, without any doubt, is on demonstration, because nothing is said there about definition.

Afterward, in text no. 69,[59] he starts to clarify the way to search for essential predicates, which are predicated by virtue of what something is and constitute that definition, which is an indemonstrable beginning-principle. Definition is no more of substance than of accident. For predicates that are called quidditative[60]

14

15

16

quo in eo quod quid est praedicantur, sive substantias, sive accidentia definienda consideremus, quemadmodum enim non possumus per causam demonstrare hominem esse rationalem vel esse
animal vel esse corpus; ita neque albedinem esse colorem vel esse
qualitatem neque eclipsim esse privationem luminis neque tonitrum[13] esse sonum. Possumus quidem per causam demonstrare
privationem luminis de luna subiecta, sed non de eclipsi; et accensionem de sanguine cordis, sed non de ira, nisi idem de seipso demonstremus.

17   Cùm igitur ea, quae praedicantur in quid tam in genere substantiae quàm in generibus accidentium, demonstrari non possint,
tamen ignota esse contingat, docet ibi Aristoteles qua via debeant
investigari an via divisiva, ut censuit Plato, an aliqua alia methodo.
Quare tota illa tractatio est de illa definitione, quam Aristoteles
antea in contextu 42 reiecerat dicens eam esse principium per se
notum in scientia, vel, si ignorari contingat, non posse per demonstrationem innotescere, sed per aliquam aliam[14] methodum, quam
in ea parte quaerit. Instrumentum igitur idoneum ad venandum
quid est, non est ipsa definitio, quandoquidem haec ignota proponitur, et quaeritur instrumentum, quo investigetur. Sed instrumentum est ipsa via divisiva vel via compositiva, per quam docet
ibi Aristoteles, quomodo eiusmodi praedicata venari debeamus et
horum venatio est venatio ipsius definitionis ignotae, quia venari
ipsum quid est et venari definitionem idem significant. Quare nihil
vanius est, nihil ab Aristotele alienius, quàm dicere definitionem
esse methodum et instrumentum, quo ipse in ea parte docet venari
praedicata in quid, ut patet tum legentibus verba Aristotelis in eo

cannot be demonstrated about that of which they are predicated in something by virtue of what it is, whether we consider defining substances or accidents. For just as we cannot demonstrate by means of a cause that man is rational or is an animal or is a body, so neither [can we demonstrate] that white is a color or a quality, nor that an eclipse is a privation of light, nor that thunder is a noise. We can, of course, demonstrate by means of a cause that privation of light is of the moon as subject but is not of the eclipse, and that boiling is of the heart's blood, but not of anger, lest we demonstrate the selfsame thing of itself.

Although, therefore, those things that are predicated in what   17
they are (*in quid*), as much in the genus of substance as in the genera of accidents, cannot be demonstrated, it nevertheless can happen that they are unknown. Aristotle there teaches by what way these ought to be investigated, whether by the divisive way, as Plato deems it, or by some other method. Therefore, that whole treatment is on the definition that Aristotle had earlier rejected in text no. 42,[61] saying that it is a beginning-principle known *per se* in a science, or, if it happens to be unknown, it cannot become known by means of demonstration, but by means of some other method that he inquires after in that part. The instrument fit for searching for what [something] is (*quid est*), therefore, is not the definition itself, since this is set out [as] unknown, and the instrument by which it can be investigated is what is being sought after. But the instrument is either the divisive way itself or the compositive way;[62] by means of it Aristotle there teaches in what way predicates of this type ought to be searched for, and the search for them is the search for the unknown definition itself, because to search for what something is and to search for the definition signify the same thing. And so nothing is more vain, nothing more alien to Aristotle, than to say that definition is a method and an instrument by which (he teaches in this part) things predicated in what they are (*in quid*),[63] are searched for — as is patent both to

loco tum rem ipsam per se considerantibus. Ergo in ea quoque parte, quae à contextu 69 usque ad 84 protenditur, manifestum est tum Aristotelem non loqui de sola definitione substantiae tum etiam non considerare definitionem ut methodum et instrumentum, quo venemur ipsum quid est.

18   Hinc sequitur nullum esse locum in eo secundo libro, in quo Aristoteles de definitione substantiae seorsum loquatur; et nullum etiam esse locum, in quo consideret definitionem ut instrumentum ducens ad cognitionem ipsius quid est; quia post illum contextum 84 usque ad finem libri nil amplius dicit de definitione, ibi enim ad demonstrationem revertitur et loquitur de scientia ipsius propter quid est per cognitionem causae usque ad centesimum contextum, in quo epilogum facit et dicit se de syllogismo ac de demonstratione docuisse, de definitione autem nihil dicit, quia revera omnia, quae usque ad illum locum docuerat, ad demonstrationem pertinuerant, ut mox clarius ostendemus.

19   Demum in postremo capite eius secundi libri loquitur Aristoteles de principiis ac de eorum cognitione et declarat qua via sit in nobis acquisita primorum principiorum notitia° et viam dicit esse inductionem.

20   Falsa est igitur et ab Aristotele alienissima sententia eorum, qui dicunt in secundo illo libro considerari definitionem ut instrumentum cognoscendi substantiam vel cognoscendi quid est generaliter accepti, nullus enim in eo libro locus apparet, in quo id fiat, hoc enim et ipsius Aristotelis testimonio et rationibus ex ipsius tractatione desumptis apertissimè, ni fallor, demonstratum est.

those reading Aristotle's words in this passage and to those considering that issue by itself. Therefore, also in the part that extends from text no. 69 up to no. 84,[64] it is manifest both that Aristotle does not speak only about the definition of substance and also that he does not consider definition as a method and an instrument by which we search for what that is (*quid est*).

And thus it follows that there is no passage in the second book 18 in which Aristotle speaks separately about the definition of substance and there is also no passage in which he considers definition as an instrument leading to knowledge *quid est*, because after that text no. 84, up to the end of the book, he says nothing more about definition; he returns there to demonstration and speaks about scientific knowledge *propter quid* by means of a knowledge of cause — [and does so] up to text no. 100.[65] There he makes a summary and says that he has taught about syllogism and about demonstration. He says nothing, however, about definition, because in truth everything that he had taught up to this point pertained to demonstration, as we will soon show more clearly.

Lastly, in the final chapter of his second book,[66] Aristotle speaks 19 about beginning-principles and about knowledge of them and makes clear in what way knowledge° of first beginning-principles is acquired by us. And he says the way is induction.

False and most alien to Aristotle, therefore, is the position of 20 those who say that, in that second book, definition is considered as an instrument for knowing substance or for knowing what [something] is (*quid*), accepted in the general sense — there appears to be no passage in the book in which this happens. And unless I am mistaken, this has been demonstrated most plainly both by the testimony of Aristotle himself and by the reasons drawn from his treatment.

: VIII :

In quo ostenditur eam sententiam
Averroi quoque adversari.

1   Posteaquam dogma illud de definitione ab Aristotele alienissimum
esse demonstravimus, nunc ab Averroe quoque alienum esse ostendamus. Sententia Averrois est in multis locis quòd primus Posteriorum liber ad secundum dirigatur et demonstratio propter definitionem tractetur, ut legere est in commentario undecimo primi
Posteriorum et in 38 secundi et in Epitome logices in capite de
demonstratione et in quaestionibus logicis, quaestione decima.
Atqui demonstratio ad definitionem substantiae non dirigitur,
cùm ad eam nihil penitus conferat, quandoquidem ad solam accidentium cognitionem conducit. Sed potius è contrario definitio
substantiae ipsi demonstrationi inservit, et ad eam dirigitur
tanquàm eius principium. Itaque hunc in modum argumentemur,
illa definitio in secundo libro Posteriorum tractatur, ad quam dirigitur tractatio de demonstratione in primo, sed ad substantiae definitionem non dirigitur demonstratio, ergo secundus liber non est
de definitione substantiae, sed est potius de definitione accidentis,
ad hanc enim dirigitur demonstratio, tamen reipsa ab ea non distinguitur, quia differunt sola terminorum positione.

2   Cùm igitur neque ad definitionem substantiae dirigatur demonstratio, neque ad definitionem latè acceptam, quae et definitionem
substantiae et definitionem accidentis complectatur, sed ad solam
accidentis definitionem, sequitur quòd Averroes non existimavit
tractari in eo secundo libro definitionem ut instrumentum cognoscendae substantiae, neque ut instrumentum cognoscendi quid est,

: VIII :

In which it is shown that this position also contradicts Averroës.

Now that we have demonstrated that this doctrine about defini-   1
tion is most alien to Aristotle, we may now show that it is alien to
Averroës too. In many passages it is Averroës' position that the
first book of the *Posterior* [*Analytics*] is directed toward the second
and that demonstration is treated for the sake of definition, as one
may read in the eleventh commentary to the first [book] of the
Posterior [*Analytics*][67] and in 38 of the second,[68] and in the *Epitome
of Logic* in the chapter on demonstration,[69] and in the logic questions, question 10.[70] But demonstration is not directed to definition of substance — it does not contribute to it at all, since it contributes only to knowledge of accidents. But instead vice versa:
definition of substance serves demonstration itself and is directed
to it as its beginning-principle. And so we may argue in this way:
That definition treated in the second book of the *Posterior* [*Analytics*] is that toward which the treatment on demonstration in the
first [book] is directed. But demonstration is not directed to the
definition of substance. Therefore, the second book is not about
the definition of substance; it is instead about the definition of accident. For demonstration is directed to this, but is not distinguished from it in reality, because they differ only by the position
of terms.

Since, therefore, demonstration is not directed toward defini-   2
tion of substance and not toward definition accepted in a broad
sense that encompasses both definition of substance and definition
of accident but only toward definition of accident, it follows that
Averroës did not judge that definition was treated in the second
book as an instrument for knowing substance or as an instrument

generaliter sumendo quid est tam substantiarum quàm accidentium; neque ut instrumentum distinctum re à demonstratione,
quia definitio accidentis re non distinguitur à demonstratione. Imò
si dictos Averrois locos legamus, vult Averroes ipsam potius demonstrationem esse instrumentum ducens ad cognoscendum quid
est, quàm definitionem; nisi dicamus definitionem ipsam accidentis eatenus esse instrumentum, quatenus est idem re, quod demonstratio, sic autem non duo erunt instrumenta, sed unum,
quemadmodum infrà declarabimus.

3   Verùm ad hoc argumentum aliqui respondent non esse ita, ut
nos putamus, intelligenda Averrois verba, quando dicit, demonstrationem dirigi ad definitionem, non enim significare voluit tractationem demonstrationis esse gratia definitionis, sed demonstrationem esse instrumentum definitione praestantius. Quando enim
plura in aliqua disciplina tractantur, quorum unum est nobilius
caeteris, dicere solemus id, quod aliorum praestantissimum est,
instar finis esse respectu aliorum, licèt aliorum tractatio non sit
propter tractationem ipsius.

4   Quae responsio quàm vana sit, clarum est legentibus et benè
considerantibus verba Averrois in omnibus memoratis locis, non
solùm enim dicit definitionem esse demonstratione nobiliorem,
sed dicit demonstrationem quaeri propter definitionem, quae ex
ipsa educitur, et illa, quae de demonstratione in primo libro dicuntur, dici propter illa, quae dicuntur in secundo de definitione; et
id, quod revera quaerunt ac desiderant illi, qui demonstrationem
construunt, esse extractionem definitionis et cognitionem quid est.
Clara igitur sunt Averrois verba et ipsius sententiam ab eorum
dogmate alienissimam esse demonstrant.

for knowing what [something] is (*quid est*), taking "what [something] is" in the general sense, as of substances as well as of accidents, or as an instrument distinct in reality from demonstration, because definition of accidents is not distinguished in reality from demonstration. Indeed, if we read the said passages in Averroës, Averroës wants demonstration itself, rather than definition, to be the instrument leading to knowing what [something] is (*quid est*), unless we say that an accident's definition itself is an instrument insofar as it is the same in reality as demonstration. But then there will not be two instruments, but one, as we will make clear below.

Now to this argument some respond that when Averroës says 3 that demonstration is directed toward definition, his words are not to be understood as we hold. For he did not want to indicate that the treatment of demonstration is for the sake of definition, but that demonstration is a more excellent instrument than definition. For when, in some discipline, many things are treated, one of which is more noble than the others, we normally say that that which is most excellent of all is, with regard to the others, like an end, even granting that the treatment of the others not be for the sake of the treatment of that one.

To what degree this response is vain is clear to those reading 4 and well considering Averroës' words in all the passages referred to. For he not only says that definition is more noble than demonstration, but he says that demonstration is inquired after for the sake of definition, which is educed from it, and that those things that are said about demonstration in the first book are said for the sake of those things that are said about definition in the second, and that in truth what those who construct a demonstration inquire after and desire is extraction of a definition and knowledge of what [something] is (*quid est*). Averroës' words, therefore, are clear and demonstrate that his position is most alien to their doctrine.

5   Praeterea in sua praefatione in primum librum Posteriorum Averroes rationem reddens cur Aristoteles in uno et eodem libro de demonstratione, ac de¹⁵ definitione tractaverit, inquit non esse inter eas magnum discrimen, quod¹⁶ diversos ac separatos tractatus postulaverit. Sed hoc Averrois dictum de definitione substantiae minimè verum est, haec enim, ut antea ostendimus, est diversa penitus à demonstratione, ergo haec non est illa definitio, de qua putat Averroes Aristotelem agere in secundo Posteriorum, neque etiam putat tractari in eo libro alterum instrumentum re distinctum à demonstratione, sed potius idem re, differens autem solo terminorum situ.

6   Alia quoque verba sunt in eo prooemio Averrois expendenda, quando dicit, postquàm Aristoteles in Prioribus Analyticis egit de forma demonstrationis, vult in Posterioribus agere de eius materia, materia autem demonstrationis sunt propositiones verae, quae quidem et in primo et in secundo libro Posteriorum Analyticorum considerantur, alia tamen et alia ratione, in primo quidem quatenus ducunt hominem ad verificationem perfectam, in secundo autem quatenus sunt definitiones et ducunt hominem ad perfectam formationem. Est autem verificatio apud Averroem cognitio quaesitorum complexorum, quae per demonstrationem, quatenus demonstratio est, innotescunt; formatio verò est conceptio naturae sive quidditatis, quae per definitionem significatur. Sententia igitur Averrois est quòd propositiones verae, id est necessariae tam in primo quàm in secundo Posteriorum libro tractentur, sed in primo quatenus ducunt ad cognitionem propter quid est, in secundo autem quatenus tradunt etiam cognitionem quid est.

7   Dicant ergo hi Averroistae an Averroes intelligat propositiones necessarias tractari in eo secundo libro prout sunt definitiones subiecti, an accidentis. Si subiecti, adversantur Averroi, qui in

In addition, in his preface to the first book of the *Posterior* [*Ana-* 5
lytics], Averroës,[71] in accounting for why Aristotle treated demonstration and definition in one and the same book, says that
between them there is not a big discriminating difference that
demanded different and separate treatises. But this dictum of
Averroës' is not at all true of the definition of substance. For this
[definition], as we showed earlier, is completely different from
demonstration. Therefore, this is not the definition that Averroës
holds that Aristotle deals with in the second [book] of the *Posterior*
[*Analytics*]; nor does he hold that in this book another instrument,
distinct in reality from demonstration, is treated; it is, instead, the
same in reality, differing only in arrangement of terms.

There are other words in Averroës' proem that have to be 6
weighed, [those] when he says:[72] After Aristotle dealt in the *Prior
Analytics* with the form of demonstration, he wants to deal in the
Posterior [*Analytics*] with its matter; but now the matter of demonstration is true premises and these are, of course, considered both
in the first and in the second book of the *Posterior Analytics*, but in
different ways—in the first indeed insofar as they lead man to
perfect verification, and in the second insofar as they are definitions and lead man to perfect formation. Now in Averroës, verification is knowledge of compound inquiries, which become known
by means of demonstration insofar as it is demonstration, and
formation is the conception of a nature or a quiddity, which is signified by means of a definition.[73] It is, therefore, Averroës' position
that true, that is, necessary, premises are treated in the first as
much as in the second book of the *Posterior* [*Analytics*], but in the
first insofar as they lead to knowledge *propter quid*, and in the second insofar as they instead convey knowledge *quid est*.

Therefore let those Averroists say whether Averroës under- 7
stands necessary premises as treated in the second book according to whether they are definitions of subject or of accident. If of
the subject, they contradict Averroës, who in the first book of the

primo libro Posteriorum commentario undecimo et in quaestione sua de medio demonstrationis confutat eorum sententiam, qui dicunt, medium esse definitionem subiecti et probat esse definitionem semper et ex necessitate maioris extremi, non minoris, nisi rarò et ex accidenti. Atqui dum dicimus propositiones demonstrationis, dicimus terminum medium, ut suprà contra Latinos ostendimus et ut isti quoque Averroistae contra eosdem disputando confitentur et sumunt. Revera enim praeter maius et minus extremum nihil est in propositionibus, nisi medium, ergo quaerere an propositiones sint definitio maioris an minoris extremi, nil aliud est, quàm quaerere an medium sit definitio huius vel illius. Quando igitur Averroes dicit propositiones, quae materia demonstrationis sunt, considerari in secundo libro Posteriorum prout sunt definitiones et ducunt ad cognoscendum quid est, intelligit accidentis, non subiecti. Ergo non est verum quòd in eo secundo libro secundùm Averroem tractetur definitio ut est instrumentum substantiae cognoscendae, neque generaliter cognoscendi quid est, quia definitionem substantiae nulla ratione respexit Averroes, in declarando Aristotelis scopo in illo secundo libro, sed solam definitionem accidentis.

8   Non est etiam verum quòd Averroes putet definitionem ibi tractari ut instrumentum re distinctum à demonstratione, cùm enim dicat easdem propositiones demonstrationis ducere ad cognoscendum et propter quid est et quid est, dicit unum instrumentum esse, quod utramque cognitionem praestat et vocatur tum demonstratio tum definitio; ut ipse quoque Aristoteles asserit in contextu 45 secundi Posteriorum et ut nos postea declarabimus.

Posterior [*Analytics*], the eleventh commentary,[74] and in his question about the middle [term] of a demonstration,[75] confutes the position of those who say that the middle [term] is the definition of the subject and proves that it is always and out of necessity the definition of the major extreme, not of the minor (except rarely and by accident). But yet when we say the premises of a demonstration, we say the middle term — as we showed above against the Latins, and as the Averroists themselves confess and accept in debating against them [i.e., the Latins], too. For in truth, in the premises there is nothing beside the major extreme and the minor except the middle [term]. Therefore, to inquire whether the premises are the definition of the major or of the minor extreme is nothing other than to inquire whether the middle [term] is the definition of the latter or the former. When Averroës, therefore, says that premises, which are the matter of demonstration, are considered in the second book of the *Posterior* [*Analytics*] in that they are definitions and lead to knowing what [something] is (*quid est*), he understands this of accident, not of subject. Therefore, it is not true that, according to Averroës, definition in the second book is treated as an instrument for knowing substance or generally for knowing what [something] is (*quid est*), because Averroës, in clarifying Aristotle's goal in that second book, gave no regard to the definition; [he regarded] instead only the definition of accident.

It is also not true that Averroës holds that definition is treated 8 there as an instrument distinct in reality from demonstration, for although he says that the same premises of a demonstration lead to knowing both what [something] is on account of (*propter quid*) and what it is (*quid est*), he says that there is one instrument that warrants both [types of] knowledge and is called both demonstration and definition, as Aristotle himself asserts also, in text no. 45 of the second [book] of the *Posterior* [*Analytics*],[76] and as we will make clear below.

: IX :

Quodnam sit Aristotelis consilium in Posterioribus Analyticis.

1   Postquàm aliorum sententias reiecimus, ad veritatis declarationem accedendum est, ut tum per ea, quae de methodis docuimus, ad cognoscendam Aristotelis mentem ducamur; tum vicissim tractatione Aristotelis intellecta sententiam nostram de methodis tanti viri testimonio mirificè comprobemus. Etenim ea instrumenta sciendi tractavit Aristoteles, quae tractare debuit, et eo modo, quo debuit.

2   Iam ostendimus duas tantùm dari methodos seu duo instrumenta sciendi, methodum demonstrativam[17] et methodum resolutivam, quae serva est methodi demonstrativae, quoniam resolutio fit propter compositionem, ut illi principia subministret. Optimè igitur fecit Aristoteles, qui in Posterioribus Analyticis duas tantùm methodos tradidit, demonstrativam et resolutivam, demonstrativam quidem primariò, resolutivam autem secundariò et prout pendet à demonstrativa. Propterea dum praecipuum Aristotelis scopum spectamus rectè dicimus unam esse utriusque Posteriorum libri subiectam materiam, demonstrationem potissimam. Ideò Aristoteles in principio primi libri Priorum Analyticorum proponere nobis volens finem omnium librorum Analyticorum, unicum proposuit, demonstrativam methodum, et in epilogo in calce secundi Posteriorum eam unam collegit. De methodo autem resolutiva nihil dixit, quia eam secundariò consideravit, quatenus dirigitur ad demonstrativam, quo fit ut appellatione demonstrativae resolutiva quoque comprehendatur, siquidem perfecta tractatio de

: IX :

What Aristotle's intent is in the Posterior Analytics.

Now that we have rejected the positions of others, we have to go 1
on to a clarification of the truth, so that both by means of that
which we taught about methods, we may be led to knowing what
Aristotle had in mind, and in turn, after understanding Aristotle's
treatment, we may confirm our position on methods wonderfully
well by the testimony of this great man. For indeed Aristotle
treated those instruments for knowing scientifically that he ought
to have treated and in the way in which he ought to have.

We now show that there are only two methods or two instru- 2
ments for knowing scientifically — demonstrative method and resolutive method, which is servant to demonstrative method, since
resolution occurs for the sake of composition, so as to furnish it
with beginning-principles. Aristotle, therefore, did it the best way:
He conveyed only two methods in the *Posterior Analytics* — demonstrative and resolutive — the demonstrative, of course, primarily
and the resolutive secondarily and in that it depends on the demonstrative. And so when we look at Aristotle's principal goal, we
correctly say that one thing, demonstration *potissima*, is the subject
matter of each book in the *Posterior* [*Analytics*].[77] Thus Aristotle in
the beginning of the first book of the *Prior Analytics*,[78] wanting to
set out for us the end of all the books of the *Analytics*, set out only
one method — the demonstrative — and gathered this one in the
summary at the end of the second [book] of the *Posterior* [*Analytics*].[79] And he said nothing about resolutive method, because he
considered it secondarily, insofar as it is directed toward the demonstrative. Hence resolutive, too, is comprehended under the
appellation "demonstrative," since indeed a perfect treatment on

methodo demonstrativa ea omnia complectitur, quae ipsi demonstrativae methodo inserviunt ac necessaria sunt.

3   Quamvis igitur non unam methodum tractet Aristoteles in Posterioribus Analyticis, unus tamen est eorum librorum scopus, quatenus aliae methodi ad unam diriguntur et ab una pendent, à qua nomen universae tractationi iure impositum est.

4   Propterea Graeci omnes rectè dixerunt, cùm de ultimo fine totius logici negotii loquentes eum unum esse asseruerunt, demonstrationem potissimam, ut Alexander in praefatione tum primi libri Priorum tum primi Topicorum, et Ammonius et Simplicius in praefatione libri categoriarum. Averroes quoque in multis locis dixit, potissimam demonstrationem esse primariò intentam in libris Analyticis, quod non dixisset, si putasset definitionem quoque ut instrumentum sciendi considerari, ea enim esset instrumentum demonstratione praestantius, ut etiam Averroistae omnes asserere videntur.

: X :

In quo ratio consilii Aristotelis exponitur.

1   Caeterum ut tum horum omnium rationem habeamus tum ea, quae dicenda manent, intelligamus, sciendum° est quòd potissima demonstratio optimam nobis rerum scientiam tradit, seriem enim rerum naturalem imitatur, et facit nos ita res cognoscere, ut sunt et ut cognoscendae sunt. Ea est rerum series naturalis ut causa prior sit effectu, natura enim ex causis effectus producit. Ex causarum igitur cognitione si effectuum scientiam consequamur, methodo

demonstrative method encompasses all the things that are necessary and serve demonstrative method itself.

Even though Aristotle, therefore, does not treat one method in the *Posterior Analytics*, nevertheless the goal of the books is one, insofar as the other methods are directed to one and depend on one; from this [method] the name is justly imposed on the tract as a universal whole.

And so the Greeks all said the correct thing when, speaking about the ultimate end of the whole business of logic, asserted that it is this one, demonstration *potissima*, as Alexander [said] both in the preface to the first book of the *Prior* [*Analytics*] and in the one to the first of the *Topics*, and as Ammonius and Simplicius[80] did in the preface to the book on the categories. Averroës too said in many passages that demonstration *potissima* is the primary intention in the *Analytics* books; he would not have said this if he had held that definition too is considered as an instrument for knowing scientifically, for then it would be an instrument more excellent than demonstration, as indeed all Averroists appear to assert.

3

4

: X :

In which the reason for Aristotle's intent is laid out.

But now, so that we may both have an account of all this and understand those things that remain to be said, it has to be understood° that demonstration *potissima* conveys to us optimal scientific knowledge of things,[81] for it imitates the natural sequence of things and makes us know things such as they are and as they have to be known. This is the natural sequence of things, that cause is prior to effect, for nature produces effects from causes. If, therefore, we gain scientific knowledge of effects from knowledge

1

utimur rerum naturae consentanea et à notis ad ignota secundùm
naturam progredimur, quia causae, ut causae sunt, naturaliter notae sunt, effectus verò, ut effectus, naturaliter ignoti. Illud enim
dicitur secundùm naturam notum, quod ut cognoscatur non pendet ex alio, ignotum autem secundùm naturam cuius cognitio
pendet necessariò ex cognitione alterius. Res namque ita debent
cognosci, uti sunt; quare illa, quae ut sint° non pendent ex alio in
cognitione quoque ex alio non pendent, huiusmodi sunt ipsae causae quatenus causae sunt. Causa enim quatenus est causa, non
pendet ab effectu ut sit,° quare nec ut cognoscatur, ipsa enim ex
sua natura cognoscibilis est, non ex effectu. Effectus verò quatenus
sunt effectus, à causis suis pendent ut sint,° ideò etiam ut cognoscantur, nunquàm enim perfectè cognosci possunt, nisi ex cognitione causarum, sicuti neque esse sine causis suis possunt. Igitur
ignoti secundùm suam naturam dicuntur.

2   Nec perturbare nos debet quòd videamus causas saepe nobis
esse incognitas, effectus verò sensiles et notos, haec enim effectuum cognitio absque causarum cognitione debilis et imperfecta
est; et plurimum interest qualisnam sit nostra cognitio et qualisnam esse deberet, est quidem imperfecta dum in statu ignorantiae
sumus constituti et effectus sine causis cognoscimus. Sed tamen
ita affecti esse deberemus, ut prius causas, quàm effectus cognosceremus, et à causis ad effectus[18] naturam imitando progrederemur.
Ideò quando dicimus causas esse naturaliter notas, effectus verò
naturaliter ignotos, non intelligimus° necessarium esse ut id semper eveniat ut prius causas, quàm effectus et ex causis effectus
cognoscamus. Sed solam spectando perfectam et distinctam cognitionem significare volumus talem esse causarum naturam quatenus

of causes, we use the method agreeing with the nature of things and we progress from what is known to what is not known, according to nature, because causes, as they are causes, are naturally known, but effects, as effects, are naturally unknown. For that is said to be known according to nature which does not, to be known, depend on anything else; and [that is] unknown according to nature whose knowledge depends necessarily on knowledge of something else.[82] Now things ought to be known such as they are, and so those things that do not depend on something else for their existence° do not depend on anything else for their knowledge either.[83] Causes themselves, insofar as they are causes, are of this type. For a cause insofar as it is a cause does not depend on an effect for its existence,° nor therefore to be known, for it is knowable itself by its own nature, not from an effect. But effects, insofar as they are effects, depend on their causes for their existence,° and therefore also to be known. For they can never be perfectly known except by knowledge of causes, just as they cannot be without their causes. They, therefore, are said to be unknown according to their nature.

It ought not to disturb us that we see that often causes are unknown to us but effects are sensible and known, for the knowledge of effects without knowledge of causes is weak and imperfect, and much lies between what sort of thing our knowledge is and what sort it ought to be. It is of course imperfect when we are situated in a state of ignorance and we know effects without causes. But we ought nevertheless to be so affected that we know causes prior to effects and by imitating nature progress from causes to effects. And so when we say that causes are naturally known but effects naturally unknown, we do not mean° that it necessarily always happens that we know causes prior to effects and effects from causes. Rather, by looking only at perfect and distinct knowledge, we want to indicate that the nature of causes insofar as they are causes and of effects insofar as they are effects is such that we

causae sunt et effectuum quatenus sunt effectus, ut prius et magis causas, quàm effectus cognoscere deberemus, quae esset optima et maximè perfecta scientia ipsique rerum naturae consentanea, siquidem distincta causarum cognitio à distincta effectuum cognitione non pendet, sed effectuum distincta cognitio haberi non potest, nisi prius causae ipsorum noscantur.°

3   Aristoteles igitur tradere nobis volens methodum et instrumentum, quo ad rerum cognitionem duceremur, perfectam cognitionem primariò respexit et primariò docere voluit instrumentum illud, quod naturalem rerum ordinem imitando perfectam nobis scientiam traderet, qua ita res cognosceremus, ut cognoscendae sunt. In qua sua primaria tractatione non respexit ingenii nostri imbecillitatem, qui causas, à quibus progrediendum esset, saepenumerò ignoramus, sed solùm quaenam sit methodus perfectam scientiam tradens consideravit.

4   Hanc sententiam optimè expressit Themistius in declaratione illorum verborum secundi capitis primi libri Posteriorum, 'Priora verò et notiora dupliciter.' Videns enim Aristotelem facta illa prioris et notioris distinctione non subiunxisse utro duorum modorum principia demonstrationis priora et notiora esse velit, an nobis an natura, hoc ipse explicare voluit inquiens principia notiora esse debere non nobis, sed natura. Quoniam autem obiicere aliquis poterat, ergo ab ignotis progrediemur, quod tamen dicendum esse non videtur, quandoquidem demonstratio instrumentum est nostri gratia inventum,° ut nos per eam cognoscamus, non ut natura cognoscat, natura enim neque demonstrat neque cognoscit, sed nos sumus, qui demonstrationem extruimus, ut scientiam adipiscamur. Ad hoc respondet Themistius demonstrationem non ingenii nostri infirmitatem, sed solam veritatem respicere, proinde

ought to know causes earlier and more than effects—this would be optimal and maximally perfect scientific knowledge, in agreement with the very nature of things, since indeed distinct knowledge of causes does not depend on distinct knowledge of effects, but distinct knowledge of effects cannot be had unless their causes are first known.°

Aristotle, therefore, wanting to convey to us the method and 3 instrument by which we are led to knowledge of things, gave regard primarily to perfect knowledge and wanted primarily to teach the instrument that, by imitating the natural order of things, conveys to us perfect scientific knowledge, by which we know things as they have to be known. In that primary treatment by him, he did not regard the feebleness of our wit—we who oftentimes do not know the causes from which we would have to progress—but considered only what is the method conveying perfect scientific knowledge.

Themistius expressed this position best in a clarification[84] of 4 those words in the second chapter of the first book of the *Posterior* [*Analytics*], "But things are prior and more known in two ways."[85] For seeing that, to the distinction made about prior and more known, Aristotle had not added in which of the two ways he wanted the beginning-principles of demonstration to be prior and more known, whether to us or by nature,[86] he [i.e., Themistius] wanted to explicate this, saying that beginning-principles ought to be more known, not to us but by nature. But now someone could object: we would then progress from unknowns; and it appears this should not be said, since demonstration is indeed an instrument invented° for our sake, so that by means of it we may know, not so that nature may know, for nature neither demonstrates nor knows; instead it is we who build up a demonstration so that we may obtain scientific knowledge. To this Themistius responds that demonstration does not give regard to the inadequacy of our wit, but only to the truth, and accordingly brings forth scientific

scientiam rerum parere prout rerum natura requirit. Quae The-mistii *sententia*, etsi aliquibus dura esse videatur, mihi tamen summoperè probatur et doctissima ac verissima esse videtur et Aristoteli maximè consentanea, qui cùm notiora distinxerit dicens alia nobis, alia natura notiora esse certè alterum membrum reiecto altero accipiendum esse voluit. Atqui illud minimè sumendum est, quòd sint priora, et notiora nobis, non natura, siquidem principia debent esse causae.[19] Ergo alterum, quòd natura, non nobis.

5   Voluit igitur significare Themistius quòd Aristoteles docens in-strumentum, quo res ignotas indagare et scire possimus,[20] non re-spexit quomodo nos affecti simus, sed quomodo affecti esse debe-remus. Neque ideò negat principia nobis nota esse debere, quia velit ea nobis esse incognita. Sed negat ea nobis notiora esse ut in statu imperfectionis et ignorantiae constitutis et vult esse nobis notiora ut naturam imitantibus in rerum cognitione pervestiganda. Hac autem ratione totam cognitionem nobis tribuit Aristoteles, non naturae, ut illius distinctionis hic sic sensus, alia dicuntur notiora non secundùm naturam, sed nobis, alia verò natura, non nobis, id est non nobis ut ordini naturae adversantibus, sed nobis ut naturam imitantibus et res cognoscentibus eo modo, quo sunt cognoscendae; id autem est quando ab iis rebus, quarum ea est natura, ut ex se sint cognoscibiles, ad eas progredimur, quarum ea est natura, ut non cognoscantur nisi ex alio, hoc est à causis ad effectus.

6   Haec igitur est ratio cur Aristoteles in Posterioribus Analyticis primariò de methodo demonstrativa tractaverit, quae à causis du-cit ad cognitionem effectuum.

knowledge of things as the nature of things requires. Although this position of Themistius' may appear coarse to some, to me it has been proven exceedingly well, and appears to be most scholarly and true, and altogether agrees with Aristotle, who certainly — since he distinguished the more known, saying that some things are more known to us, others by nature — if one of the branches [of the division] was rejected, wanted the other to be accepted. But yet it does not at all have to be taken that they are prior and more known to us and not by nature, since beginning-principles ought to be causes; therefore, it is the other [branch], that [they are prior and more known] by nature, not to us.

Themistius, therefore, wanted to indicate that Aristotle, teach- 5 ing the instrument by which we can track down and scientifically know things unknown, gave regard not to what way we are affected but to what way we ought to be affected. And so he does not deny that beginning-principles ought to be known to us, because he wants them to be unknown to us. Rather he denies that they are more known to us as situated in a state of imperfection and ignorance, and wants them to be more known to us as [people] imitating nature in thoroughly investigating knowledge of things. And so by this reasoning, Aristotle ascribes knowledge as a whole to us, not to nature, with the result that the sense of the distinction is this: Some things are said to be more known, not according to nature but to us, others by nature and not to us — that is, not to us as opposing the order of nature but rather to us as imitating nature and knowing things in the way in which they have to be known,[87] that is, when from those things whose nature it is that they are knowable from themselves we progress to those whose nature it is that they are not known unless from another, that is, from causes to effects.

This, therefore, is the reason why Aristotle, in the *Posterior Ana-* 6 *lytics*, primarily treated demonstrative method, which leads from causes to knowledge of effects.

7   Eadem est ratio cur primaria Aristotelis tractatio in iis libris sit de instrumento, quo accidentia cognoscantur, non de instrumento, quo cognoscantur substantiae. Accidentium enim ea est natura, ut per se sint ignota, quia ab externis causis pendent, id est extra eorum essentiam positis, proinde sciri non possunt nisi ex causarum cognitione. Substantia verò ab externa causa non pendet, cùm essentiam à causa essentiae distinctam non habeat, ut asserit Aristoteles in contextu 42 secundi libri Posteriorum, ideoque dixit definitionem substantiae non posse demonstrari, quia quod causam non habet, non est demonstrabile. Substantia igitur secundùm naturam et secundùm essentiam suam nota dicitur, quia ab externa causa non pendet. Propterea Aristoteles in toto primo Posteriorum libro, quando de definitione substantiae loquitur, semper eam numerat in principiis per se notis, quae secundùm naturam demonstratione non egent. Itaque vanissima est eorum sententia, qui praecipuum Aristotelis scopum esse arbitrantur docere instrumentum cognoscendae substantiae, quandoquidem dum praecipuum eius scopum respicimus, substantia proponitur, ut principium notum et nullo instrumento eget, ut cognoscatur.

8   Sed quamvis substantia secundùm propriam naturam instrumento non egeat, tamen habita infirmitatis et ignorantiae nostrae ratione saepe contingit ut ad eam cognoscendam instrumento egeamus. Ideò Aristoteles, cuius praecipuum consilium fuit loqui de potissima demonstratione, quae est instrumentum eorum scientiam tradens, quae secundùm suam naturam sunt ignota, non omninò spernere voluit considerationem nostrae infirmitatis, sed ad humani ingenii imbecillitatem oculos convertens voluit etiam secundaria tractatione docere nos viam, qua in eorum notitiam° ducamur, quae licèt secundùm propriam naturam sint nota, nobis

And it is the same reason why Aristotle's primary treatment in   7
those books is on the instrument by which accidents are known,
not on the instrument by which substances are known. For it is
the nature of accidents that they are unknown *per se*, because they
depend on causes external, that is, located outside their essence;
accordingly they cannot be known scientifically except by knowledge of [their] causes. But a substance does not depend on an external cause, since it does not have an essence distinct from the
cause of the essence, as Aristotle asserts in text no. 42 of the second book of the *Posterior* [*Analytics*],[88] and so he said that the definition of a substance cannot be demonstrated, because what does
not have a cause is not demonstrable. Substance, therefore, is said
to be known according to nature and according to its own essence,
because it does not depend on an external cause. Therefore Aristotle, in the whole first book of the *Posterior* [*Analytics*], when he
speaks about the definition of a substance, always counts it among
the beginning-principles known *per se*, which do not, according to
nature, need demonstration. And so the position is most vain of
those who think Aristotle's principal goal is to teach an instrument
for knowing substance, since when we give regard to his principal
goal, substance is put forward as a known beginning-principle; to
be known, it needs no instrument.

But even though substance does not, according to its proper   8
nature, need an instrument, nevertheless, by reason of our ignorance and inadequacy, it often happens that we need an instrument for knowing it. And so Aristotle, whose principal intent was
to speak about demonstration *potissima*, which is an instrument
conveying scientific knowledge of those things that are unknown
according to their own nature, did not want to entirely leave aside
consideration of our inadequacy, but, turning [his] eyes to the
feebleness of human wit, he also wanted in a secondary treatment,
to teach us the way in which we are led to knowledge° of those
things, which—granted that they may be known according to

tamen ignota esse contingit, ideò de methodo quoque resolutiva loqui voluit, quae nos ad principiorum cognitionem perducit. Nam de demonstratione ab effectu in primo libro ita locutus est, ut eam tractationem secundariam esse significaverit, eam enim conferens cum demonstratione potissima et ostendens quantum ab eius perfectione declinet, separare potius eam à potissima demonstratione, quàm de ipsa primariò sermonem facere videtur. Nam qui modum tradit perfectè sciendi, de modo quoque imperfectè cognoscendi agere debet, secundaria tamen intentione, et potissimae demonstrationis usus penitus necessarius est iis, qui scientiam consequi volunt, at usus methodi resolutivae non est simpliciter necessarius, sed solùm constituta° nostra ignorantia et imbecillitate.

9   De inductione autem egit Aristoteles in ultimo capite secundi libri Posteriorum et ostendit quomodo per eam ad primorum principiorum cognitionem ducamur. Quam tractationem secundariam esse nemini dubium est, cùm eam fecerit Aristoteles post communem epilogum omnium Analyticorum librorum, postquàm enim in contextu centesimo eius secundi libri omnia collegerat, quae in libris Analyticis docuerat, ut significaret, se ea omnia tractasse, quae in prooemio primi libri Priorum pollicitus fuerat, proponit in capite ultimo declarandum modum, quo principia cognoscuntur, quasi extra primariam intentionem caput illud in calce libri collocans, tanquàm secundaria tractatione ad artis perfectionem et ad doctrinae abundantiam adiectum.

10   De eadem locutus est etiam in eodem secundo libro, quando modum docuit venandi ea, quae praedicantur in quid et definitionem essentialem constituunt, ut mox declarabimus.

their proper nature — nevertheless happen to be unknown to us. And so he wanted to speak also about resolutive method, which leads us through to knowledge of beginning-principles. He spoke thus about demonstration *ab effectu* in the first book,[89] in a way that indicated this treatment was secondary. For comparing it with demonstration *potissima* and showing how much it diverges from the other's perfection, he appears to separate it from demonstration *potissima* rather than make a discourse about it primarily. For whoever conveys the way of perfectly knowing scientifically, ought also to deal with the way of knowing imperfectly, though with a secondary intention. And use of demonstration *potissima* is completely necessary to those who want to gain scientific knowledge. Use of resolutive method is not necessary absolutely but only from the ignorance and feebleness with which we are endowed.°

Aristotle did however deal with induction in the last chapter of   9
the second book of the *Posterior* [*Analytics*],[90] and showed in what way we are led by means of it to knowledge of first beginning-principles. It is doubtful to no one that this treatment is secondary, since Aristotle made it after the general summary of all the books of the *Analytics*. For after he had gathered, in text no. 100 of his second book,[91] everything that he had taught in the *Analytics* books, so that he might indicate that he had treated everything that he had promised [to treat] in the proem of the first book of the *Prior* [*Analytics*], he sets out in the last chapter to clarify the way by which beginning-principles are known, including, at the end of the book, as if outside [his] primary intention, this chapter, added on like a secondary treatment, for perfection of the art and the full furnishing of the teaching.

In the same second book, he spoke about the same thing also   10
when he taught the way of searching for those things that are predicated by what they are and that establish the essential definition, as we will soon make clear.[92]

11   Haec dicuntur ab Aristotele in Posterioribus Analyticis de utraque methodo resolutiva. Reliqua omnia, quae tum in primo tum
in secundo libro tractantur, ad demonstrativam methodum pertinent, quae in utroque libro primaria est, ut memoratum prooemium et epilogus manifestè declarant.

: XI :

Solutio dubii de noto et ignoto
secundùm naturam.

1 Adversus ea, quae diximus, dubium non leve oritur, antea enim
cùm discrimen inter demonstrationem ab effectu et inductionem
declararemus, diximus illa principia inductione cognosci, quae
sunt nota secundùm naturam, hoc est quae ex seipsis, non ex alio
cognoscibilia sunt; demonstratione autem ab effectu illa principia
declarari, quae ignota secundùm naturam sunt et per alia demonstrantur, quoniam sensilia non sunt, nec ex se possunt innotescere.
Itaque aliqua principia esse asservimus, quae sunt ignotae secundùm naturam. Nunc verò contrarium pronunciasse videmur dicentes omnia principia et omnes causas esse secundùm naturam
nota quatenus principia et causae sunt, nam quod alicui competit
quatenus est tale, omni competat necesse est. Quomodo igitur
aliqua possunt esse principia ignota secundùm naturam?

2   Solvitur dubium hoc distinguendo id, quod dicitur secundùm
naturam notum vel ignotum, aut enim intelligimus secundùm

These things are said about each resolutive method[93] by Aristotle in the *Posterior Analytics*. Everything else that is treated, both in the first book and in the second, pertains to demonstrative method; this is primary in each book, as the proem referred to and the summary manifestly make clear.

: XI :

Solution to a doubt regarding known
and unknown according to nature.

Against the things we have said, there arises a doubt — and not a light one. For earlier,[94] when we clarified the discriminating difference between demonstration *ab effectu* (from effect) and induction, we said that those beginning-principles that are known according to nature, that is, that are knowable in themselves and not from something else, are known by induction, and then that those beginning-principles that are unknown according to nature are made clear by demonstration *ab effectu* and are demonstrated by means of others, since they are not sensible and cannot become known from themselves. And so we asserted that there are some beginning-principles that are unknown according to nature. But we now appear to have stated the contrary, saying that all causes and all beginning-principles are known according to nature, insofar as they are causes and beginning-principles. For what appertains to something insofar as it is such and such, necessarily appertains to all. How, therefore, can there be any beginning-principles unknown according to nature?

This doubt is done away with by distinguishing that which is said to be, according to nature, known or unknown. For either we understand "according to the nature proper to things" [(a)] without

propriam rerum naturam, non considerando conditionem et naturam nostrorum cognoscentium, qua quidem ratione omnes causae omniaque principia dicuntur secundùm propriam naturam nota[21] quatenus principia et causae sunt, quoniam ab alio non pendent ut cognoscantur, quemadmodum neque ut sint.° Ita intellexit Aristoteles in secundo capite primi libri Posteriorum et in[22] prooemio primi Physicorum, quando notiora secundùm naturam dixit esse notiora simpliciter, hoc est sine ullo respectu virium humani ingenii, sed ipsam secundùm se rerum naturam considerando, haec enim est significatio illius vocis, simpliciter.

3   Aut intelligimus secundùm naturam nostrorum cognoscentium seu secundùm naturam rerum ipsarum respectu[23] conditionis humani ingenii, qua ratione causae aliquae dicuntur nobis ignotae secundùm naturam, quia insensiles sunt. Cùm enim ea sit nostra naturalis conditio, ut omnis nostra cognitio à sensu originem ducat, illud omne, quod sensu cognosci non potest, etiam si principium et causa aliqua fuerit, ignotum nobis secundùm naturam dicitur et ex alio cognoscitur, licèt simpliciter, et secundùm propriam naturam à nullo pendeat et ex se cognoscibile sit.

4   Eundem sensum aliis verbis referemus, si distinguamus cognitionem nostram distinctam et confusam, distincta enim cognitione omnia principia sunt nota secundùm naturam, quamvis abscondita et insensilia[24] sint; et omnes effectus secundùm naturam sunt ignoti, etiam si sensiles sint.

5   At si de confusa nostra cognitione loquamur, quae in sensus perceptione consistit, aliqui effectus sunt noti secundùm naturam et aliquae causae secundùm naturam ignotae.

considering what is characteristic of and the nature of our know-ing. By this reasoning, of course, all causes and all beginning-principles are said to be known according to their proper nature, insofar as they are beginning-principles and causes, since, to be known, they do not depend on something else, just as, to exist,° they do not [depend on something else]. This is how Aristotle understood it in the second chapter of the first book of the *Posterior [Analytics]*[95] and in the proem of the first [book] of the *Physics*,[96] when he said that things more known according to nature are more known absolutely, that is, without any regard to the powers of human wit and instead by considering the very nature of things in and of itself. For this is the signification of the word "absolutely."[97]

Or [(b)] we understand [it] according to the nature of our 3 knowing or according to the nature of things themselves with regard to what is characteristic of human wit. By this reasoning some causes are said to be, to us, unknown according to nature, because they are insensible. For since it is a natural characteristic of ours that all our knowledge draws its origin from sense, every-thing that cannot be known by sense, even if it were a beginning-principle and some cause, is said to be, to us, unknown according to nature, and is known from another, granted that absolutely and according to [its] proper nature, it would depend on nothing else and be knowable from itself.

We will recount the same sense in other words if we distin- 4 guish our distinct and confused knowledge. For in distinct knowl-edge, all beginning-principles are known according to nature, even though they are hidden and insensible; and all effects are unknown according to nature, even if they are sensible.

But if we speak about our confused knowledge, which consists 5 in perception by sense, some effects are known according to nature and some causes unknown according to nature.

: XII :

Quomodo definitio à primo philosopho et quomodo à logico consideretur.

1   Quoniam autem in Posterioribus Analyticis multa ab Aristotele de definitione dicuntur, declarandum nobis est quonam consilio id egerit et quomodo ad eos libros et ad tractationem de methodis definitio pertineat, hoc enim declarato, plenè cognita erit ipsa methodorum natura.

2   Quòd igitur in iis libris tractetur definitio omnes utriusque sectae interpretes concessere, sed de ratione huius tractationis ac de intentione Aristotelis dubium est, cùm alii eam ut demonstrationis medium, alii ut instrumentum sciendi à demonstratione distinctum tractari existimaverint. Tractari etiam à primo philosopho definitionem manifestum est, cuius tractationis differentiam cognoscere oportet, ut hanc ab illa discernamus.

3   Videntur omnes concedere quòd definitio ea ratione à metaphysico consideretur, quatenus significat quidditates rerum, idque apud Averroem apertè legimus in commentario 42 septimi libri Metaphysicorum. Est igitur cognoscendum quidnam sit definitionem considerari ut significatricem quidditatis. Mihi quidem videtur nil aliud esse, quàm considerari ut definitio est, nam definitio ab Aristotele ita definitur, definitio est oratio significans quid est,[25] quare si à metaphysico consideratur ut significans quidditates rerum, certè consideratur ut est definitio. Quoniam autem definitio est oratio significans quid est et oratio omnis, quemadmodum et omne nomen et omne verbum, est instrumentum, quo

: XII :

In what way definition is considered by the philosopher of metaphysics and in what way by the logician.

Now, since in the *Posterior Analytics*, many things are said by Aristotle about definition, it has to be made clear by us, with what intent he did this and in what way definition pertains to these books and to the treatment on methods. Once this is made clear, the very nature of methods will be fully known.

That definition, therefore, is treated in these books, all commentators of each sect have conceded. But there is a doubt about the reason for this treatment and about Aristotle's intention, since some judged that it [i.e., definition] was treated as the middle [term] of a demonstration and others as an instrument, distinct from demonstration, for knowing scientifically. It is manifest that definition is treated even by the philosopher of metaphysics, and the way his treatment differs must be understood so that we may discern the latter one from the former.

All appear to concede that definition is considered in the way it is by the metaphysician insofar as it signifies the quiddities of things, and we read this plainly in Averroës, in commentary 42 to the seventh book of the *Metaphysics*.[98] It has to be known, therefore, what it is for a definition to be considered as a signifier of quiddity. It appears to me, of course, to be nothing other than to be considered as it is the definition, for definition is defined by Aristotle as follows: Definition is speech signifying what [something] is (*quid est*).[99] If therefore it is considered by the metaphysician as signifying quiddities of things, then certainly it is considered as it is a definition. Moreover, since a definition is speech signifying what something is, and all speech, as also every noun and every verb, is an instrument by which we signify the

conceptiones animi significamus, sequitur quòd definitio ut est definitio et ut est oratio significans quidditatem, est instrumentum significandi[26] quidditatem, quam animo concipimus. Et definitionem considerari à primo philosopho ut significantem quidditates rerum, est eam considerari ut instrumentum significandi ipsas quidditates.

4   Nec propterea divina scientia est instrumentalis, non enim ab huiusmodi instrumentis disciplinae vocantur instrumentales et logicae, sed ab instrumentis ratiocinantibus et discurrentibus à noto ad ignotum, haec enim sunt instrumenta logica, ut antea demonstravimus. Divinus quidem philosophus agere vult de quidditatibus rerum et docere in quonam cuiusque rei quidditas consistat, at de quidditate loquens non potest non definitionem considerare, qua ipsam quidditatem significet. Ideò docere in quonam quidditas rei consistat, est docere ex quibusnam definitio constituenda sit. Quippe philosophi omnes, dum in rerum contemplatione versantur, vocibus utuntur res ipsas significantibus. Sic philosophus naturalis per voces significat res naturales, tanquàm per instrumenta significandi conceptiones animi, nec tamen est artifex instrumentalis. Dicimus itaque definitionem considerari à metaphysico prout est instrumentum significans rerum quidditates, neque hoc dicentes in absurditatem ullam incidimus.

5   Hac igitur eadem ratione si etiam à logico definitionem considerari dixerimus, absque dubio logicam cum metaphysica considerationem confundemus, quod minimè faciendum est.

6   Sed neque dicere possumus eam à logico considerari ut instrumentum notificandi rem ignotam, à metaphysico autem ut instrumentum significandi quidditatem, quandoquidem definitionem instrumentum notificandi non esse iam in praecedentibus fusè demonstravimus.

conceptions of our soul, it follows that a definition, as it is a definition and as it is speech signifying quiddity, is an instrument for signifying a quiddity that we conceive in our soul. And for the definition to be considered by the philosopher of metaphysics as signifying the quiddities of things, is for it to be considered as an instrument for signifying the quiddities themselves.

And so the science of divinity is not instrumental, for disciplines are not called instrumental and logical on account of instruments of this type, but on account of instruments for ratiocinations and discursive movements from known to unknown, for these are logical instruments, as we demonstrated earlier. The philosopher of divinity, of course, wants to deal with the quiddities of things and to teach in what the quiddity of any thing consists. But speaking about a quiddity, he cannot not consider the definition, by which he may signify the quiddity itself. And so to teach in what the quiddity of something consists is to teach from what things the definition has to be constituted. And so of course all philosophers, when they are concerned with contemplation of things, use words signifying the things themselves. So the natural philosopher signifies natural things by means of words, as by means of instruments for signifying the conceptions of man's soul. Nevertheless, he is not an instrumental practitioner. And so we say that definition is considered by the metaphysician in that it is an instrument signifying the quiddities of things, and saying this, we do not fall into any absurdity.

If, therefore, we say that definition is considered in the same way by the logician, without doubt we conflate logical consideration with metaphysical; this should not be done at all.

But we cannot say it is considered by the logician as an instrument for making known something unknown, and by the metaphysician as an instrument for signifying quiddity, since we already demonstrated at length in the preceding that a definition is not an instrument for making known.

4

5

6

7   Quoniam igitur non considerat logicus nisi instrumenta notificandi et tale instrumentum non est definitio, relinquitur eam nulla
ratione posse à logico considerari, nisi prout ad logica instrumenta
relationem aliquam habet vel ad eorum cognitionem vel traditionem aliqua ratione confert. Cùm enim logica sit disciplina instrumentalis, nil debet considerare, nisi instrumenta sciendi vel ea,
quae ad instrumenta sciendi aliquo modo referantur atque perti-
neant. Id autem Aristoteles ipse manifestissimè pronunciavit tum
in primo tum in secundo Posteriorum libro, quando dixit omnem
definitionem aut principium esse demonstrationis aut conclusionem demonstrationis aut ipsammet demonstrationem solo terminorum situ ab ea discrepantem. Nullam igitur definitionem in
iis libris considerare voluit, nisi cum respectu ad demonstrationem, idque ex utriusque libri lectione manifestè colligitur, nam in
fine primi capitis secundi libri quando proponit Aristoteles se diligentem tractationem de definitione facturum, proponit inter caetera declarandum quomodo definitio ad demonstrationem pertineat. Postea in contextu 47 epilogum faciens dicit se docuisse
quomodo definitio ad demonstrationem referatur.

8   Qualis autem sit haec relatio facilè intelligemus, si et instrumenti logici et definitionis naturam consideraverimus. Logicum
enim instrumentum est progressus et discursus ab hoc ad illud
tanquàm à noto ad ignotum. Definitio verò simplex quoddam est
ac individuum, quod omni discursu caret. Igitur eam rationem
habet ad logica instrumenta, quam punctum habet ad lineam et
terminus individuus ad id, cuius terminus est, etenim in linea
partes notare possumus et progressum ab una ad aliam, punctum
verò individuum est et partibus caret et est vel principium vel finis

Since, therefore, the logician does not consider [anything] ex- 7
cept instruments for making known, and definition is not such an
instrument, it remains that it can be considered by the logician
in no way except in that it is somehow relative to logical instruments or contributes in some way to knowledge of them or to
conveying them. For while logic is an instrumental discipline, it
ought to consider nothing except instruments for knowing scientifically or those things that are referred and pertain in some way
to instruments for knowing scientifically. Moreover, Aristotle himself stated this most manifestly, both in the first and in the second
book of the *Posterior* [*Analytics*], when he said that every definition
is either a beginning-principle of demonstration or the conclusion
of a demonstration or a demonstration itself, being different from
it only in arrangement of terms.[100] He wanted, therefore, to consider no definition in these books except with regard to demonstration; this is gathered manifestly from a reading of either book.
For at the end of the first chapter of the second book,[101] when
Aristotle sets out to make a careful treatment on definition, he
sets out to clarify, among other things, in what way definition
pertains to demonstration. Afterward, in making the summary in
text no. 47,[102] he says that he has taught in what way definition is
referred to demonstration.

What sort of relation this is we will easily understand, once we 8
have considered the nature of both logical instrument and definition. For a logical instrument is a progression and discursive
movement from this to that, as from known to unknown. But
definition is something simple and indivisible; it lacks all discursive movement. It has, therefore, the same relation to logical instrument that point has to line and an indivisible terminus to that
of which it is the terminus. For indeed in a line we can note parts
and a progression from one to another, but a point is indivisible
and lacks parts and is either the beginning (*principium*) or the end

lineae. Sic definitio vel principium, vel finis logicarum methodorum dicenda est, principium quidem quando est nota, finis verò
quando est ignota et per logica instrumenta investigatur, nam
quaerere quid res sit, nil aliud est, quàm eius definitionem indagare. Quomodo autem haec omnia sese habeant manifestum
fiet, si omnes definitionis species singillatim consideraverimus et
singularum respectum ad demonstrationem et ad logicas methodos declaraverimus.

∶ XIII ∶

Qua ratione definitio substantiae in
libris Analyticis consideretur.

1 Ut hac in re à rebus definitis tanquàm à notioribus progrediamur,
omne, quod definitur, aut est substantia aut accidens, ut nomine
substantiae illa omnia complectamur, quae in scientiis ut alteri
substrata et subiecta considerantur. Et nomine accidentis ea, quae
considerantur ut alteri inhaerentia, quae solent vocari passiones
sive affectiones. Propterea magnitudines et numeri, quae in scientiis mathematicis habent locum subiecti, de quo affectiones demonstrantur, licèt simpliciter accidentia sint, tamen substantiae
nomine comprehendentur. Ideò Aristoteles in contextu 178 primi
libri Posteriorum unitatem° et punctum substantias vocavit, quia
in toto eo libro ea semper ut alteri subiecta, non ut accidentia no-
minaverat.

2 Ut autem à definitione substantiae exordiamur, huius duplex
est in Posterioribus Analyticis consideratio, una primaria, altera

(*finis*) of a line. Thus definition has to be said to be either the beginning-principle (*principium*) or the end (*finis*) of logical methods — the beginning-principle when it is known, of course, and the end when it is unknown and investigated by means of logical instruments.[103] For to inquire what something is, is nothing other than to track down its definition. And in what way all this is so will be made manifest, once we have considered all species of definition one by one and have clarified the relationship of each to demonstration and to logical methods.

: XIII :

*In what way the definition of a substance
is considered in the Analytics books.*

Now on this issue, so that we may progress from things defined as   1
from things more known: Everything that is defined is either substance or accident. Under the name of substance is encompassed everything that, in the sciences, is considered as subject and substrate of another. Under the name of accident [is encompassed everything] that is considered as inhering in something else; this is normally called a passion or affection. And so magnitudes and numbers, which in mathematical sciences have the place of subject and about which affections are demonstrated, even granted that absolutely they may be accidents, are nevertheless comprehended under the name of substance. That is why Aristotle, in text no. 178 of the first book of the *Posterior* [*Analytics*],[104] called unit° and point substances — because in the whole book he had always named them as subjects of something else, not as accidents.

Let us begin then with the definition of substance. Con-   2
sideration of it in the *Posterior Analytics* is of two types — one

secundaria, primaria quidem prout est principium notum methodi demonstrativae, secundaria verò prout est finis ignotus methodi resolutivae. Cùm enim primaria tractatio sit de potissima demonstratione, definitio substantiae ad hanc non refertur, nisi ut eius principium notum, nam haec definitio non traditur ex causis extra rei definitae essentiam positis, sed ex internis, quae essentiam ipsam constituunt; propterea Aristoteles indemonstrabilem talem definitionem esse dixit in contextu 42 secundi libri Posteriorum, definitur enim substantia per formam et materiam et eius definitionis nulla causa afferri potest, quia illa forma in materia inest sine[27] ulla media causa, siquidem ipsa est et essentia rei et causa, qua res esse dicitur. Ideò secundùm ipsam rerum naturam haec definitio locum habet causae non pendentis ab alia causa, proinde habet locum principii noti secundùm propriam naturam. Propterea in primaria Aristotelis tractatione haec definitio non consideratur nisi ut principium demonstrationis per se notum, ut patet legentibus Aristotelem in primo Posteriorum libro, ubi passim definitionem hanc numerat in principiis demonstrationis per se notis et indemonstrabilibus.

3   De hac definitione Latini rectè sentiunt, cùm dicunt eam non rectè[28] considerari nisi ut medium demonstrationis, dicentes enim prout est medium dicunt prout est principium, nam quomodo est principium demonstrationis, nisi quatenus est medius terminus in syllogismo demonstrativo? In eo tamen decepti sunt, quòd putarunt hanc definitionem prout est medium in secundo tantùm Posteriorum libro tractari, cùm hac ratione consideret eam Aristoteles in toto primo libro, nec video quomodo hoc inficiari audeant,

primary, the other secondary: the primary, in that it is the known beginning-principle of demonstrative method; and the secondary, in that it is the unknown end of resolutive method. Now although the primary treatment is on demonstration *potissima* and the definition of a substance is not referred to this except as its known beginning-principle — for this definition is not conveyed from causes lying outside the essence of the thing defined, but from [those] internal, which constitute the essence itself — so Aristotle, in text no. 42 of the second book of the *Posterior* [*Analytics*],[105] said that such a definition is indemonstrable. For a substance is defined by means of form and matter, and no cause of its definition can be brought forward, because that form belongs to matter without any middle cause, since indeed that is both the essence of the thing and the cause by which the thing is said to be. And so, according to the very nature of things, this definition has the place of a cause not depending on another cause, and correspondingly, has the place of a known beginning-principle, according to its proper nature. And so in Aristotle's primary treatment this definition is not considered except as a demonstration's beginning-principle known *per se*, as is patent to those reading Aristotle in the first book of the *Posterior* [*Analytics*], where in various passages he counts this definition among the beginning-principles of demonstration indemonstrable and known *per se*.

Regarding this definition, the Latins take the correct position, 3 since they say that it is not considered correctly except as the middle [term] of a demonstration. For saying "in that it is a middle [term]," they say, "in that it is a beginning-principle." Now in what possible way is it the beginning-principle of a demonstration except insofar as it is the middle term in a demonstrative syllogism? Nevertheless they were deceived in that they held that this definition, in that it is a middle [term], is treated only in the second book of the *Posterior* [*Analytics*], although Aristotle considers it in this way in the whole first book. I do not see how they dare

cùm Aristoteles in primo libro non alia ratione eam consideret, quàm ut principium demonstrationis. In contextu etiam 42 secundi libri dixit Aristoteles[29] hanc definitionem esse immediatam et indemonstrabilem, et habere locum principii per se noti vel suppositi.

4   Averroistae quoque decepti sunt dum dixerunt hanc definitionem considerari ut instrumentum cognoscendae substantiae, cùm enim definitio subiecti duos habeat respectus, unum ad subiectum, cuius est definitio, alterum ad affectiones subiecti, quarum est causa, Aristoteles hoc tantùm secundo modo eam in libris Analyticis consideravit, nimirum prout est medium demonstrationis, quod nobis perfectam accidentium scientiam praebet. Sed respectu subiecti non consideratur in iis libris definitio, siquidem longè diversus est hic respectus ab eo, quem ad affectiones subiecti habet. Affectiones quidem declarat per discursum eas notificando, naturam verò et quidditatem subiecti declarat non notificando, sed significando, sicuti nomen quoque rem significare dicitur. Quae quidem consideratio non ad logicum, sed ad metaphysicum pertinet, quemadmodum ostendimus. Dum igitur primariam Aristotelis tractationem et intentionem spectamus, non alia ratione definitionem substantiae consideravit, quàm ut principium notum methodi demonstrativae, hoc est ut medium potissimae demonstrationis, ut benè Latini dixerunt, eaque tractatio tota de demonstratione nominanda est, non de definitione, quia cùm aliquid ut alterius principium et eius gratia tractatur, de illo altero tractatio esse dicitur, ut Aristoteles testatur in principio libri duodecimi Metaphysicorum dicens, 'de substantia est contemplatio, quoniam substantiarum principia, et causae quaeruntur.'

deny this, since Aristotle considers it in the first book in no way other than as the beginning-principle of demonstration. In text no. 42 of the second book,[106] Aristotle said that this definition is immediate and indemonstrable and has the place of a beginning-principle known *per se* or supposed.

Averroists too were deceived, when they said that this defini- 4 tion is considered as an instrument for knowing substance. For although the definition of a subject has two relationships, one to the subject whose definition it is, the other to the subject's affections of which it is the cause, Aristotle considered it only in this second way in the *Analytics* books, namely, of course, in that it is the middle [term] of a demonstration; this provides us with perfect scientific knowledge of accidents. But definition is not considered in relation to the subject in these books, since indeed this relation is very different from that which it has to affections of the subject. It makes the affections clear, of course, by making them known by means of discursive movement. But it makes the nature and quiddity of the subject clear, not by making known, but by signifying, just as a name too is said to signify something. This consideration, of course, pertains not to the logician but to the metaphysician, as we showed. When, therefore, we look at Aristotle's primary treatment and intention, [we see that] he considered the definition of a substance in no way other than as a known beginning-principle of demonstrative method, that is, as the middle [term] of a demonstration *potissima*, as the Latins said well. And the whole treatment has to be named with regard to demonstration, not definition, because when something is treated as the beginning-principle of something else and for its sake, the treatment is said to be with regard to that other thing, as Aristotle attests in the beginning of the twelfth book of the *Metaphysics*, saying, "The contemplation [here] is with regard to substance, since the beginning-principles and causes of substances are being inquired after."[107]

5   Primaria[30] igitur huius definitionis tractatio non alia est, quàm ea, quam diximus.

6   Dum verò secundariam in iis libris tractationem respicimus, in qua Aristoteles viam tradit, qua ad principiorum notitiam,° si latuerint, ducamur, definitio substantiae consideratur ut finis ignotus methodi resolutivae et de hac loquitur Aristoteles in secundo libro Posteriorum Analyticorum à contextu 69 usque ad 84 ibi namque docet qua via eam definitionem venari debeamus, quae est principium indemonstrabile et refutat opinionem Platonis, qui divisione definitionem venabatur, non quidem quòd penitus reiiciat divisionis usum, sed quia solam non sufficere arbitratur, ut eam, quae vim syllogisticam non habet. Ipse igitur praefert viam compositivam, quae vim habet illatricem ignoti ex noto, si enim proponatur investiganda definitio generis, ut animalis, vult Aristoteles spectandas esse species omnes illius generis et ea, quae de ipsis speciebus essentialiter praedicantur, his namque praedicatis omnibus collectis, exceptisque illis, quae singularum specierum propria sunt, reliqua generis definitionem constituent. Simili ratione si speciei definitionem quaeramus, individua considerare debemus et sumere omnia, quae de ipsis praedicantur in eo quod quid est, ex iis enim speciei definitionem conflabimus.

7   Hanc viam vocavit Aristoteles compositivam, quia ex specierum collectione et attributorum compositione definitionem generis colligimus et ex individuorum collectione definitionem speciei. Attamen si benè rem consideremus, haec via resolutiva potius est, quàm compositiva, quia ab inferioribus ad superiora procedere est à posterioribus ire ad priora, ut antea demonstravimus. Quare nil aliud est via haec, quàm inductio, ex eo enim quòd illa praedicata

The primary treatment of this definition, therefore, is nothing 5
other than what we said it was.

Now when we give regard to the secondary treatment in these 6
books, in which Aristotle conveys the way by which we are led to
knowledge° of beginning-principles, if they were hidden, [we see
that] the definition of a substance is considered as the unknown
end of resolutive method. And Aristotle speaks about this in the
second book of the *Posterior Analytics* from text no. 69 up to no.
84.[108] For there he teaches in what way we ought to search for a
definition that is an indemonstrable beginning-principle, and he
confutes the opinion of Plato, who searched for definition by division — not of course because he completely rejects the use of division, but because he thinks that, as something that does not have
syllogistic power, it is not sufficient alone. He prefers, therefore, a
compositive way, which has the power to infer unknown from
known. For if the definition of a genus, say animal, is set out to be
investigated, Aristotle wants all species of that genus to be looked
at as well as those [attributes] that are predicated essentially of the
species themselves. Once everything predicated has been gathered[109] and whatever is proper to each of the species has been ex-
cepted, the things remaining constitute the definition of the genus.
In a similar way, if we inquire after the definition of a species, we
ought to consider the individuals and take everything that is predicated of them, by virtue of what each is; from these we will as-
semble the definition of the species.

Aristotle called this way compositive, because we gather the 7
definition of the genus from a gathering of the species and a composition of attributes, and the definition of the species from a
gathering of individuals. But nevertheless, if we consider the issue
well, this way is resolutive rather than compositive, because to
proceed from the lower to the higher is to go from posteriors to
priors, as we demonstrated earlier. This way, therefore, is nothing
other than induction. For from the fact that these predicates

omnibus speciebus competunt, colligimus eadem competere generi; et ex eo quòd omnibus individuis competunt, probamus ea
competere speciei. Haec igitur via compositiva vim inductionis
habet et voluit Aristoteles significare neque sola divisione utendum
esse neque sola via compositiva absque divisione in venandis definitionibus, quia si absque divisione utamur via compositiva, illa
omnia praedicata inordinatè et confusè colligemus, facit enim divisio ut rectum ordinem in partibus definitionis exponendis serve-
mus. Sed ratione efficacitatis probandi id, quod quaeritur, praefert
Aristoteles viam compositivam primasque ei tribuit, cùm enim
vim syllogisticam habeat, totam rei ignotae declarationem ipsi acceptam ferimus, à divisione verò nullam probationem, sed solam
partium dispositionem habemus. Itaque si definitionem ita venemur, ut divisione utendo singulas definitionis partes ordinatè
sumamus, easque rei definiendae inesse, eò quòd omnibus particularibus competant, ostendamus per inductionem, optimus erit modus venandi talem definitionem, qui Aristoteli mirificè placuit.

8   De demonstratione autem à signo quòd ad venandas partes definitionis utilis sit, nihil in ea parte dixit Aristoteles, satis enim
habuit se de illa in primo libro locutum esse. At in ea parte
ostendere volens infirmitatem viae Platonicae divisivae, satis eam
declaravit ostendens divisionem ad eas definitiones venandas inutilem esse, quae secundùm naturam notae sunt et leviore indigent
instrumento logico. Si namque ad has venandas divisio inefficax
est, multò minùs ad eas indagandas, quae sunt naturaliter ignotae,
efficax erit. Sed quòd demonstratione quoque ab effectu demonstrentur partes definitionis naturaliter ignotae, testatur ipse
Aristoteles in contextu undecimo primi libri de Anima dicens

appertain to all the species, we gather that the same [predicates] appertain to the genus, and from the fact that they appertain to all individuals, we prove that they appertain to the species. This compositive way, therefore, has the power of induction,[110] and Aristotle wanted to indicate that in searching for definitions, neither is division alone to be used, nor a compositive way alone without division. Because if we use a compositive way without division, we will gather all those predicates confusedly and without order, for division makes it that we maintain the correct order in laying out the parts of a definition. But for reason of efficacy in proving that which is inquired after, Aristotle prefers a compositive way and ascribes first place to it. For since it has syllogistic power, we get, thanks to it, the whole clarification of the unknown thing, but with division we have no proof, only a disposition of parts. And so if we thus search for a definition — such that by using division we take up all of the parts of the definition in order, and we show by means of induction that they belong to what is being defined because they appertain to all the particulars[111] — this will be the best way of searching for such a definition; it pleased Aristotle wonderfully well.

Now regarding demonstration *a signo*,[112] that it might be useful 8 for searching for parts of a definition, Aristotle said nothing in that part. For it was enough for him to have spoken about it in the first book. But wanting to show in that part the inadequacy of Plato's way of division, he made it clear enough, showing that division is useless for searching for those definitions that are known according to nature and need a lighter logical instrument. And if division is ineffectual in searching for these, it will be much less effectual for tracking down those that are naturally unknown. But that naturally unknown parts of a definition may be demonstrated by demonstration *ab effectu* also, Aristotle himself attests in text no. 11 of the first book of *On the Soul*,[113] saying that

accidentia plurimùm conferre ad cognoscendum quid est, nam ex iis, quae rei definiendae accidunt, duci in cognitionem eorum, quae eius rei essentiam constituunt, est ab effectibus ea venari et demonstrare à posteriori.

9 Ex his patet definitionem substantiae nulla alia ratione considerari ab Aristotele in libris Analyticis, quàm ut principium methodi demonstrativae et ut finem methodi resolutivae, si eam ignotam esse et quaeri contingat. Prior quidem illa consideratio primaria est, haec autem secundaria, quod significavit Aristoteles in contextu 42 libri secundi Posteriorum de hac definitione loquens, quando dixit eam esse principium notum vel, si ignorari contingat, non demonstratione, sed alia via manifestum fieri. De qua alia via sermonem postea facere voluit in ea parte, quae inscribitur de venatione definitionis.

10 Has duas tractationes de definitione substantiae in iis libris invenio et praeterea nullam. In quarum neutra dicere possumus definitionem ut instrumentum substantiae cognoscendae considerari ab Aristotele, sed solùm ut principium vel ut finem instrumenti. Quando enim eam considerat ut principium demonstrationis, non ad substantiae, sed ad accidentis cognitionem ipsam dirigit. Quando autem docet viam venandi praedicata essentialia et partes definitionis, non videmus Aristotelem dicere definitionem esse methodum vel instrumentum, quo talia praedicata venemur, id enim si dixisset, ridiculum se omnibus praebuisset. Sed viam divisivam et viam compositivam ut instrumenta huius venationis consideravit, definitionem verò non ut instrumentum neque ut viam, sed ut finem viae, nam quaerere ea, quae in eo quod quid est praedicantur, est ipsam definitionem quaerere, ut saepius dictum est in praecedentibus.

accidents greatly contribute to knowing what [something] is (*quid est*), for to be led from those things that are accidents of the thing being defined into knowledge of those that constitute the essence of the thing, is to search for them from effects and to demonstrate [them] *a posteriori*.

From all this it is patent that in the *Analytics* books the defini- 9 tion of a substance is considered by Aristotle in no other way than as the beginning-principle of demonstrative method and as the end of resolutive method, if it happens to be unknown and inquired after. That first consideration is, of course, primary and the latter is secondary, as Aristotle indicated in text no. 42 of the second book of the *Posterior [Analytics]*,[114] speaking about this definition, when he said, it is a known beginning-principle, or, if it happens to be unknown, it is made manifest not by demonstration but in some other way. He wanted to discourse on this other way afterward, in that part which is titled *On the Search for Definition*.[115]

In these books, I find these two treatments on the definition of 10 a substance, and no other. In neither of them can we say that definition is considered by Aristotle as an instrument for [coming to] know a substance, but only [instead] as an instrument's beginning-principle or end. For when he considers it as the beginning-principle of a demonstration, he directs it [i.e., the demonstration] toward the knowledge of an accident but not of a substance. And when he teaches the way of searching for essential predicates and parts of a definition, we do not see Aristotle say that definition is a method or an instrument by which we search for such predicates. For if he had said this, he would have surrendered himself to the ridicule of everyone. Instead he considered the divisive way and the compositive way as instruments of this search, but definition not as an instrument or a way, but as the end of a way. For to inquire after those things that are predicated by virtue of what something is, is to inquire after the definition itself, as was frequently said in the preceding.

: XIV :

Quotuplex sit definitio accidentis.

1   Accidentis autem definitio multiplex est, alia enim est perfecta et omnibus numeris absoluta, quae est demonstratio positione differens, alia imperfecta et manca, quae tum conclusio tum principium demonstrationis esse potest. Quae omnia facilè intelligentur, si quaenam sit perfecta accidentis definitio declaraverimus, quod enim in quoque genere summum perfectumque est, norma esse solet, qua caetera eius generis dignoscuntur.

2   Accidentia omnia per formam et materiam definienda sunt, quemadmodum substantiae, alia tamen ratione, nam accidentis genus formae locum tenet, proprium autem subiectum est loco materiae et in definitione pro differentia sumitur, ut definitio tonitrus est, sonus in nube et sonus quidem est genus et forma, quia omne accidens forma quaedam est, nubes autem est subiecta materia, quae tanquàm differentia sumitur separans tonitrum ab aliis sonis. Contra quàm in substantiis accidat, in quibus genus habet locum materiae, differentia verò formae. Et in substantiis materia interna est, ex qua res constat; in accidentibus verò externa, in qua inhaerent.

3   Definitio haec ex materia et forma constans dicitur essentialis sive quidditativa, quia tota accidentis essentia per eam exprimitur, scilicet per genus restrictum à subiecto proprio fungente officio differentiae, etenim quidnam aliud est eclipsis, quàm privatio quaedam luminis lunae? Quid aliud est tonitrus, quàm sonus quidam in nube factus?

: XIV :

Of how many types is the definition of accident.

Now the definition of accident is of multiple types. One is perfect 1
and on all counts complete; this is a demonstration differing in
position. The other is imperfect and deficient; this can be either
the conclusion or the beginning-principle of a demonstration. All
this will be easily understood if we clarify what a perfect definition
of accident is. For what is highest and perfect in any genus is,
normally, the norm by which others of the genus are distinguished.

As with substances, though in another way, all accidents have to 2
be defined by means of form and matter. Now the genus of an accident holds the place of form and the proper subject is in the
place of matter and is taken in the definition as the differentia, as,
for instance, the definition of thunder is noise in a cloud; noise, of
course, is the genus and form—because every accident is some
sort of form—and then cloud is the subject matter that is taken as
the differentia separating thunder from other noises. It would happen otherwise with substances; in them genus has the place of
matter and the differentia [the place] of form. And in substances
the matter out of which the thing is composed is internal, but in
accidents the matter in which [they] inhere is external.

This definition, composed out of matter and form, is said to be 3
essential or quidditative,[116] because the whole essence of the accident is expressed by means of it, that is, by means of the genus
restricted by the proper subject performing the function of differentia. For indeed, what is an eclipse other than some sort of privation of the light of the moon? What is thunder other than some
sort of noise made in a cloud?

4   Sed aliud est magni momenti discrimen inter hanc accidentis definitionem et definitionem substantiae, quod ad propositam nobis contemplationem maximè pertinet, quòd definitio substantiae per materiam et formam perfecta est et nihil ei deest, haec enim hominis definitio, animal rationale et essentialis est et perfecta definitio, quia essentiam hominis ita declarat, ut nil quaerendum maneat, huius autem ratio est, quoniam illa forma inest in illa materia sine medio, neque ab ulla externa causa pendet, sed est simul essentia et causa rei.

5   Ast illa accidentis definitio, quam diximus, essentialis quidem est, non tamen perfecta, quia essentia accidentis pendet ab externa causa diversa ab ipsa essentia, ideò quando ad quaestionem quid est respondetur talis definitio essentialis, animus quaerentis non conquiescit, praesertim animus viri sapientis, ut doctissimè notavit Averroes in calce sexti commentarii libri secundi Posteriorum, quando enim quidditas pendet ab alia externa causa, animus in cognitione quidditatis non quiescit, nisi causam quoque noscat° eius quidditatis. Propterea ut definitio accidentis perfecta reddatur, tertia particula praeter formam et materiam requiritur, nempè causa inhaerentiae illius formae in illa materia. Perfecta igitur definitio eclipsis est, privatio luminis in luna ob terrae interpositionem. Et perfecta tonitrus definitio est, sonus in nube propter extinctionem ignis, si modò haec ipsius causa sit, ut supponit Aristoteles.

6   Ob hanc igitur rationem in contextu 42 et 43 secundi libri Posteriorum dicit Aristoteles definitionem accidentis quidditativam demonstrabilem esse, non quòd tota definitio de definito[31]

But there is another discriminating difference of great moment 4
between this definition of an accident and the definition of a substance. It pertains especially to the contemplation set out before
us, that the definition of a substance by means of matter and form
is perfect and lacks nothing — as the definition of man as rational
animal is a definition both essential and perfect — because it makes
the essence of man so clear that nothing remains to be inquired
after. And the reason for this is that since that form belongs in
that matter without a middle, it does not depend on any external
cause, but is at the same time the essence and cause of the thing.

But now that definition of accident that we spoke of is essen- 5
tial, of course, but nevertheless not perfect, because the essence of
an accident depends on an external cause different from the essence itself. And so, when such an essential definition is given in
response to the question of what [something] is (quid est), the soul
of the one inquiring, especially the soul of a wise man, does not
rest, as Averroës noted in a most learned way at the end of the
sixth commentary to the second book of the Posterior [Analytics].[117]
For when the quiddity depends on another, external cause, our
soul does not rest in knowledge of the quiddity, unless it also
knows° the cause of that quiddity. And so for the definition of an
accident to be rendered perfect, a little third part, besides matter
and form, is required; namely the cause of that form's inherence in
that matter. The perfect definition of an eclipse, therefore, is: a
privation of light in the moon on account of interposition of the
earth. And the perfect definition of thunder is: a noise in a cloud
on account of the extinction of fire — if in reality this is its cause,
as Aristotle supposes.[118]

For this reason, therefore, Aristotle says, in texts no. 42 and no. 6
43 of the second book of the Posterior [Analytics],[119] that the quidditative definition[120] of an accident is demonstrable, not because

demonstretur, ut si colligatur eclipsim esse privationem luminis lunae et tonitrum esse sonum in nube, quem sensum multi perperam accipiunt; sed quòd altera definitionis pars de altera parte demonstretur, nempè forma de materia per aliam tertiam partem adiectam,[32] quae est causa inhaerentiae formae in materia. De luna enim demonstratur privatio luminis per mediam terrae interpositionem et de nube sonus per mediam ignis extinctionem. Idcircò haec perfecta accidentis definitio apud Aristotelem est ipsamet demonstratio et solo terminorum situ ab ipsa demonstratione discrepat, quia ex iisdem tribus terminis demonstratio accidentis et eiusdem definitio constituitur, genus quidem accidentis est in demonstratione maius extremum, subiectum est minus extremum, causa verò medius terminus.

7   Haec perfecta definitio in duas imperfectas dividitur, ut notat Averroes in commentario 64 primi libri Posteriorum, quarum altera est principium demonstrationis, altera verò conclusio. Ut enim utraque simul sumpta facit integram demonstrationem, quia simul conclusionem et principia continet; ita si in duas dividatur, altera erit principium demonstrationis tantùm, altera verò erit solùm conclusio. Illa enim, quae traditur per solam causam, est principium demonstrationis, ut si dicamus, eclipsis est interpositio terrae, quae quidem impropriè dicitur definitio, cùm dicat potius propter quid, quàm quid, ideò vocatur definitio causalis et ad demonstrationem refertur non alia ratione, quàm ut eius principium, ut asserit etiam Themistius interpretans contextum 46 secundi

the whole definition of what is defined is demonstrated — as if [for example] it were gathered that an eclipse is a privation of the light of the moon and thunder is noise in a cloud — a sense [of perfect definition] that many wrongly accepted, but because one part of the definition is demonstrated of another part, that is, the form [is demonstrated] of the matter by means of another, third part, added in, which is the cause of the form's inhering in the matter. For privation of light is demonstrated of the moon by means of the interposition of the earth in the middle and the noise [is demonstrated] of the cloud by means of extinction of fire. And accordingly, this perfect definition of an accident is in Aristotle the very demonstration itself; it is different from the demonstration itself only in arrangement of the terms, because from the same three terms the demonstration of the accident and the definition of the same are constituted; in the demonstration, of course, the genus of the accident is the major extreme, the subject is the minor extreme, and the cause is the middle term.[121]

This perfect definition is divided into two imperfect ones, as Averroës notes in commentary 64 to the first book of the *Posterior* [*Analytics*].[122] Of the two, one is the beginning-principle of a demonstration, and the other the conclusion. For just as both taken together make an intact demonstration — since it contains at the same time the conclusion and the beginning-principles — so, if it is divided into two, one will be only the beginning-principle of the demonstration and the other will be only the conclusion. For the former, which is conveyed by means of cause alone, is the beginning-principle of the demonstration, as if [for example] we were to say, an eclipse is the interposition of the earth; this, of course, is said to be the definition [only] improperly, since it says what [something] is on account of (*propter quid*) rather than what [something] is (*quid est*). And so the definition is called causal and is referred to demonstration in no other way than as its beginning-principle, as even Themistius asserts, commenting on text no. 46

libri Posteriorum, est enim principium, quia est demonstrationis medium. Ideò quando Aristoteles dicit aliquam definitionem esse principium demonstrationis, intelligit° quidem praecipuè definitionem subiecti, sed ad illud membrum redigitur illa quoque affectionis definitio, quae vocatur causalis, ut vult eo in loco Themistius.

8 Altera verò perfectae definitionis pars, quae formam et materiam sine causa continet, illa est, quae dicitur demonstrationis conclusio, ut si dicamus, eclipsis est privatio luminis lunae, tonitrus est sonus in nube, et vocatur conclusio, quia cùm extra se causam habeat, demonstrari potest. Hanc diximus vocari quidditativam sive essentialem sive etiam formalem definitionem; internam enim rei naturam, et essentiam exprimit. Hanc eandem Aristoteles in 44 contextu secundi libri Posteriorum nominalem definitionem appellavit, quia essentiam ab externa causa pendentem proferens sine expressione causae non benè ipsam essentiam declarat, id enim, quod causam aliquam habet propter quam est, non benè cognoscitur, nisi per illam. Ut igitur hanc distingueret ab illa, quae essentiam cum essentiae causa declarat et est perfecta definitio, eam vocavit nominalem, quia nominis tantùm significationem exprimit, essentiam verò non benè declarat. Postea verò in contextu 46 ex iis, quae prius dixerat, colligens omnes definitionis species eam vocat conclusionem demonstrationis, quia causam extra se habet, per quam demonstrari potest. Tot igitur numero sunt accidentis definitiones.

of the second book of the *Posterior* [*Analytics*].[123] For it is the beginning-principle because it is the middle [term] of a demonstration. And so when Aristotle says that some definition is the beginning-principle of a demonstration, he, of course, means° principally the definition of the subject; but that definition of the affection that is called causal is directed to that branch [of the division] also, as Themistius would have it in this passage.

But now the other part of a perfect definition, which contains form and matter without cause, is that which is said to be the conclusion of a demonstration, as if [for example] we were to say an eclipse is the privation of the light of the moon, thunder is noise in a cloud. And it is called a conclusion because, since it has a cause outside itself, it can be demonstrated. We said this is called a quidditative or an essential or also a formal definition,[124] for it expresses the internal nature and essence of the thing. Aristotle, in text no. 44 of the second book of the *Posterior* [*Analytics*],[125] called this same thing a nominal definition, because, advancing as it does an essence that depends on an external cause without expression of the cause, it does not well clarify the essence itself, for that which has some cause on account of which it is, is not known well except by means of that. To distinguish, therefore, this latter from the former, which makes the essence clear using the cause of the essence and is a perfect definition, he called it nominal,[126] because it expresses only the signification of the name; it does not well clarify the essence. But afterward, in text no. 46,[127] gathering all species of definition from that which he had said earlier, he calls it the conclusion of a demonstration, because it has a cause outside itself, by means of which it can demonstrated. That is, therefore, the number of definitions of accident there are.

8

: XV :

De aliorum erroribus in ea definitione,
quae est demonstrationis conclusio.

1   Non sunt hîc silentio praetereundi duo aliorum errores in intelligenda illa definitione, quae secundùm Aristotelem est demonstrationis conclusio. Unus error est quòd multi putant definitionem hanc non esse eandem, quam Aristoteles in contextu 44 eius libri nominalem vocaverat, cùm tamen ea ipsa sit, ut patet legentibus totam illam partem et ut nos alio in loco opportunius demonstrabimus. Alter autem error est et is quidem maximus, in quem omnes penitus interpretes et antiquiores et posteriores inciderunt, uno Averroe excepto et fortasse etiam Themistio, quòd non intellexerunt quomodo ea definitio dicatur ab Aristotele demonstrationis conclusio, putaverunt enim eam definitionem totam in conclusione praedicari de definito, ut demonstrationis conclusio talis sit, ergo eclipsis est privatio luminis in luna, ergo tonitrus est sonus in nube. Sic etiam definitionem irae, quam materialem vocant, de ipsa ira demonstrant per alteram irae definitionem formalem, nam per appetitum vindictae demonstrant iram esse accensionem sanguinis in corde, ut in conclusione accensio sanguinis in corde sit praedicatum, ira verò subiectum.

2   Qua quidem sententia nihil absurdius, nihil ab Aristotele alienius excogitari potest. Sic enim ponitur modus ille demonstrandi definitionem, quem prorsus reiecit Aristoteles in contextu 36 et 37 secundi libri Posteriorum, nam definitionem demonstrare de

: XV :

On the errors of others in the definition that is the conclusion of a demonstration.

Two errors of others in understanding that definition that is, ac- 1
cording to Aristotle, the conclusion of a demonstration are not to
be passed over in silence here. One error is that many hold that
this definition is not the same one that Aristotle had called nominal in text no. 44 of the book,[128] although it is [in fact] that very
one, as is patent to those reading that whole part and as we will
demonstrate with better opportunity in another place. The other
error, and of course the greatest one, into which absolutely all the
commentators, both ancient and later, fell, except Averroës and
perhaps also Themistius, is that they did not understand in what
way the definition is said by Aristotle to be the conclusion of a
demonstration. For they held that in the conclusion the whole
definition is predicated of what is being defined, so that the
conclusion of the demonstration is something like "Therefore an
eclipse is the privation of light in the moon" or "Therefore thunder
is noise in a cloud." So too, regarding anger, they even demonstrate the definition of anger that they call material by means of
the other, formal, definition of anger:[129] By means of "the eagerness for revenge" they demonstrate that anger is a boiling of blood
in the heart, so that in the conclusion, boiling of blood in the
heart is the predicate, and anger the subject.

Of course, nothing more absurd than this position, nothing 2
more alien to Aristotle, can be imagined. For what is being set out
is the way of demonstrating definition that Aristotle utterly rejected in texts no. 36 and no. 37 of the second book of the *Posterior [Analytics]*.[130] To demonstrate the definition of what is being

definito vanum penitus est et naturae demonstrationis repugnan-
tissimum, quod multifariam ostendere possumus, primùm quidem
id nobis constituendum est, tum de subiecto demonstrationis tum
de affectione demonstranda praecognoscendum esse quid nomen
significet et significationem nominis intelligi non posse nisi cognito genere aliquo illius rei vel propinquo vel saltem remoto, quomodo enim nomen eclipsis intelligemus, nisi sciamus eclipsim significare privationem quandam luminis? Quomodo intelligemus
nomen tonitrus, nisi noscamus° esse sonum quendam? Dum igitur
demonstramus eclipsim esse privationem luminis lunae, eclipsim
facimus in ea demonstratione subiectum, ergo praecognoscimus
eclipsim significare privationem luminis, idem igitur demonstramus, quod praecognoscimus, quod certè absurdissimum esse quilibet sanae mentis assereret, nam significatio nominis eclipsis nulla
alia est, quàm privatio luminis in luna, ideò si hanc praecognitam
esse oportet, eadem demonstrari non potest, nisi idem de seipso
demonstretur, cùm nihil aliud sit eclipsis, quàm privatio luminis
lunae. Similiter non est demonstrandum iram esse accensionem
sanguinis in corde neque tonitrum esse sonum in nube, quia haec
est significatio nominis penitus indemonstrabilis de ipso nomine.

3   Neque aliquid roboris habet id, quod multi dicunt, definitionem de definito non demonstrari ea ratione, qua est definitio, sed
quatenus est quaesitum ignotum, hoc enim ridiculum est, quia si
est quaesitum ignotum, significatio igitur nominis subiecti ignota
est, quomodo igitur demonstrare aliquid possumus de subiecto,
cuius nomen non intelligimus? Quòd si dicant praecognosci quidem quòd eclipsis significat privationem luminis, non tamen eam,

defined is completely vain and most incompatible with the nature of definition. We can show this in more ways than one. First, of course, it has to be established by us that, regarding both the subject of a demonstration and the affection being demonstrated, what the name signifies has to be known beforehand, and the signification of the name cannot be understood unless some genus of the thing, either near or at least remote, is known. For how can we understand the name "eclipse," unless we know scientifically that eclipse signifies some sort of privation of light? How can we understand the name of thunder unless we know° it is some sort of noise? When we demonstrate, therefore, that an eclipse is the privation of the light of the moon, we make eclipse the subject in the demonstration, and therefore we know beforehand that eclipse signifies a privation of light. We demonstrate, therefore, the same thing we know beforehand. Certainly, anyone of sound mind would assert that this is most absurd. For the signification of the name "eclipse" is nothing other than privation of the light in the moon. And so if this must be known beforehand, the same thing cannot be demonstrated, unless the selfsame thing is demonstrated of that thing itself, since an eclipse is nothing other than privation of the light of the moon. Similarly, it is not to be demonstrated that anger is a boiling of blood in the heart or thunder a noise in a cloud, because this signification of the name is completely indemonstrable of the name itself.

Nor does what many say have any weight: That the definition 3 is not demonstrated of what is being defined in the way in which it is a definition, but insofar as it is an unknown inquired after. For this is ridiculous. Because if it is the unknown inquired after, then the signification of the name of the subject is unknown. How, therefore, can we demonstrate anything of a subject whose name we do not understand? Now if they say that it is, of course, known beforehand that eclipse signifies a privation of light but not

quae est in luna et hoc esse id, quod demonstratur, in aliud absurdum incident non minùs grave, sic enim subiectum demonstrationis erit ipsum accidens, ut eclipsis, praedicatum verò, quod
quaeritur, erit subiectum accidentis, in illa enim demonstratione
nihil manebit ignotum, quod dicatur quaeri, nisi luna. Hoc autem
est inconveniens, debet enim accidens de subiecto monstrari, ut sit
praedicatio naturalis, non subiectum de accidente.

4   Praeterea certum est eam esse potissimae demonstrationis
conditionem ut medium sit causa inhaerentiae maioris extremi in
minore, ut rationale est causa ut risibile inhaereat homini et interpositio terrae est causa ut lunae insit eclipsis. Hoc tamen in illa
puerili demonstratione non videmus, nam interpositio terrae non
est causa ut eclipsi competat privatio luminis lunae, sed potius ut
privatio luminis insit lunae. Sic appetitus vindictae non est causa
ut irae insit accensio sanguinis cordis, sed potius ut sanguini cordis insit accensio; siquidem propria vocis significatio ei sine causa
competit, nulla enim est causa cur eclipsis sit privatio luminis vel
cur ira sit accensio sanguinis. Illi itaque, qui ita demonstrant, non
demonstrant conclusionem per suam causam, quoniam eiusmodi
conclusionis nulla causa afferri potest.

5   Nos igitur dicimus alia ratione definitionem illam esse conclusionem demonstrationis. Non enim quòd tota definitio de definito
praedicetur, sed quòd altera definitionis pars de altera, nempè
forma de materia concludatur, ideò[33] Aristoteles talem definitionem conclusionem demonstrationis esse dixit. Per mediam enim
interpositionem terrae demonstramus in luna esse privationem

the one that is in the moon, and that this is what is demonstrated, they fall into another absurdity no less grave. For then the subject of the demonstration will be the accident itself, such as eclipse, and the predicate — what is inquired after — will be the subject of the accident. For in that demonstration nothing that is said to be inquired after will remain unknown, except the moon. This, however, is inconsistent, for accident ought to be demonstrated of subject — as natural predication would be — not subject of accident.

Moreover, it is certainly a characteristic of demonstration *potissima* that the middle [term] is the cause of the inherence of the major extreme in the minor, as rational is the cause by which risible inheres in man, and interposition of the earth is the cause by which an eclipse belongs to the moon. Nevertheless, we do not see this in that puerile demonstration. For interposition of the earth is not the cause by which privation of the moon's light applies to an eclipse, but rather [the cause] by which privation of light belongs to the moon. And thus eagerness for revenge is not the cause by which boiling of the heart's blood belongs to anger, but rather [the cause] by which the boiling belongs to the heart's blood, if indeed the word's proper signification applies to it without the cause — for there is no cause why eclipse is privation of light or why anger is boiling of blood. And so those who demonstrate this way, do not demonstrate the conclusion by means of its cause, since no cause for a conclusion of this type can be brought forward.

We say, therefore, that that definition is the conclusion of a demonstration in another way. For it is not because the whole definition is predicated of what is being defined, but because one part of the definition [is predicated] of another part, namely the form is concluded of the matter, that Aristotle therefore said that such a definition is the conclusion of a demonstration. For by means of the middle [term] "interposition of the earth" we demonstrate that privation of light is in the moon, and by means of the

luminis et per medium appetitum vindictae demonstramus in sanguine cordis fieri accensionem. Quare hae definitiones sunt con-
clusiones demonstrationis, privatio luminis in luna et accensio
sanguinis in corde. Si etenim has demonstremus, verè causam
conclusionis afferimus, id est causam inhaerentiae praedicati in
subiecto. Et quia conclusio illa formam et materiam accidentis
continet, subiectum enim est materia, praedicatum autem forma,
ideò est illa definitio imperfecta et demonstrabilis, quae causam rei
non continet. Ipsi igitur si causam addamus, conclusioni medium
terminum addimus et definitionem perfectam facimus, quae reipsa
à demonstratione non differt, sed solo terminorum situ.

6 Haec est sententia Aristotelis, si benè eius verba perpendantur
in secundo libro Posteriorum à contextu 36 usque ad 47 eaque rationi maximè consona est, nam haec imperfecta definitio ad conclusionem demonstrationis eam rationem habere debet, quam
perfecta definitio habet ad demonstrationem totam, ut enim haec
dicitur integra demonstratio, sic illa dicitur demonstrationis conclusio. Sed perfecta definitio idem est re quod demonstratio, modo
autem et ratione differt à demonstratione et differentia in duobus
consistit, ut inquit Averroes in commentario 45 secundi libri Posteriorum,[34] in forma et in ordine terminorum, nam demonstratio
formam syllogismi habet, quae definitioni non convenit; et, cùm ex
iisdem tribus terminis utraque constituatur, alio tamen et alio ordine in hac et in illa dispositi sunt, nam in definitione primùm
profertur genus, ultimo loco causa, dicimus enim, eclipsis est privatio luminis lunae ob terrae interpositionem. At in demonstratione nunquàm à genere auspicamur, imò postremum est, quia est
id, quod quaeritur, ideò si minorem propositionem anteponamus,

middle [term] "eagerness for revenge," we demonstrate that boiling happens in the blood of the heart. And so these definitions are conclusions of demonstration: Privation of light in the moon and boiling of blood in the heart. For indeed if we demonstrate these, we truly bring forward the cause of the conclusion, that is, the cause of the inhering of the predicate in the subject. And because that conclusion contains the form and matter of the accident — for the subject is the matter and the predicate the form — that definition, therefore, which does not contain the cause of the thing is imperfect and demonstrable. If, therefore, we add a cause to it, we add a middle term to the conclusion and we make a perfect definition that differs from demonstration, not in reality but only in arrangement of the terms.

This is Aristotle's position, if his words in the second book of 6 the *Posterior* [*Analytics*], from text no. 36 up to no. 47,[131] are examined well, and it is [the position] most consonant with reason. For this imperfect definition ought to have the relationship to the conclusion of the demonstration that the perfect definition has to the whole demonstration. For as the latter is said to be an intact demonstration, so the former is said to be the conclusion of a demonstration. But a perfect definition is the same in reality as what a demonstration is; it differs, however, from demonstration in mode and in reason, and the difference consists in two things, as Averroës says in commentary 45 to the second book of the *Posterior* [*Analytics*][132]: in form and in the order of terms. For demonstration has the form of a syllogism, which is not appropriate for definition, and though each is constituted from the same three terms, they are nevertheless disposed using one order in one and using another order in the other. In a definition, genus is advanced first, cause in the last place — for we say eclipse is the privation of light in the moon on account of the interposition of the earth. But in a demonstration we never commence with the genus; indeed it is final, because it is that which is inquired after. And so if we

primo loco subiectum dicimus, deinde causam, postremo genus. Quòd si anteponamus maiorem, à causa incipimus, deinde genus, tandem subiectum proferimus. Omninò igitur alius est in definitione perfecta, alius in demonstratione ordo, ac situs terminorum.

7   Idem dicendum est de illa imperfecta definitione et de conclusione demonstrationis, sunt enim idem re, quia ex iisdem duobus terminis constant, genere et subiecto, ratione autem differre debent, id est et forma et ordine terminorum, forma quidem quoniam conclusio ex necessitate est enunciatio affirmans vel negans, definitioni verò forma enunciationis minimè convenit, sicuti neque forma syllogismi, definitio enim, ut ait Aristoteles, neque rem esse neque non esse dicit, sed solùm quid sit. Ordine autem terminorum, quoniam in conclusione necessarium est subiectum anteponere generi praedicato, in definitione autem necesse est anteponere genus subiecto locum differentiae obtinenti, haec enim est definitio, privatio luminis in luna, haec autem conclusio, ergo in luna est privatio luminis. Ita haec est definitio, accensio sanguinis cordis, haec autem conclusio, igitur in sanguine cordis fit accensio et sic de caeteris. Oportet enim easdem voces alio modo proferre ut definitionem, alio modo ut enunciationem et conclusionem demonstrationis.

8   Sententiam hanc legere possumus clarè apud Averroem in commentario 42 secundi libri Posteriorum, ubi dicit potissimam demonstrationem esse potestate° definitionem, cùm unam partem definitionis nominalis demonstret alteri parti inesse per causam externam, quae est perfectio et complementum definitionis, quo

place the minor premise first, we say the subject in the first place, then the cause, and finally the genus. Now if we place the major first, we start from the cause, then we advance the genus and finally the subject. The order and arrangement of the terms, therefore, is one way in a perfect definition, thoroughly another in a demonstration.

The same has to be said about that imperfect definition and about the conclusion of a demonstration; for they are the same in reality, because they are composed out of the same two terms — genus and subject. They ought to differ, however, in reason, that is, both in form and the order of terms. [They differ] in form, of course, since a conclusion is out of necessity a proposition affirming or denying, but the form of a proposition is not at all appropriate for a definition, just as the form of a syllogism is not; for a definition, as Aristotle says, does not say that the thing is or is not, but only what it is. And [they differ] in the order of terms, since in a conclusion it is necessary to place the subject before the predicated genus, but in a definition it is necessary to place the genus before the subject occupying the place of the differentia. For this is a definition: "Privation of light in the moon." But this is a conclusion: "Therefore in the moon there is a privation of light." And this is a definition: "Boiling of the heart's blood." But this is a conclusion: "Therefore in the heart's blood boiling happens." And so on. So it must be that the same words are advanced in one way as a definition and in another way as a proposition and the conclusion of a demonstration.

We can read this position clearly in Averroës in commentary 42 to the second book of the *Posterior* [*Analytics*],[133] where he says that demonstration *potissima* is in its force° a definition, since it demonstrates that one part of a nominal definition belongs to another part by means of an external cause; this is the perfection and

7

8

fit ut tota demonstratio in definitionem perfectam accidentis convertatur, quae est ultimus, ac praestantissimus demonstrationis finis.

9   Haec itaque et Aristotelis et Averrois sententia est et vera. Qui verò aliter sentiunt, crassa nimis ignorantia ac caecitate laborant.

10   Quaerere hîc aliquis posset cur in maiore extremo demonstrationis non ipsum affectionis nomen posuerimus, ut eclipsim, sed potius eius genus, privationem luminis. Verùm ne nimia cum prolixitate in praesentia digredi et à proposito recedere videamur, id alio in loco seorsum expendendum relinquemus. Non est autem hîc praetermittendum, sed summoperè annotandum quòd definitio illa affectionis, quae dicitur esse demonstrationis conclusio, est etiam alia ratione principium demonstrationis, idque ut primo aspectu mirabile fortasse videtur, ita, si benè intelligatur, verissimum esse comperietur, aliud namque est cognoscere rem esse, aliud est quid nomen significet intelligere, definitione quidem nominali quid significet nomen affectionis apprehendimus, sed non propterea ipsam esse cognoscimus. Ideò Aristoteles inter praecognitiones demonstrationi necessarias illam nominavit, qua et subiecti et affectionis quaesitae cognitionem habemus quid utriusque nomen significet. Tota igitur conclusio praecognita est, quia ante demonstrationem debet esse intellecta, licèt ipsius veritas nondum cognoscatur, ut si eclipsim dari ignoremus, eamque demonstrare velimus, necessarium nobis est antequàm demonstremus intelligere quid nomen eclipsis significet, significat autem privationem luminis lunae, sed licèt hoc praecognoscamus, attamen an detur in luna

completion of the definition. Hence the whole definition is converted into a perfect definition of the accident; this is the ultimate and most excellent end of demonstration.

This then is the position both of Aristotle and of Averroës, and 9 it is true. Those who take another position labor under excessively thick ignorance and blindness.

Someone could here inquire why in the major extreme of the 10 demonstration we did not set out the very name of the affection, such as eclipse, but rather its genus, privation of light. Now lest we digress at present with a great prolixity and appear to back away from what has been set out, we will leave this matter to be weighed separately in another place. It is, however, not to be overlooked here, but is to be very diligently noted, that that definition of the affection that is said to be the conclusion of a demonstration is also, in another way, the beginning-principle of the demonstration. And while this perhaps appears wonderfully strange on first glance, still, if it is well understood, it will be found to be most true. For it is one thing to know that something is; it is another to understand what the name signifies. In a nominal definition, of course, we grasp hold of what the name of an affection signifies, but we do not on that account know that it exists. Because of this, Aristotle named that among the prior knowledge necessary to demonstration, by which we have knowledge of both the subject and the affection inquired after, [knowledge of] what the name of each signifies.[134] The whole conclusion, therefore, is known beforehand, because it ought to be understood before the demonstration, even granted that the truth of it is not yet known. As, for instance, if we do not know that there is an eclipse, and we want to demonstrate it, it is necessary for us, before we demonstrate it, to understand what the name "eclipse" signifies. Now it signifies privation of the light of the moon; but granted that we know this beforehand, nevertheless, whether in the moon there is

haec luminis privatio adhuc non cognoscimus, idque per demonstrationem ostendere volumus.

11   Quoniam igitur omne, quod ante demonstrationem praecognoscendum est, principium demonstrationis vocari potest, definitio haec nominalis tanquàm nomen affectionis demonstrandae declarans, et tanquàm conclusio antequàm demonstretur intellecta, principium quoddam est demonstrationis. Non est tamen principium scientiam agens,° sed potius ad scientiam dirigens, non enim in propositionibus demonstrationis assumitur, sed ante demonstrationem praecognoscitur, quia sine illo vana esset demonstrationis extructio.

12   Cùm igitur eadem sit definitio nominalis et conclusio demonstrationis, ea quatenus est definitio nomen affectionis declarans, praecognoscitur, quatenus verò est enunciatio dicens inhaerentiam alterius partis in altera, eatenus est conclusio ignota, quae per demonstrationem innotescit.

: XVI :

Quomodo omnis definitio accidentis ad
demonstrationem referatur et ad demonstrativae
methodi traditionem[35] *pertineat.*

1   Ex his, quae diximus, facilè colligere possumus quomodo omnis definitio affectionis sive perfecta sive imperfecta ad demonstrationem pertineat. Ea enim, quae imperfecta est, vel est principium vel conclusio demonstrationis. Quae verò perfecta et ex iis

this privation of light, we still do not know; we want to show it by
means of demonstration.

Since everything, therefore, that has to be known before dem- 11
onstration can be called a beginning-principle of demonstration,
this nominal definition, as making clear the name of the affection
being demonstrated and as the conclusion understood before it is
demonstrated, is some sort of beginning-principle of demonstration. Nevertheless it is not a beginning-principle producing° scientific knowledge but is rather [one] directing toward scientific
knowledge. For it is not assumed in the premises of the demonstration, but is known before the demonstration, because without
it, building up of the demonstration would be vain.

Therefore, although the nominal definition is the same as the 12
conclusion of the demonstration, insofar as it is the definition
clarifying the name of the affection, it is known beforehand, but
insofar as it is a proposition stating the inherence of one part in
another, it is an unknown conclusion that becomes known by
means of demonstration.

: XVI :

*In what way every definition of an accident is
referred to demonstration and pertains to the
conveying of demonstrative method.*

From all the things that we have said, we can easily gather in what 1
way an affection's definition, either perfect or imperfect, pertains
to demonstration. For that which is imperfect is either a beginning-
principle or the conclusion of a demonstration. And that which is

duabus conflata, demonstratio ipsa est solo ab ipsa differens terminorum situ. Haec ut finis praecipuus demonstrationis atque ut demonstrantium ultimus scopus ab Aristotele consideratur in secundo libro Posteriorum ab initio libri usque ad 47 contextum; quia revera finis ultimus demonstrantium est è demonstratione definitionem colligere omnibus numeris absolutam et quiescere in cognitione quid est. Veritatem hanc docet in ea parte Aristoteles. Proponit enim in principio eius libri investigandum instrumentum ducens ad cognoscendum quid est, quando est naturaliter ignotum, cuiusmodi est definitio omnium affectionum, quoniam ipsarum essentia pendet ab externa causa. Postea ostendit non alio instrumento nos duci ad hanc cognitionem, quàm illo eodem, quo ducimur ad scientiam propter quid est, nempè demonstratione, ad quod declarandum voluit in primo capite hoc fundamentum constituere, quòd idem est cognoscere quid est et propter quid est, hinc enim fit ut demonstratio, cùm faciat cognitionem propter quid est, tradat simul cognitionem quid est. Declarat etiam modum Aristoteles, quo per demonstrationem ducimur ad cognoscendam accidentium quidditatem et ostendit non posse definitionem de definito demonstrari, in lucem tamen prodire per demonstrationem, quatenus tota demonstratio in definitionem perfectam accidentis convertitur, neque alia via, quàm demonstratione, posse talem definitionem innotescere.

2   Debemus autem eo in loco notare summum Aristotelis artificium, qui ut ostenderet idem esse demonstrationem ac definitionem, utitur ratione commutata et demonstrationi tribuit officium

perfect, and assembled from these two, is the demonstration itself differing from it only in arrangement of terms. This is considered by Aristotle, in the second book of the *Posterior [Analytics]*, from the start of the book up to text no. 47,[135] as the principal end of demonstration and as the ultimate goal of those who make demonstrations, because the ultimate end of those who make demonstrations is, in truth, to gather from a demonstration a definition complete on all counts and to rest in the knowledge *quid est*.[136] Aristotle teaches this truth in that part. For in the beginning of the book, he sets out to investigate an instrument that leads to knowing what [something] is (*quid est*), when it is naturally unknown. The definition of all affections is of this type, since their essence depends on an external cause. He afterward shows that we are led to this knowledge by no instrument other than that same one by which we are led to scientific knowledge (*propter quid*), namely, by demonstration. To clarify this, he wanted, in the first chapter,[137] to establish this foundation: to know what [something] is (*quid est*) and [to know] what [something] is on account of (*propter quid*) are the same. For thus it happens that demonstration, although it brings about knowledge *propter quid*, conveys at the same time knowledge *quid est*. Aristotle also makes clear the way by which we are led by means of demonstration to knowing the quiddity of accidents, and shows that the definition of what is being defined cannot be demonstrated, and that, nevertheless, it comes to light by means of demonstration, insofar as the whole demonstration is converted into a perfect definition of the accident. In no way other than by demonstration can such a definition be made known.

We ought to note Aristotle's great skill in this passage. So that he might show that demonstration and definition are the same, he uses converse reasoning and ascribes the function of definition to

2

definitionis et definitioni officium demonstrationis, est enim definitionis officium significare quid est, demonstrationis verò notificare propter quid est. Nam si demonstratio declarat etiam quid est, id facit quatenus est definitio, non quatenus est demonstratio. Et definitio si declarat propter quid est, id facit quatenus demonstratio, non quatenus definitio, haec autem manifesta sunt ex ipsa vocabulorum significatione.

3   Aristoteles igitur in contextu 42 et 43 asserit demonstrationem ostendere quid est, deinde in 45 definiens definitionem dicit eam esse orationem significantem propter quid est et orationem demonstrantem, perfecta enim definitio significat quidditatem cum causa quidditatis. Sunt autem in illis locis benè perpendenda Aristotelis verba, qui artificiosissimè locutus est ad denotandum demonstrationem esse methodum et instrumentum logicum, definitionem verò nequaquàm, sed potius finem methodi et instrumenti logici. Quamvis enim demonstratio et definitio reipsa sint idem, tamen ratione formae logicae distinguuntur, definitio enim omni discursu, omnique enunciatione caret; demonstratio autem et enunciationem et discursum habet, quoniam igitur instrumentum logicum non est nisi cum hac forma logica, cum ratiocinio, sic enim dicitur methodus, quae ab hoc ad illud progreditur, hinc fit ut demonstratio sit logicum instrumentum et methodus, definitio verò neque instrumentum neque methodus dici possit, sed finis methodi, sicuti punctum est finis lineae et sicuti quies gravium et levium in suis locis est finis motuum ipsorum ad eadem loca. Demonstratio enim motus quidam est intellectualis, in quo tria notare possumus, unum quidem[36] subiectum et duo contraria,

demonstration and the function of demonstration to definition. For it is the function of definition to signify what [something] is (*quid est*), and that of demonstration to make known what [something] is on account of (*propter quid*). But if demonstration also makes clear what [something] is (*quid est*), it does so insofar as it is a definition, not insofar as it is a demonstration. And if definition makes clear what [something] is on account of (*propter quid*), it does so insofar as it is a demonstration, not insofar as it is a definition. And these things are manifest from the very signification of the words.

Aristotle, therefore, in texts no. 42 and no. 43,[138] asserts that demonstration shows what [something] is (*quid est*). Then, in no. 45,[139] defining definition, he says that it is speech that signifies what [something] is on account of (*propter quid*) and speech that demonstrates. For perfect demonstration signifies the quiddity with the cause of the quiddity. Now Aristotle's words in those passages have to be examined well. He spoke very skillfully to point out that demonstration is a method and a logical instrument but definition is neither and is rather the end of a method and of a logical instrument. For even though demonstration and definition are in reality the same, nevertheless they are distinguished by reason of logical form. For a definition lacks any discursive movement and any proposition, while demonstration has both proposition and discursive movement. Since, therefore, it is not a logical instrument unless it has this logical form, [that is,] ratiocination — such is said to be the method that progresses from this to that — it thus happens that demonstration is a logical instrument and a method, and definition can be said to be neither an instrument nor a method, but the end of a method, just as a point is the end of a line, and just as the resting of heavy and light things in their places is the end of their motion to those places. For demonstration is some sort of motion of [the] understanding in which we can note three things — one subject, of course, and two contraries,

3

alterum à quo, reliquum ad quod fit motus. Definitio verò est simplex essentiae conceptio cum quiete et sine ullo animi motu, cuius quietis gratia extruitur ab omnibus demonstrantibus demonstratio.

4   Haec omnia significavit ibi Aristoteles dum munus definitionis demonstrationi attribuit et munus demonstrationis definitioni, tribuit enim demonstrationi ut ostendat et notificet ipsum quid est, tanquàm methodo et instrumento, nam ostendere et notificare quid est significat motum et discursum à noto ad ignotum, quae est propria logici instrumenti conditio. In contextu autem 45 tribuit definitioni officium demonstrationis, non tamen amplius cum discursu et motu, sed cum quiete, dicit enim, definitio est oratio significans propter quid, non dicit, notificans, sed significans, notificare enim est cum discursu, significare autem sine discursu.

5   Voluit igitur denotare Aristoteles, demonstrationem esse instrumentum cognoscendi quid est, id est extrahendae definitionis, definitionem verò esse fructum et finem illius instrumenti. Accidentium enim causas adducere est eorum demonstrationem afferre, ex causis autem cognitis definitionem constituere est definitionem educere è demonstratione, ac veluti lineam retrahere, et redigere ad punctum et ex re dividua facere conceptum simplicem et individuum, id enim totum, quod per discursum invenimus, postea ut simplex et omni discursu carens comprehendimus.

one from which, the other to which motion occurs. But definition is a simple conception of essence, with rest and without any motion of our soul; for the sake of this rest, demonstration is built up by everyone who makes demonstrations.

Aristotle indicated all this there when he attributed the job of   4
definition to demonstration and the job of demonstration to definition. For he ascribed to demonstration that it shows and makes known what something is (*quid est*), just as [he ascribed the same] to method and instrument; for showing and making known what [something] is (*quid est*) indicates motion and discursive movement from known to unknown and this is a characteristic proper to a logical instrument. And then, in text no. 45,[140] he ascribed to definition the function of demonstration, but not so much with discursive movement and motion but with rest. For he says definition is speech signifying what [something] is on account of (*propter quid*). He does not say "making known" but "signifying." For to make known is with discursive movement, but to signify is without discursive movement.

Aristotle, therefore, wanted to point out that demonstration is   5
an instrument for knowing what [something] is (*quid est*), that is, for extracting the definition. And definition is the fruit and end of that instrument. For to adduce the causes of accidents is to bring forward a demonstration of them, and to establish a definition from known causes is to educe a definition from a demonstration; it is just like retracting a line and reducing it to a point, and making from something divisible a simple and indivisible concept. For the whole that we discover by means of discursive movement we afterward comprehend as simple and lacking all discursive movement.

꞉ XVII ꞉

In quo ea, quae dicta sunt, confirmantur ex
multis Averrois atque Aristotelis locis.

1 Sententiam hanc, quam exposuimus, apertè legimus apud Averroem in pluribus locis, nam in commentario undecimo primi libri
Posteriorum dicit potissimam demonstrationem potestate° complecti quaestionem quid est, quae est definitio, quam naturaliter
desideramus et propter eam quaerimus cognitionem causarum,
dicit igitur demonstrationem ducere nos ad cognoscendum quid
est, hoc est, ad definitionem et hanc esse ultimum omnium demonstrantium finem, cuius gratia causas rerum quaerunt et demonstrationes construunt.

2   In commentario etiam 38 secundi libri dicit demonstrationem
esse potestate° definitionem, et ideò quicquid in primo libro dictum est de demonstratione, dirigi ad illa, quae dicuntur in secundo
de definitione.

3   In Epitome quoque in capite de demonstratione et in 10 quaestione logica saepe eandem sententiam profert Averroes quòd finis
demonstrationis et praecipua intentio demonstrantium est eductio
definitionis, quaerunt enim propter quid est, ut tandem inventa
causa cognoscant quid est.

4   Et in praefatione sua in primum librum Posteriorum, cuius
verba etiam antè perpendimus, inquit Averroes easdem propositiones demonstrationis considerari ab Aristotele et in primo et in
secundo Posteriorum libro, in primo quidem ut ducunt ad cognitionem propter quid est, in secundo autem ut tradunt cognitionem

: XVII :

In which the things that have been said are confirmed
by many passages in Averroës and Aristotle.

This position that we have laid out we plainly read in Averroës in   1
many passages. In the eleventh commentary to the first book of
the *Posterior* [*Analytics*][141] he says that demonstration *potissima* encompasses in its force° the question of what [something] is (*quid
est*); this is the definition that we naturally desire, and on account
of it we inquire after knowledge of causes. He says, therefore, that
demonstration leads us to knowing what [something] is (*quid est*),
that is, to a definition, and that this is the ultimate end of all those
who demonstrate; for its sake they inquire after the causes of
things and construct demonstrations.

In commentary 38 to the second book,[142] also, he says that   2
demonstration is in its force° definition, and so whatever was said
in the first book about demonstration is directed to that which is
said in the second about definition.

Also, in the *Epitome* [*of Logic*], in the chapter on demonstra-   3
tion[143] and in the tenth of the *Questions on Logic*,[144] Averroës often
advances the same position, that the end of demonstration and the
principal intention of those who demonstrate is to educe a definition. For they inquire after what [something] is on account of
(*propter quid*), so as, once the cause is finally discovered, to know
what [the thing] is (*quid est*).

And in his preface to the first book of the *Posterior* [*Analytics*],[145]   4
the words of which we also examined earlier, Averroës says that
the same premises of a demonstration are considered by Aristotle in both the first and second book of the *Posterior* [*Analytics*] —
in the first, of course, as they lead to knowledge *propter quid*, and
in the second as they convey knowledge *quid est*. According to

quid est, idem igitur instrumentum demonstratio secundùm Averroem ducit ad cognoscendum propter quid est et quid est.

5   Hoc idem in tota Aristotelis naturali philosophia observare possumus, plurimorum enim accidentium naturalium definitiones investigandas proponit, ut quid motus, quid locus, quid tempus, quid generatio, quid mistio, quid putredo, quid coctio, et alia eiusmodi. Dicant igitur Averroistae, Si demonstratio est proprium instrumentum, quo accidentia cognoscantur, nonne debuit Aristoteles eorum accidentium demonstrationem potius indagare, quàm definitionem? Quid ipsi ad hoc respondere possint, non video. Nos autem dicimus definitiones illas esse demonstrationes situ differentes et in omnibus iis locis Aristotelem è demonstratione definitionem colligere. Cùm enim demonstratio satisfaciat omnibus quatuor quaestionibus de accidente factis, ultima tamen et omnium praecipua est quaestio quid est, quae non potest declarari, nisi alia omnes declarentur. Hanc igitur saepe proposuit Aristoteles et ei per demonstrationem satisfecit, putredinis enim definitionem invenire volens sciensque eam non esse perfectam nisi cum causae cognitione, causam prius eius accidentis inquirit, quod nil aliud est, quàm demonstrationem inquirere, nam causam rei adducere, quatenus causa est, demonstrare rem est. Ex causae autem cognitione definitionem constituere est è demonstratione definitionem extrahere.

6   Quod in definitione illa putredinis ostendere possumus, Aristoteles enim primo loco definitionem nominalem putredinis accipit, quae ante omnia cognoscenda est, dicens eam esse corruptionem proprii et innati caloris in corpore humido. Mox causam putredinis quaerit, eamque dicit esse calorem ambientis. Allata causa

Averroës, therefore, the same instrument, demonstration, leads to
knowing what [something] is on account of (*propter quid*) and what
[something] is (*quid est*).

We can observe the same thing in Aristotle's whole natural phi- 5
losophy. For he sets out to investigate definitions of many natural
accidents, such as what motion is, what place is, what time is,
what generation is, what mixture is, what putrefaction is, what
digestion is, and others of this type. The Averroists might then say
[in response]: if demonstration is the proper instrument by which
accidents are known, ought not Aristotle to have tracked down the
demonstration, rather than the definition, of these accidents? I do
not see what they can respond to the following. For we say that
those definitions are demonstrations differing in arrangement and
that in all those passages Aristotle gathers definition from demonstration. And although demonstration answers all four questions[146] made about an accident, nevertheless the ultimate and
principal of all is the question of what [something] is (*quid est*);
this cannot be made clear unless all others are made clear. Aristotle, therefore, often set out this [question] and answered it by
means of demonstration. For wanting to discover the definition of
putrefaction and knowing scientifically that it is not perfect without knowledge of the cause, he first asks about the cause of the
accident; this is nothing other than to ask about a demonstration.
For to adduce the cause of a thing, insofar as it is a cause, is to
demonstrate the thing. And moreover, to establish the definition
from knowledge of the cause is to extract the definition from a
demonstration.

We can show this in the definition of putrefaction. For Aris- 6
totle, in the first place, accepts a nominal definition of putrefaction, which has to be known before all [else], saying that it is the
corruption of the proper and innate heat in a moist body.[147] Next,
he inquires after the cause of putrefaction and says that it is the
heat of what is surrounding. A cause tendered is said to be a

dicitur demonstratio allata esse, nam per calorem ambientis tanquàm per medium terminum possumus demonstrare cur in unoquoque humido corpore fiat corruptio proprii, et innati caloris et ita definitio nominalis fiet demonstrationis conclusio. Ex hac igitur demonstratione Aristoteles perfectam putredinis definitionem educit dicens, putredo est corruptio proprii et innati caloris in unoquoque humido à calore ambientis. Alias quoque accidentium naturalium definitiones considerare possemus, in quibus omnibus illud inspiceremus, nempè definitionem nominalem traditam ex genere et subiecto accidentis et causam inharentiae illius in hoc, quare omnes tales definitiones sunt demonstrationes situ terminorum differentes. Ut autem alii apud Aristotelem hoc non animadvertant causa est[37] quòd non vident ab eo formatum syllogismum demonstrativum, quem Aristoteles, si cum pueris tantùm ac tironibus sermonem haberet, utique ad formam redigere semper deberet; sed eruditioribus satis esse debet investigatio et allatio causae, haec enim demonstratio ipsa est, quam levi negotio potest quisque ad syllogismi formam redigere, quemadmodum in putredine declaravimis.

7   Sunt etiam summoperè perpendenda verba Aristotelis in prooemio primi libri Priorum Analyticorum et alia eis similia in epilogo in calce secundi Posteriorum, necnon verba Averrois in utriusque loci interpretatione. Aristoteles in eo prooemio quaerit circa quid et cuius sit consideratio, quam quaestionem Averroes ita intelligit, circa quid tanquàm circa rem consideratam; et cuius, id est cuiusnam[38] finis gratia. Ideò cùm postea Aristoteles ad utramque quaestionem respondens ait, circa demonstrationem et scientiae demonstrativae gratia, Averroes putat philosophum nobis proponere tum[39] instrumentum, in quo est versaturus, tum fructum et utilitatem, quam ex illo instrumento sumus percepturi. Instrumentum

demonstration tendered: by means of the heat of what is surrounding, just as by means of a middle term, we can demonstrate why corruption of the proper and innate heat occurs in any moist body whatever, and thus the nominal definition will become the conclusion of a demonstration. From this demonstration, therefore, Aristotle educes a perfect definition of putrefaction, saying: Putrefaction is corruption of the proper and innate heat in any moist [body] whatever by the heat of what is surrounding. We could also consider other definitions of natural accidents; in all of them we observe this, namely the nominal definition conveyed from the genus and the subject of the accident, and the cause of the former inhering in the latter.[148] All such definitions are demonstrations differing in arrangement of terms. The cause of others not noticing this in Aristotle is that they do not see the demonstrative syllogism formed by him. If he discoursed just with children and beginners, by all means Aristotle ought always to reduce it to [this] form, but for the more learned the investigation and tendering of the cause ought to be enough. For this is the demonstration itself, what anyone can, with light effort, reduce to the form of a syllogism, just as we made clear with putrefaction.

Now Aristotle's words in the proem of the first book of the   7
Prior Analytics[149] and others similar to these in the summary at the end of the second [book] of the *Posterior [Analytics]*,[150] as well as Averroës' words in the commentary on each passage, have to be examined very diligently. In that proem Aristotle inquires what [his] consideration is about, and what it is of. Averroës understands the question this way: "about what" is "about the thing under consideration," and "of what" is "for the sake of what end." And so since Aristotle, responding to each question, afterward says "about demonstration" and "for the sake of demonstrative science," Averroës holds that the philosopher is setting out before us both the instrument with which he will concern himself and the fruit and utility that we will receive from this instrument. Of course,

quidem unicum proponitur demonstratio, finis autem illius instrumenti scientia demonstrativa, id est perfectissima et absolutissima.

8   Verba autem Averrois[40] in huius partis declaratione sunt haec,
'et dicamus etiam quòd utilitas, quae nobis ex demonstratione
proveniet, erit ut assequamur cognitionem demonstrativam de
omni re, quantum potest homo naturaliter consequi,' si ergò ab
Averroistis petamus quaenam sit haec praestantissima cognitio,
quam per demonstrationem assequimur, certum est eos esse re-
sponsuros illam, quae in duabus quaestionibus complexis constituta est, quòd est et propter quid est. At non sunt boni Averroistae, imò neque boni Aristotelici, quandoquidem Averroes in
omnibus antea memoratis locis clarè dicit ultimum finem et supremam perfectionem demonstrationis esse extractionem definitionis
et cognitionem quid est. Igitur cognitio propter quid est non est ea
tota et absolutissima cognitio, quam à demonstratione consequi
possumus, sed potius cognitio quid est. Aristoteles quoque in
principio secundi libri Posteriorum ostendit, quòd idem demonstrationis medium tradit nobis cognitionem et propter quid est et
quid est, quo fit ut demonstratio ducat nos ad cognoscendum quid
est, ut ipse in eo libro postea declarat.

9   Ergo secundùm Averroem et Aristotelem dicere cogimur, scientiam praestantissimam, quam ex demonstratione percipimus, esse
cognitionem omnium quatuor quaesitorum, praesertim autem ipsius quid est, in quo summa demonstrationis perfectio et utilitas
consistit, quia quaestio haec est finis omnium aliarum quaestionum. Hanc igitur intellexit Averroes quando dixit utilitatem demonstrationis esse scientiam praestantissimam de omni re, quantam potest homo naturaliter consequi.

the only one instrument set out is demonstration, and the end of that instrument, demonstrative science, that is, the most perfect and complete [scientific knowledge].

Now Averroes' words in clarification of this part are these: "And 8 now we say that the utility that will arise for us from demonstration, will be that we secure demonstrative knowledge about everything, to the extent man can gain [it] naturally."[151] If then we ask of the Averroists, "What is this most excellent knowledge that we secure by means of demonstration?" it is certain they will respond that it is that which is constituted in two compound questions — that [something] is the case (*quòd*) and on account of what [something] is (*propter quid*). But these are not good Averroists; indeed they are not good Aristotelians. For Averroes, in all passages referred to earlier, clearly says that the ultimate end and highest perfection of demonstration is the extraction of a definition and the knowledge *quid*. Knowledge *propter quid*, therefore, is not the whole and most complete knowledge that we can gain from demonstration; rather, knowledge *quid* is. Aristotle too, in the beginning of the second book of the *Posterior* [*Analytics*],[152] shows that the same middle [term] of a demonstration conveys to us both knowledge *propter quid* and [knowledge] *quid*. Hence demonstration leads us to knowing what [something] is (*quid est*), as he himself makes clear afterward in that book.[153]

Therefore, according to Averroes and Aristotle, we are forced to 9 say that the most excellent scientific knowledge, what we grasp by demonstration, is knowledge of all four things inquired after, but especially of what something is (*quid est*), in which the highest perfection and utility of demonstration consist, because this question is the end of all the other questions. This, therefore, is what Averroes understood when he said that the utility of demonstration is the most excellent scientific knowledge about everything, to the extent that man can gain [it] naturally.

10   Similiter quando in calce secundi Posteriorum in epilogo colligit Aristoteles se de syllogismo ac de demonstratione ac demonstrativa scientia tractasse, scientiam demonstrativam intelligit Averroes scientiam quaesitorum, quae in scientiis speculativis locum habent, nec dicit aliquorum quaesitorum, sed simpliciter quaesitorum, omnia igitur quaesita intelligit, quaeritur autem in scientiis etiam quid est, imò haec est maximè praecipua quaestio, quare hanc quoque complectitur Averroes nomine scientiae demonstrativae.

11   Ex his colligimus, definitionem non esse instrumentum sciendi apud Aristotelem et Averroem, qui tametsi de definitione plura[41] dixerunt, tamen solam demonstrationem ut instrumentum nominarunt, quia definitionem finem potius atque utilitatem instrumenti esse arbitrati sunt. Idcircò eam nomine demonstrativae scientiae comprehenderunt, finem enim demonstrationis praecipuum esse diximus cognitionem[42] definitionis, quae è demonstratione colligitur. Propterea inquit ibi Aristoteles demonstrationem ac scientiam demonstrativam idem esse, quoniam idem reipsa est definitio ac demonstratio, solùm in forma discrimen est, quae forma, cùm sit logicorum instrumentorum conditio necessaria, facit ut demonstratio sit instrumentum logicum, minimè verò definitio, nisi impropriè et per translationem loquamur et quod demonstrationis est, tribuamus definitioni, sic enim definitionem instrumentum appellare possumus ea ratione, qua est idem re[43] quod demonstratio. Attamen non erit instrumentum à demonstratione distinctum, sed idem. Dum igitur definitionem consideramus quatenus à demonstratione distinguitur, finis est ultimus instrumenti, non instrumentum. Dum verò ipsam sumimus quatenus idem est re, quod demonstratio, instrumentum vocari potest,

Similarly, when, in the summary at the end of the second 10
[book] of the *Posterior* [*Analytics*], Aristotle gathers that he has
treated the syllogism and demonstration and demonstrative science,[154] Averroës understands demonstrative science as the scientific knowledge of the things inquired after that have a place in the
speculative sciences, and he does not say "of some things inquired
after," but "of things inquired after, absolutely"; he, therefore, understands "all things inquired after." But now in the sciences what
[something] is (*quid est*), is also inquired after; indeed this is the
most important question of all. And so Averroës encompasses
this, too, under the name of demonstrative science.

From all this we gather that definition is not an instrument for 11
knowing scientifically in Aristotle and Averroës, who, although
they said many things about definition, nevertheless named only
demonstration as an instrument, because they thought that definition is rather the end and utility of an instrument. And accordingly they comprehended it under the name of demonstrative science. For we said that the most important end of demonstration is
knowledge of a definition that is gathered from demonstration.
And so Aristotle there[155] says that demonstration and demonstrative science are the same, since definition and demonstration are in
reality the same. The discriminating difference is only in form.
This form, since it is a necessary characteristic of logical instruments, makes it that demonstration is a logical instrument, but
definition not at all, unless we speak improperly and figuratively
and ascribe to definition what is of demonstration. For then we
can call definition an instrument for the reason that it is in reality
the same as what demonstration is. But even then it will not be an
instrument distinct from demonstration; instead it will be the
same. When we consider definition, therefore, insofar as it is distinguished from demonstration, it is the ultimate end of an instrument, not the instrument. But when we take it insofar as it is in
reality the same as demonstration, it can be called an instrument,

siquidem demonstratio instrumentum est, nec propterea duo sunt instrumenta, sed unum, quod cùm actu sit demonstratio, vocatur etiam definitio, quia potestate° est definitio. Dummodò enim ipsam rei veritatem intelligamus, haec omnia vera sunt, neque ullis aliorum cavillis perturbari possumus.

: XVIII :

Quomodo universae logicae finis definitio sit et omnis methodus possit appellari definitiva.

1 Manifestum est igitur unicum esse logicum instrumentum, quod praecipuè ab Aristotele libris Analyticis consideratur, methodum demonstrativam, cuius gratia reliqua omnia tractantur et ad quam omnia referuntur.

2   Quòd si definitionis gratia omnia tractari in logica et definitionem esse universi logici negotii finem asseramus, necnon methodum omnem esse definitivam, idque sano modo accipiamus, nil falsi, vel absurdi dicemus. Omnia namque logica instrumenta ad definitionem ducunt, quoniam omnium rerum cognitio in definitione consistit, quaelibet enim res tunc plenè cognoscitur sive substantia fuerit sive[44] accidens, quando perfecta ipsius definitio habetur, ut ait Ammonius in principio suae praefationis in libellum Porphyrii, Averroes in commentario 97 secundi libri Posteriorum.[45] Atqui methodo demonstrativa definitiones accidentium venamur, methodo autem resolutiva definitiones substantiarum. Omnium igitur methodorum finis est definitio, proinde totius logicae finis est, non quidem ut instrumentum, sed ut finis logicorum instrumentorum. Sumpta igitur nominatione à fine potest

since indeed demonstration is an instrument. But there are not on account of this two instruments, but one, which is actually demonstration, even though it is called definition, because it is definition in its force.° As long as we understand the very truth of the issue, all these things are true, and we cannot be disturbed by any caviling of others.

: XVIII :

In what way definition is the end of logic as a universal whole and every method can be called definitive.

It is manifest, therefore, that there is only one logical instrument, demonstrative method. It is principally considered by Aristotle in the *Analytics* books; for its sake all others are treated; and to it all are referred.

Now if we assert that everything in logic is treated for the sake of definition and that definition is the end of the business of logic as a universal whole, and also that every method is definitive, and we accept this in a sound way, we will say nothing false or absurd. For every logical instrument leads to definition, since knowledge of all things consists in definition. For anything, whether it is substance or accident, is fully known whenever a perfect definition of it is had, as Ammonius says in the beginning of his preface to the little book by Porphyry[156] and Averroës in commentary 97 to the second book of the *Posterior* [*Analytics*].[157] But we search for definitions of accidents by demonstrative method and definitions of substances by resolutive method. Definition, therefore, is the end of all methods; accordingly, it is the end of logic as a whole—not, of course, as an instrument but as the end of logical instruments. Naming being taken from the end, therefore, every method can be

omnis methodus vocari definitiva, quatenus ad definitionem ducit. Quare haec utraque simul vera sunt, omnem methodum esse definitivam et nullam dari methodum definitivam, nulla enim datur praeter demonstrativam et resolutivam.

: XIX :

De ordine servato ab Aristotele
in methodorum traditione.

1 Ex his, quae hactenus dicta sunt, manifestum est non aliud fuisse Aristotelis consilium in Posterioribus Analyticis, quàm de methodis agere, ipsumque id optimè praestitisse, cùm et naturam et utilitatem methodorum diligentissimè declaraverit. Relinquitur ut pauca quaedam dicamus de ordine, quo eam tractationem disposuit, huius enim declaratio ad plenam consilii Aristotelis intelligentiam plurimùm conferet.

2 Illud in primis in memoriam revocandum est, quod suprà declaravimus, primariam esse in iis libris tractationem de methodo demonstrativa, secundariam verò de resolutiva. Hinc factum est ut totam tractationis ordinationem sumpserit Aristoteles à methodo demonstrativa, quam solam in prooemio proposuit considerandam et quam solam in epilogo collegit declaratam fuisse, patet enim tum in prooemio tum in epilogo iam memoratis nullam factam esse mentionem methodi resolutivae, quoniam igitur huius consideratio non fuit principalis sed potius ut connexa methodo demonstrativa et ab illa pendens, tota tractationis omniumque theorematum dispositio sumenda fuit à methodo demonstrativa.

called definitive, insofar as it leads to definition. And so both these are at the same time true: every method is definitive, and there is no definitive method, for there is none besides demonstrative and resolutive.

: XIX :

On the order maintained by Aristotle in the conveying of methods.

From all this that has been said up until now, it is manifest that 1 Aristotle's intent in the *Posterior Analytics* was nothing other than to deal with methods, and he really did this the best way, since he clarified most carefully both the nature and utility of methods. It remains that we say some little bit about the order in which he disposed that treatment, for a clarification of this will contribute greatly to a full understanding of Aristotle's intent.

In the first place, what was made clear above has to be remem- 2 bered: in these books the primary treatment is on demonstrative method and the secondary is on resolutive. From this it happened that Aristotle took the whole ordering of the treatment from demonstrative method, which in the proem he set out as the only thing to consider and which in the summary he concluded was the only thing that had been clarified.[158] For it is patent in both the proem and the summary just referred to that no mention was made of resolutive method. Since, therefore, consideration of this was not primary but [was made] rather as this is connected to demonstrative method and dependent on it, the whole disposition of the treatment and of all the theorems had to be taken from demonstrative method.

3   Dicimus itaque methodi demonstrativae traditionem scriptam esse ab Aristotele ordine resolutivo neque alio ordine scribi potuisse, ut manifestum est tum ex iis, quae suprà de ordinibus diximus, tum etiam ex iis, quae aliàs de natura logicae scripsimus. Certum est enim Aristotelem auspicatum esse à notione finis et ab eo ad principiorum inventionem per resolutionem processisse, finis autem duplex ei proponebatur, unus internus, ipsa demonstrativa methodus, alter externus, nempè huius instrumenti utilitas, quam in prooemio primi libri Priorum dixerat esse demonstrativam scientiam. Sic etiam aedificatori duo huiusmodi fines proponuntur, domus quidem ut instrumentum ab ipso fabricandum, proinde ut internus finis, habitatio verò, sive conservatio supellectilis tanquàm externus finis et utilitas illius instrumenti. Ab utriusque igitur finis praenotione exordiendum fuit.

4   Quod quidem ipse Aristoteles fecit, nam in primo capite primi libri sine ulla probatione constituit tum scientiam dari tum dari per demonstrationem. Deinde in principio secundi capitis etiam quid sit et scientia et demonstratio declaravit. Ex qua finis notione ad conditiones principiorum demonstrationis indagandas ascendit et in iis declarandis per satis longam ac diligentem tractationem versatus est. Non sunt autem in praesentia ipsae principiorum conditiones declarandae, quandoquidem non sententias Aristotelis explanare, sed solum ordinem ab eo servatum expendere constituimus. Credimus autem de principiorum conditionibus Aristotelem egisse usque ad contextum 56 primi libri, quo in loco agere incipit de proprietatibus, quae ipsa principia inventa et

And so we say that the conveying of demonstrative method was   3
written by Aristotle using resolutive order and could not have
been written using any other order, as is manifest both from the
things we said above about orders and also from the things we
wrote elsewhere about the nature of logic.[159] For it is certain that
Aristotle commenced with a notion of the end and proceeded
from it to discovery of beginning-principles by means of resolution. And the end that was being set out for him was of two types
— the one internal, the demonstrative method itself; the other external, namely the utility of this instrument, what in the proem of
the first book of the *Prior [Analytics]*[160] he said was demonstrative
science. Thus also two ends of this type are set out for the
builder — first the house, as the instrument being fabricated by
him and accordingly as the internal end; and the second, habita-
tion or the conservation of its furnishings as the external end and
utility of that instrument. He [i.e., the builder] had to begin,
therefore, from a prenotion of each end.

    Aristotle himself, of course, did this. For in the first chapter of   4
the first book [of the *Posterior Analytics*][161] he established without
any proof both that there is scientific knowledge and that there is
[such] by means of demonstration. And then in the beginning of
the second chapter, he made clear both what scientific knowledge
is and what demonstration is. He ascended from this notion of
an end to tracking down the characteristics of the beginning-
principles of demonstration and concerned himself with making
these clear by means of a sufficiently long and careful treatment.
(Now, at present, the very characteristics of the beginning-
principles do not have to be made clear, since we have decided not
to explain Aristotle's positions, but only to weigh the order maintained by him.) We believe Aristotle then dealt with the characteristics of beginning-principles up to text no. 56 of the first
book,[162] in which passage he starts to deal with properties that
ensue from the discovered beginning-principles themselves and

demonstrationem ex iis constructam consecuntur, de quibus agit in tota reliqua eius libri parte, idque egregiè animadvertit Averroes in commentario centesimo secundi libri Posteriorum.

5   Inter ipsas autem demonstrationis proprietates principem locum tenere videtur ipsa demonstrationis utilitas, quam declarat Aristoteles partim in primo libro, partim in secundo. Cùm enim utilitas ea sit, quam superius diximus, nimirum omnibus quaestionibus de eadem re factis satisfacere tum complexis tum simplicibus, de complexis quidem loquitur Aristoteles in primo libro declarans quomodo per talem demonstrationem ex talibus principiis constructam assequamur scientiam non solùm quòd sit, verùm etiam propter quid sit; simulque in ea⁴⁶ parte declarat imperfectionem illius demonstrationis, quae ob alicuius conditionis defectum non declarat propter quid, sed solùm quòd; ideò ibi manifestum est quomodo Aristoteles agat secundariò de methodo resolutiva, omnes enim conditiones tribuit primariò demonstrationi potissimae tanquàm eius proprias; demonstrationi autem à signo non tribuit eas nisi per quandam participationem, participat enim quibusdam conditionibus potissimae demonstrationis, sed non omnibus. Videtur itaque Aristoteles de hac sermonem facere potius ut eam separet à potissima demonstratione, quàm per se; vel per se quidem, sed secundariò.

6   Sed ne credamus summam demonstrationis perfectionem, totamque eius utilitatem consistere in declarando propter quid est, ideò in principio secundi libri quaestiones omnes et omnia ea, quae sub scientiam cadunt, enumerat, ut nominatione quaestionis quid

demonstration constructed from them. He deals with these things in the whole remaining part of the book. And Averroës notes this very well in the hundredth commentary to the second book of the *Posterior* [*Analytics*].[163]

The very utility of demonstration, moreover, appears to hold 5 the foremost place among the very properties of demonstration, [a utility] that Aristotle makes clear partly in the first book, partly in the second. For since this utility is what we said above, namely of course, to answer all questions, both compound and simple, made about the same thing—of course, Aristotle speaks of the compound ones in the first book, clarifying in what way, by means of such a demonstration constructed from such beginning-principles, we secure not only scientific knowledge *quòd* but indeed also [scientific knowledge] *propter quid*, and at the same time he clarifies in that part the imperfection of that demonstration, which, on account of the absence of some characteristic, clarifies not what something is on account of (*propter quid*), but only that [something] is the case (*quod*)—it is, therefore, manifest there in what way Aristotle deals with resolutive method secondarily. For he ascribes all the characteristics primarily to demonstration *potissima* as proper to it; he does not ascribe them to demonstration *a signo* except by means of some sort of participation, for it participates in some characteristics of demonstration *potissima*, but not in all. And so Aristotle appears to discourse on this so as to separate it from demonstration *potissima* rather than [to discourse on it] *per se*—or *per se*, of course, but only secondarily.

Now lest we believe that the highest perfection of demonstra- 6 tion and its whole utility consists in making clear what [something] is on account of (*propter quid*), in the beginning of the second book he enumerates all the questions and everything that falls under scientific knowledge.[164] In naming the question, "what

est, occasionem sumat declarandi quonam instrumento ducamur ad cognoscendum quid est, quando est naturaliter ignotum. Deinde longa facta disputatione, in qua omnia, quae excogitari possunt, instrumenta refellit, tanquàm inutilia ad declarandum quid est, inter quae reiicit et demonstrationem et definitionem, incipit postea ex animi sententia° loqui et ipsam rei veritatem declarare et ostendit quòd, etsi non potest demonstrari ipsum quid est, tamen in lucem prodit per demonstrationem, quando primam definitionis partem inesse in secunda demonstramus per tertiam, quemadmodum antea declaravimus. Ea autem tractatio de demonstratione est tanquàm de instrumento, de definitione verò tanquàm de summa ipsius instrumenti utilitate atque ultimo fine.

7   Est autem animadvertendum quòd quando cum Averroe dicimus Aristotelem in secundo Posteriorum libro considerare demonstrationem ut instrumentum ducens ad cognitionem ipsius quid est, non totam secundi libri tractationem intelligimus, sed primam tantùm et praecipuam eius libri partem, quandoquidem alia quoque multa praeter hoc tractantur ab Aristotele in eo secundo libro, is enim liber in duas principes partes divisus est, quarum quae prior est, ad contextum usque 47 extenditur, in qua docet Aristoteles quomodo demonstratio in definitionem convertatur et ducat ad cognoscendum quid est. Eamque dicimus esse praecipuam eorum librorum partem et scopum ultimum totius Analyticae tractationis, cùm in ea Aristoteles ad apicem usque scientiae perfectae et ad supremam demonstrationis utilitatem cognoscendam nos perducat.

8   In reliqua verò eius libri parte quaedam alia considerat ad perfectam demonstrativae methodi traditionem necessaria, qua in re videtur egregios pictores imitari, qui prius totam figuram depingere volunt, postea ad singulas eius partes melius considerandas

[something] is" (*quid est*), he takes the occasion to make clear by what instrument we are led to knowing what [something] is (*quid est*), when it is naturally unknown. Then in a long debate he refutes as useless all instruments that can be imagined for making clear what [something] is (*quid*), among which he rejects both demonstration and definition, and then[165] he starts to speak from what he has firmly in mind° and to make the very truth of the issue clear. He shows that, although what [something] is (*quid est*) cannot itself be demonstrated, nevertheless it is brought to light by means of demonstration, when we demonstrate that the first part of a definition belongs to the second by means of the third, as we made clear earlier. And so this treatment is on demonstration as on an instrument and on definition as the highest utility and ultimate end of instrument itself.

But now it has to be noted that when we say, with Averroës, 7 that in the second book of the *Posterior* [*Analytics*], Aristotle considers demonstration as an instrument leading to knowledge *quid est*, we are not to understand the whole treatment in the second book, but only the first and the principal part of the book, since indeed much else besides this is also treated by Aristotle in the second book. For this book is divided into two main parts, the first of which extends up to text no. 47,[166] in which Aristotle teaches in what way demonstration is converted into definition and leads to knowing what [something] is (*quid est*). And we say this is the principal part of the books and the ultimate goal of the treatment in the whole *Analytics*, since in it Aristotle leads us all the way through to the apex of perfect scientific knowledge and to knowing the highest utility of demonstration.

In the remaining part of the book he considers some other 8 things necessary for perfect conveyance of demonstrative method. In this issue he appears to imitate excellent painters, who want to depict the whole figure first and afterward go back to better

redeunt, ut si quid omissum fuit, aut si quid imperfectum mansit, instauretur ac perficiatur. Ipse etenim totam demonstrativam disciplinam continuato sermone perscribere voluit et ad apicem usque ipsius pervenire, declarans tum demonstrationis naturam, tum totam utilitatem, quam ex ea sumus percepturi, quod quidem facit in primo libro et in secundo usque ad 47 contextum. Postea considerans plura à se antea fuisse omissa consultò et data opera, ne nimis prolixa digressione tractationem praecipuam interrumperet, revertitur ad quaedam aliqua declaranda, quae ad doctrinae illius necessitatem vel abundantiam explicanda erant.

9   Primo quidem loco animadvertit se de demonstratione loquentem dixisse eam fieri ex proxima rei causa, ob quam res esse dicitur et per quam assequimur perfectam rei scientiam tum propter quid sit tum etiam quid sit, sed nomen causae communiter ac generaliter accepisse, sine ulla causarum distinctione, ut in mentibus nostris dubium manere possit an omne causae genus, an aliquod solùm ad tradendam nobis perfectam illam cognitionem idoneum sit. Ideò genera causarum distinguere constituit et ostendere quomodo singulum in potissima demonstratione medium esse queat. Sic enim ab universali ad particularia progreditur, à causa latè accepta ad singula causarum genera. Ibique multa de causis dicit, quae in primo libro, vel in prima parte secundi absque tractationis interruptione dicere non potuerat.

10   Deinde loquitur de modo venandi illam definitionem, quam dixerat esse indemonstrabilem et demonstrationis principium, primaria enim consideratione haec ut per se nota usque ad eum locum habita fuerat, ibi verò vult secundariam quoque facere

consider each of the parts, so that if something was omitted or if something remained imperfect, it may be redone and perfected. For indeed he wanted, in a continuous discourse, to write out the whole demonstrative discipline and go all the way through to the apex of it, clarifying both the nature of demonstration and the whole utility that we will receive from it. He does this, of course, in the first book and in the second up to text no. 47.[167] Afterward, considering that many things had earlier been intentionally and deliberately omitted by him so that he would not have to interrupt the principal treatment with an excessively prolix digression, he returns to clarify some other things that had to be explicated for him to meet or exceed the needs of his teaching.

He notes, of course, in the first place, that, when he was speaking of demonstration, he said it occurred from something's proximate cause, on account of which that thing is said to be, and by means of which we secure perfect scientific knowledge of that thing, both knowledge *propter quid* and [knowledge] *quid est*. But [he said] he had accepted the name of cause in the common and general sense, without any distinction of causes, so in our minds there could remain a doubt whether every kind of cause or only some are fit for conveying to us that perfect knowledge. And so he decided to distinguish kinds of causes and to show in what way each can be the middle [term] in a demonstration *potissima*. For thus does he progress from universal to particulars, from cause accepted in a broad sense to each kind of cause. And he says many things about causes that he could not say in the first book or in the first part of the second without interrupting the treatment.

He then speaks about the way of searching for that definition that he said is indemonstrable and a beginning-principle of demonstration.[168] In the primary consideration, all the way to this passage, this had been taken as known *per se*. But here he wants

9

10

eiusdem tractationem, quatenus eam à nobis ignorari et quaeri contingit.

11   Eiusmodi sunt reliqua omnia, quae in calce eius libri tractantur ab Aristotele, pertinent enim ad maiorem eorum, quae in primo libro dicta erant, intelligentiam, ut per se quisque considerare potest. Nobis enim in praesentia satis est methodum et artificium Aristotelis, atque eorum librorum fabricam exposuisse, ut tractatio illa Aristotelis de methodis pro praesenti occasione optimè intelligatur.

: XX :

Solutio argumentorum prioris sectae
Grecorum et Latinorum.

1   Declarato consilio Aristotelis in Posterioribus Analyticis reliquum est ut aliorum argumenta expendamus et quantum roboris habeant, consideremus; et prius quidem argumenta Latinorum et Graecorum quibus ostendere nitebantur Aristotelem in secundo libro Posteriorum Analyticorum agere de medio et de definitione prout est medium demonstrationis. Primum horum argumentum sumptum fuit ex tractatione de syllogismo in primo libro Priorum, quemadmodum enim ibi Aristoteles docet et constructionem ratiocinationis et facilem inventionem propositionum ad ratiocinandum, ita etiam in Posterioribus Analyticis agere debet tum de constructione demonstrationis tum de facili inventione, constructionem quidem docet in primo libro, facilem autem constructionem in secundo, dum docet facilem inventionem medii, medio namque invento facilis est demonstrationis constructio.

also to make a secondary treatment of the same thing insofar as it happens to be unknown and inquired after by us.

All the other things that are treated by Aristotle at the end of the book are of this type. For they pertain to greater understanding of those things that were said in the first book, as anyone can consider by himself. At present it is enough for us to have laid out Aristotle's method and skill and the fabric of these books, so that this treatment by Aristotle on method may be understood in the best way for the present occasion.

: XX :

Solution to the arguments of the earlier
sect of the Greeks and Latins.

Aristotle's intent in the *Posterior Analytics* having been made clear, it remains that we weigh others' arguments and consider how much weight they have — and, of course, first the arguments of the Latins and Greeks, by which they endeavored to show that Aristotle, in the second book of the *Posterior Analytics*, dealt with the middle [term] and definition in that it is the middle [term] of a demonstration. The first of these arguments was taken from the treatment on the syllogism in the first book of the *Prior* [*Analytics*]: For just as Aristotle there teaches both construction of ratiocination and easy discovery of the premises for ratiocinating, [they claim,] so also he ought to deal in the *Posterior Analytics* both with construction of a demonstration and with easy discovery. And indeed he does teach construction in the first book, and easy construction in the second, [they claim,] when he teaches easy discovery of the middle [term], for once the middle [term] is discovered, construction of a demonstration is easy.

2   Ad hoc dicimus illud quidem verissimum esse, quòd Aristoteles, ut de syllogismo, ita et multo magis de demonstratione debuit
et constructionem et facilem principiorum inventionem docere,
neutrum tamen asserimus in secundo libro factum esse, sed
utrumque in primo, quam veritatem ante nos nemo cognovit praeter Themistium.

3   In secundo quidem libro quòd Aristoteles non tradiderit facilem principiorum inventionem manifestum est eum locum considerantibus, nam tractatio, quae est ab eius libri initio usque ad 47
contextum, adeò obscura est et intellectu difficilis, ut à nemine
post Averroem tum sententia tum intentio Aristotelis in ea parte
fuerit intellecta, credimusque nos primos hac tempestate locum
illum vera interpretatione illustrasse. Igitur si in ea parte facilem
medii inventionem reddere voluit Aristoteles, spe atque opinione
sua egregiè frustratus est, quippe cùm difficillima sit medii inventio, si in illius partis intellectione constituta est.

4   Sed dicant, quaeso, Latini quonam modo putent Aristotelem
ibi facilem medii inventionem docere, mihi quidem dicere videntur, Aristotelem id facere docendo medium esse definitionem, sed
hoc quomodo facilem reddat medii inventionem equidem non
video, facilis enim fit inventio medii quando conditio aliqua
maximè aliarum conspicua declaratur, per quam reliquae igno-
tiores inveniuntur, quemadmodum si in magno hominum numero
unum aliquem nobis de facie non cognitum quaereremus et aliquis
diceret; ille, quem quaeritis, est vir sapiens, prudens, iustus, tem-
perans, hic certè illius inventionem facilem nobis non redderet,
quoniam conditiones absconditas nobis proponeret. At si aetatem,
colorem ac formam faciei aliasque corporis conditiones declararet,
facilem utique nobis eius inventionem redderet.

To this we say that it is of course quite true that Aristotle, as 2
regards the syllogism and so also even more so demonstration,
ought to have taught both construction and easy discovery of
beginning-principles. Nevertheless we assert that neither is done
in the second book, but both in the first — a truth no one before
us, besides Themistius, knew.

To those who consider the passage carefully, it is manifest that 3
Aristotle, of course, did not convey easy discovery of beginning-
principles in the second book. For the treatment that is from the
start of the book up to text no. 47[169] is so obscure and difficult to
understand that both Aristotle's position and intention in that
part were understood by no one after Averroës. And we believe we
are the first in these times to elucidate that passage with the true
interpretation. If in that part, therefore, Aristotle wanted to render
easy the discovery of a middle [term], he was inordinately frus-
trated in his hope and opinion, since discovery of a middle [term]
will be most difficult, of course, if it is established in an understanding of that part.

But let the Latins say, I pray, in what way they hold that Aris- 4
totle there teaches easy discovery of the middle [term]. To me they
appear to be saying that Aristotle does it by teaching that the
middle [term] is a definition. But for my part I do not see in what
way this renders discovery of a middle [term] easy. For discovery
of a middle [term] is made easy when some characteristic is made
clear that, among the others, is most plainly apparent and by
means of which others less known are discovered — just as if in a
great number of men, we inquired after someone not known to us
by appearance, and someone said, "He whom you inquire after is a
man wise, prudent, just, temperate"; this would certainly not render discovery of him easy to us, since it sets out for us hidden
characteristics. But if [someone] made clear the age, color, shape
of the face and other characteristics of the body, that would by all
means render his discovery easy for us.

5   Medium autem demonstrationis notius nobis est ut causa, quàm ut definitio, ipsum enim esse proximam rei causam satis in primo libro Posteriorum novimus et Aristoteles ex hoc quòd medium est causa declarans propter quid est, ostendit in secundo libro ipsum esse definitionem et notificare etiam quid est. Conditionem igitur ignotiorem ex notiore deducit, neque ullam novam conditionem evidentiorem in medium affert, per quam facilis nobis fiat ipsius medii inventio. Cuius rei argumentum manifestum ex tota Aristotelis philosophia desumere possumus, quando enim alicuius accidentis cognitionem nobis tradere vult et eius demonstrationem extruere, non quaerit ante omnia medium, quod sit definitio, quod quidem secundùm Latinorum opinionem facere deberet, postea verò ex tali medio invento demonstrationem construere. Sed contra quaerit primùm ipsam causam rei, ex qua inventa demonstrationem facit, postmodum causa rei cognita definitionem assignat, in qua animus noster conquiescit. Postremum igitur, quod cognoscimus, est quòd medium sit definitio rei, tantum abest ut sit conditio, per quam facilis nobis reddatur inventio medii antequàm demonstremus.

6   Sed ipsimet Latini satis superque dogmatis sui falsitatem declarant, quomodo enim medium sit definitio adhuc nemo ipsorum plenè intelligere potuit, ut in libro nostro de medio demonstrationis fusè demonstravimus, fuit namque inter eos altercatio maxima qualisnam definitio sit ipsum medium et utrius extremi sit definitio, eaque controversia inter eorum sectatores ad haec usque tempora perdurat, ut adhuc sub iudice lis esse videatur. Quomodo igitur Aristoteles facilem nobis reddidit inventionem

Moreover, the middle [term] of a demonstration is more known 5
to us as cause than as definition, for it is the proximate cause of
something, as we sufficiently came to know in the first book of the
Posterior [Analytics]. And in the second book, Aristotle shows, from
the fact that the middle [term] is the cause clarifying what [some-
thing] is on account of (propter quid), that it is a definition and that
it also makes known what [something] is (quid est). He deduces,[170]
therefore, the more unknown characteristic from the more known,
and brings up no other new characteristic more evident, by means
of which discovery of the middle [term] itself may be made easy
for us. We can draw an obvious argument for this from Aristotle's
whole philosophy. For when he wants to convey to us knowledge
of some accident and to build up a demonstration of it, he does
not inquire after, before all [else], the middle [term], which is the
definition   which, of course, he ought to do, according to the
opinion of the Latins — and then afterward construct the demonstration from such and such a middle [term] once it has been
discovered. On the contrary, he first inquires after the cause itself
of something, from which, once discovered, he makes the demonstration. After the cause of the thing is known, he assigns a definition in which our soul rests. Finally, therefore, what we know is
that the middle [term] is the definition of the thing — so far is it
from being a characteristic by means of which discovery of the
middle [term] is rendered easy for us before we demonstrate.

But the Latins all by themselves make abundantly clear the fal- 6
sity of their doctrine. For in what way a middle [term] is a definition, none of them could yet fully understand, as we demonstrated
at length in our book on the middle [term] of a demonstration.[171]
For there was the greatest wrangling between them over what sort
of definition the middle [term] is and of which extreme it is the
definition. And this controversy among their sectarians continues
all the way to our own times, so that it still appears to be a lawsuit
before a judge. How, therefore, did Aristotle render discovery of

medii, dum docuit ipsum esse definitionem, si nondum per eius verba cognoscimus quam definitionem pro medio sumere debeamus? Non est igitur secundus ille liber de inventione medii, sed potius primus, quanquàm ante me nemo id vidit, nisi Themistius, cuius verba me ad hoc inveniendum excitarunt et illuminarunt.

7   Credo itaque Aristotelem de constructione demonstrationis egisse in secundo capite primi libri, de facili autem principiorum inventione seu de abundantia propositionum in capite quarto aliisque duobus sequentibus usque ad contextum 56 in secundo enim capite docuit constructionem demonstrationis, dum tradidit conditiones principiorum, ex quibus construenda est demonstratio, neque aliae conditiones praeter illas requiruntur, ut fiat praestantissima demonstratio, quia si demonstremus ex principiis veris, primis, medioque carentibus et prioribus et notioribus et causis conclusionis, ea est potissima et omnibus numeris absolutissima demonstratio, in qua nulla amplius conditio desiderari possit. Quando igitur Aristoteles in quarto capite et aliis sequentibus docet principia demonstrationis debere esse de omni, per se et universalia, non profert has tanquàm alias conditiones, quae in demonstratione requirantur praeter illas antea memoratas, quandoquidem illae sine his esse non possunt, proinde ad demonstrationem praestantissimam construendam illas nominasse satis fuit. Sed has in medium affert tanquàm conditiones evidentiores et ad facilem illarum inventionem et ad propositionum abundantiam pertinentes.

8   Propterea pars illa secundae sectioni primi libri Priorum proportione respondet et utraque cum primo libro Topicorum, imò cum omnibus Topicis libris comparari potest. Ut enim in illa secunda sectione agit Aristoteles de abundantia propositionum ad

the middle [term] easy for us when he taught that it is the definition itself, if we do not yet know by means of his words which
definition we ought to take for the middle [term]? It is not the
second book, therefore, but rather the first that is about discovery
of the middle [term]. But no one saw this before me except Themistius, whose words enlightened me and stimulated me to discover it.

And so, I believe that Aristotle dealt with construction of a   7
demonstration in the second chapter of the first book,[172] and with
easy discovery of beginning-principles or with the supply of premises in chapter four[173] and in the two following chapters, up to text
no. 56.[174] For in the second chapter he taught construction of a
demonstration when he conveyed the characteristics of beginning-
principles, from which a demonstration has to be constructed. For
there to be a most excellent demonstration, no other characteristics besides these are required, because if we demonstrate from
beginning principles that are true, first, lacking a middle [term],
prior, more known, and causes of the conclusion, this is a demonstration *potissima* and on all counts most complete; in it, no further
characteristic could be desired. When Aristotle teaches, therefore,
in chapter four and the others following, that the beginning-
principles of a demonstration ought to be *de omni*, *per se*, and *universale*, he does not advance these as other characteristics that are
required in the demonstration, besides those referred to earlier,
since those cannot exist without these.[175] Accordingly, for constructing a most excellent demonstration it was enough to have
named those. But he brings these up as more evident characteristics, pertaining both to their easy discovery and to the supply of
premises.

And so that part corresponds in comparative relation to the   8
second section of the first book of the *Prior* [*Analytics*],[176] and
both with the first book of the *Topics*. Indeed, they can be compared with all the books of the *Topics*. For as Aristotle deals in that

simpliciter ratiocinandum; et in libris Topicis de abundantia propositionum ad disputandum; ita in illa parte primi libri Posteriorum loquitur de abundantia propositionum ad demonstrandum.
Fit autem nobis alicuius rei copia et abundantia, quando locus
ostenditur, ubi ea res posita est, ut optimè dicebat Marcus Tullius
in principio Topicorum suorum declarans eorum librorum inscriptionem et utilitatem, cuius verba sunt haec, 'Ut earum rerum, quae
absconditae sunt, demonstrato et notato loco facilis, inventio est;
sic cùm pervestigare argumentum aliquod volumus, locos nosse
debemus, sic enim appellatae sunt ab Aristotele hae quasi sedes, è
quibus argumenta promuntur,' de abundantia igitur propositionum loqui nil aliud est, quàm locos declarare, ubi inveniri possint
quando iis uti voluerimus.

9   Ideo Aristoteles in tertio capite primi libri Topicorum, de locis
argumentorum tractationem instituens asserit eam ad abundantiam syllogismorum pertinere; et in secunda sectione primi libri
Priorum proponens se locuturum de abundantia propositionum
declarat locos, unde propositiones sumendae sunt. Ita in quarto
capite primi Posteriorum docet locos propositionum demonstrativarum, qui sunt primus atque secundus modus dicendi per
se, ut nos aliàs declaravimus dum librum illum publicè interpretaremur, cuius loci° clara est similitudo cum secunda illa sectione
primi libri Priorum, ut enim ibi Aristoteles locos syllogismorum
docet regulas tradens de iis, quae simpliciter praedicantur et subiiciuntur, sic enim facilis fit medii et propositionum inventio, ita
hîc locos propositionum demonstrativarum declarat per regularum
traditionem de iis, quae essentialiter praedicantur ac subiiciuntur.
Sed manifesta res est considerantibus verba Aristotelis in fine

second section with the supply of premises for ratiocinating absolutely and in the *Topics* books with the supply of premises for debating, so in that part of the first book of the *Posterior* [*Analytics*] he speaks about the supply of premises for demonstrating. Moreover, the supply and abundance of some thing occurs to us when the locus where that thing is located is shown, as M. Tullius [Cicero] finely said in the beginning of his *Topics*, while making the title[177] and utility of those books clear. His words are these, "Just as discovery of those things that are hidden is easy once the locus is demonstrated and noted, so when we want to thoroughly investigate some argument, we ought to know the loci—for thus are they called by Aristotle, as if they are dwelling places from which arguments are pulled out."[178] To speak, therefore, of the supply of premises is nothing other than to make clear the loci where they can be discovered whenever we want to use them.

And so Aristotle, in the third chapter of the first book of the *Topics*,[179] putting together a treatment on the loci of arguments, asserts that it pertains to the supply of syllogisms, and in the second section of the first book of the *Prior* [*Analytics*],[180] intending to speak about the supply of premises, makes clear the loci from which the premises have to be taken. Thus, in the fourth chapter of the first [book] of the *Posterior* [*Analytics*],[181] he teaches the loci of demonstrative premises that are *per se* in the first and second way of saying [so], as we made clear elsewhere when we made public comments on that book;[182] the resemblance of the locus passage°[183] to that second section of the first book of the *Prior* [*Analytics*] is clear. For as Aristotle there teaches the loci of syllogisms, conveying rules for the things that are made predicates and subjects absolutely—for that is how the discovery of a middle [term] and of premises is made easy—so here he makes the loci of demonstrative premises clear by means of conveyance of rules for the things that are made predicates and subjects essentially. And the issue is manifest to those considering Aristotle's words at the

contextus 28, quibus illam de locis tractationem proponit, nam si-
milibus verbis usus est in tertio capite primi Topicorum proponens
tractationem de locis et de abundantia argumentorum.

10   Themistius autem in eorum verborum declaratione clarè dicit
Aristotelem ibi agere de abundantia propositionum demonstrativarum. Et Averroes ibidem ita loquitur, ut ostendat se huiusce
rei veritatem saltem sub nube vidisse. Sed ea nos aliis consideranda et expendenda relinquimus.

11   Ad secundum argumentum dicimus Aristotelem et in primo et
in secundo Posteriorum libro agere de medio, de principiis enim
agere est de medio agere, ut antè ostendimus. Sed in primo libro
consideratur medium, eiusque conditiones inveniuntur, prout ducit ad scientiam propter quid est; in secundo autem declaratur
quòd idem medium et eisdem conditionibus praeditum tradit
etiam cognitionem quid est. In hoc igitur sensu non negamus secundum librum esse de medio, quemadmodum illi negare non de-
berent primum quoque librum de medio esse. Hanc autem differentiam indicat° Aristoteles in primo capite secundi libri dum
docet quaestionem propter quid et quaestionem quid unius et eiusdem medii esse, propterea quòd unum, et idem medium, una et
eadem causa utrique simul quaestioni satisfacit. Sic enim primum
librum cum secundo connectit, in primo enim de medio egerat ut
tradente scientiam propter quid est, in secundo autem de eodem
medio demonstrationis acturus est ut ducente etiam ad cognitionem quid est.

12   Tertium argumentum peccat ex insufficiente enumeratione,
nam definitio non solùm potest considerari ut definitio et ut medium demonstrationis, verùm etiam ut finis demonstrationis. À

end of text no. 28,[184] in which he sets out that treatment on loci. For he used similar words in the third chapter of the first [book] of the *Topics*,[185] setting out a treatment on loci and the supply of arguments.

Themistius too, in clarification of these words,[186] clearly says   10
that Aristotle deals there with the supply of demonstrative premises. And in the same place,[187] Averroës speaks in a way that shows that he too had seen — at least in a cloudy way — the truth of this issue. But we leave this for others to consider and weigh.

To the second argument we say that Aristotle deals with the   11
middle [term] in both the first and the second book of the *Posterior [Analytics]*. For to deal with beginning-principles is to deal with the middle [term], as we showed earlier. But in the first book the middle [term] is considered and, in that it leads to scientific knowledge *propter quid*, its characteristics are discovered; in the second [book] it is made clear that the same middle [term], and endowed with the same characteristics, also conveys knowledge *quid*. In this sense, therefore, we do not deny that the second book is about the middle [term], just as they ought not to have denied that the first book too is about the middle [term]. Moreover, Aristotle indicates° this difference in the first chapter of the second book,[188] when he teaches that the question of what [something] is on account of (*propter quid*) and the question of what [something] is (*quid est*) are of one and the same middle [term], and so [indicates] that one and the same middle [term], one and the same cause, answers both questions at the same time. For this is how the first book connects with the second: in the first he had dealt with the middle [term] as conveying scientific knowledge *propter quid*, and in the second he will deal with the same middle [term] of a demonstration as leading to knowledge *quid est*.

The third argument goes astray from an insufficient enumera-   12
tion. For a definition can be considered not only as a definition and as the middle [term] of a demonstration, but also as the end

primo quidem philosopho consideratur ut est definitio, et oratio significans quidditatem; aliis autem duobus modis à logico, in secundo enim Posteriorum libro consideratur[47] praecipuè ut est demonstrationis finis, ut autem est medium, seu principium demonstrationis tum in primo libro tum in secundo, medium enim est definitio causalis, quae ducit nos ad definitionem accidentis perfectam, quae est finis demonstrationis et demonstratio positione differens.

13   Similiter ad quartum dicimus verum esse id, quod ipsi sumunt, definitionem non considerari à logico nisi prout ad demonstrationem refertur. At relatio haec potest esse duplex, aut enim ut servi ad dominum aut ut domini ad servum, nam potest definitio dirigi ad demonstrationem, nempè quando ut medium demonstrationis accipitur. Potest etiam demonstratio ad definitionem dirigi, quae est praecipua definitionis consideratio in secundo libro, quod isti ignoraverunt. Ideò Aristoteles quando hanc relationem significare voluit, non solùm dixit definitionem esse demonstrationis principium, sed etiam conclusionem et demonstrationem situ terminorum differentem. His namque omnibus modis definitio ad demonstrationem refertur, non solùm ut medium.

14   Iure itaque Aristoteles de definitione agens in secundo libro etiam demonstrationis saepe mentionem facit, ibi namque de demonstratione agit prout est instrumentum ducens ad talis definitionis eductionem.

15   Ex ultimo autem argumento, quod fuit Eustratii adversus Alexandrum, colligimus Alexandri fuisse eam sententiam, quam nos in praesentia sequimur. Argumentum autem ab inscriptione eorum librorum sumitur, quos dicit Eustratius inscribi resolutorios à re considerata, à demonstratione, quae est resolutio, unde

of a demonstration. Of course, it is considered by the philosopher of metaphysics as it is a definition and speech signifying quiddity, and by the logician in the two other ways. In the second book of the *Posterior* [*Analytics*], it is considered principally as it is the end of a demonstration, but in both the first book and the second [it is considered] as it is a middle [term] or a beginning-principle of demonstration. For a middle [term] is a causal definition that leads us to an accident's perfect definition, which is the end of demonstration and a demonstration differing by position.

Similarly, to the fourth [argument] we say that what they are assuming is true, [namely] that definition is not considered by the logician except in that it is referred to demonstration. But this relation can be twofold, for it is either like servant to lord or like lord to servant. For definition cannot be directed to demonstration, namely, when it is accepted as the middle [term] of a demonstration. But demonstration can be directed to definition, which is the principal consideration of definition in the second book, they ignored this. And so Aristotle, when he wanted to indicate this relation, not only said that definition is a beginning-principle of demonstration, but also that it is a conclusion and a demonstration differing in arrangement of the terms.[189] For definition is referred to demonstration in all these ways, not only as a middle [term]. 13

And so Aristotle, in dealing with definition, often makes mention of demonstration in the second book also, and rightly so. For there he deals with demonstration in that it is an instrument leading to the eduction of such a definition. 14

Furthermore, from the last argument, which was Eustratius' against Alexander,[190] we gather that the position that we follow at present was Alexander's. The argument is taken from the title of the books, which Eustratius says are titled "Analytics (*Resolutoria*)" from the issue being considered, [i.e.] from demonstration, which 15

infert secundum librum non esse de definitione, ut putavit Alexander, sic enim illi ea inscriptio non competeret, quandoquidem definitio non est resolutio.

16   Sed facile est ad hoc argumentum pro Alexandro respondere, primùm quidem negare possemus inscriptionem Analyticorum librorum sumptam esse à re considerata et demonstrationem esse resolutionem, hoc enim non parum in se difficultatis habet. Eo tamen in praesentia concesso, quia id excutere ad propositam nobis contemplationem non pertinet, dicimus secundum quoque Posteriorum librum esse de demonstratione, ut Aristoteles in epilogo in calce eius libri significavit et ut nos efficaciter contra Averroistas ostendimus, idque significavit Alexander, si eius verba ab Eustratio relata benè perpendamus, dixit enim agi in secundo libro de definitione prout ad demonstrationem pertinet, non dixit prout est instrumentum à demonstratione distinctum, ut Averroistae dicunt, quare liber ille ita est de definitione, ut sit etiam de demonstratione secundùm Alexandrum.

17   Sed quonam modo intellexit Alexander definitionem ibi considerari prout ad demonstrationem pertinet? Certè altero duorum modorum pertinet, vel quòd ad demonstrationem dirigatur vel quòd demonstratio ad eam. Si primum, non rectè carpitur ab Eustratio Alexander, ea namque est ipsa Eustratii sententia quòd definitio tractetur ut medium demonstrationis, non enim alia ratione ad demonstrationem dirigitur, nisi ut ipsius medium. Cùm igitur sententiam Alexandri refutet Eustratius, indicat° se cognovisse aliam fuisse Alexrandri mentem, scilicet definitionem in secundo libro tractari quatenus ad eam tanquàm ad finem dirigitur demonstratio, sic enim definitionem pertinere ad demonstrationem intellexit Alexander, proinde ea fuit ipsius de secundo illo libro sententia, quam nos Averrois quoque fuisse demonstravimus et quam solam veram esse arbitramur.

is resolution; from this he infers that the second book is not about definition, as Alexander held. For then this title would not apply to it, since definition is not resolution.

But it is easy to respond to this argument in favor of Alexan- 16 der.[191] First, of course, we could deny that the title of the *Analytics* books is taken from the issue being considered and that demonstration is resolution; for this has no small amount of difficulty to it. Nevertheless, let this be conceded at present, because to examine it does not pertain to the contemplation set out before us. We say that the second book too of the *Posterior* [*Analytics*] is about demonstration, as Aristotle indicated in the summary at the end of the book[192] and as we effectually showed against the Averroists. And Alexander indicated it, if we examine well his words, as reported by Eustratius. For he said that, in the second book, definition is dealt with in that it pertains to demonstration; he did not say "in that it is an instrument distinct from demonstration," as Averroists say. And so, according to Alexander, that book is as much about definition as it is also about demonstration.

But in what way did Alexander understand definition to be 17 considered there as pertaining to demonstration? Certainly, it pertains in one of two ways — either because it is directed to demonstration, or demonstration to it. If the first, Alexander is not correctly rebuked by Eustratius. For it is the very position of Eustratius that definition is treated as the middle [term] of a demonstration, for it is directed to demonstration in no other way than as its middle [term]. Since, therefore, Eustratius criticizes Alexander's position, he indicates° that he himself knew that Alexander had something else in mind, namely that definition is treated in the second book insofar as demonstration is directed to it as to an end. For thus did Alexander understand definition to pertain to demonstration. And accordingly, this was his position on that second book; we demonstrated this was Averroes' also and we think it is the only true one.

: XXI :

Solutio argumentorum posterioris sectae Averroistarum.

1 Sequitur ut ad Averroistas seu potius Pseudoaverroistas respondeamus; in primo quidem argumento id, quod assumunt, duo esse in rebus cognoscenda, substantiam et accidens, concedimus. Sed cùm postea inferunt, ergo de definitione agendum in logica ut de instrumento substantiae cognoscendae, neganda est consequentia. Nam debet quidem logicus agere de instrumento, quo substantia ignota notificetur, sed illud non est definitio, sed est methodus resolutiva, ut nos superius declaravimus. Hoc igitur argumento nil aliud probant, quàm agendum fuisse in logica tum de methodo demonstrativa, qua accidentia cognoscantur, tum de resolutiva, qua substantiae ignotae notificentur.

2 Unde manifestum est, quàm debilis ac vana sit argumentatio quorundam, qui de nobiliore instrumento logico disputantes et ostendere volentes[48] definitionem esse instrumentum nobilius demonstratione, ita argumentantur, definitione substantiam cognoscimus, demonstratione accidentia, at substantia accidentibus nobilior est, ergo et definitio est nobilius instrumentum demonstratione, quandoquidem illud est instrumentum nobilius, quod ad rei nobilioris cognitionem dirigitur.

3 Sed hi in eo primùm peccant, quòd comparationem absurdam faciunt, solemus enim dicere, comparativum supponit positivum, ergo non poterant disputare utrum sit nobilius instrumentum, demonstratio, an definitio, nisi prius ostendissent definitionem

: XXI :

Solution to the arguments of the later school of Averroists.

It follows that we may respond to the Averroists, or rather, the   1
pseudo-Averroists [as follows]: In the first argument, of course, we
concede that which they assume, that two [things] are to be
known in things: substance and accident. But when they afterward
infer that therefore definition has to be dealt with in logic as an
instrument for knowing substance, the consequence has to be denied. For the logician, of course, ought to deal with the instrument by which an unknown substance is made known, but this is
not definition; it is instead resolutive method, as we made clear
earlier. By this argument, therefore, they prove nothing other than
that in logic, both the demonstrative method, by which accidents
are known, and the resolutive, by which unknown substances are
made known, had to be dealt with.

From this it is manifest just how weak and vain is the argu-   2
mentation of those who, debating about the more noble logical
instrument and wanting to show that definition is an instrument
more noble than demonstration, make the following argument:
We know substance by definition, accidents by demonstration; but
substance is more noble than accidents; therefore definition is an
instrument more noble than demonstration, since the nobler instrument is the one that is directed to knowledge of something
more noble.

But they go astray, first, in this, that they make an absurd com-   3
parison. For we normally say that the comparative presupposes the
positive. Therefore, they could not debate whether the nobler instrument is demonstration or definition, unless they had first

instrumentum esse notificandi, hoc enim negato, ut nos omninò negamus, tota eorum disputatio inanis redditur.

4   Sed etiam si definitio instrumentum sciendi esset, argumentum ipsorum, quod memoravimus, nihil prorsus haberet efficacitatis ad probandum eam esse instrumentum praestantius demonstratione. Si quid enim roboris haberet, probaret etiam methodum resolutivam demonstrativa praestantiorem esse, siquidem resolutiva substantias cognoscimus, demonstrativa verò accidentia. Id tamen est manifestè falsum, cùm apud omnes constet potissimam demonstrationem demonstratione à signo atque inductione nobiliorem esse.

5   Solvitur autem argumentum, negando propositionem illam, quam pro axiomate sumunt, illud est praestantius instrumentum, quod ad rei praestantioris cognitionem dirigitur, haec enim falsa est et multa eam sequerentur absurda, veluti idem demonstrationis genus nobilius esse in demonstrandis accidentibus hominis, quàm in demonstrandis accidentibus asini, cùm tamen eadem utrobique sit potissimae demonstrationis natura, et vis; et demonstrationem à signo, qua primus aeternus motor ostenditur, praestantius instrumentum esse ea demonstratione à signo, qua demonstratur prima materia, imò et omni[49] potissima demonstratione, quoniam Deus omnibus rebus nobilitate praestat. Itaque pro rerum cognoscendarum varietate innumeri erunt demonstrationum gradus inter se nobilitate et ignobilitate dissidentes, quae omnia vana sunt et à nullo philosopho pronuncianda.

6   Nos igitur dicimus maiorem vel minorem instrumenti logici nobilitatem non esse sumendam ex ipsa rerum, quae notificantur, nobilitate, sed ex maiore vel minore notificandi vi et efficacitate, cùm enim Aristoteles in principio primi libri de Anima dicat duas

shown that definition is an instrument for making [something] known. If this is denied, as we altogether do deny, their whole debate is rendered inane.

But even if definition were an instrument for knowing scien- 4
tifically, their argument, which we just referred to, would have no efficacy at all for proving that it is an instrument more excellent than demonstration. For if this had weight, it would also prove that resolutive method is more excellent than demonstrative, since indeed we know substances by the resolutive and accidents by the demonstrative. Nevertheless, this is manifestly false, since every-one agrees that demonstration *potissima* is more noble than dem-onstration *a signo* and than induction.

Furthermore, the argument is done away with by denying that 5
premise which they take to be an axiom: The more excellent in-strument is the one directed to knowledge of the more excellent thing. This is false and many absurdities would follow from it, such as that the same kind of demonstration is more noble in demonstrating the accidents of man than in demonstrating the ac-cidents of ass, even though the nature and power of the demon-stration *potissima* is in both cases the same, and that the demon-stration *a signo*, by which the first, eternal mover is shown, is a more excellent instrument than the demonstration *a signo*, by which first matter is demonstrated—indeed [more excellent too] than every demonstration *potissima*, since God excels everything in nobility. And so, on account of variety of things to be known, there will be countless degrees of demonstrations, differing among themselves in nobility and ignobility. All this is vain and no phi-losopher should say it.

We say, therefore, that greater or lesser nobility of a logical in- 6
strument has to be taken, not from the nobility itself of the things that are made known, but from its greater or lesser power and ef-ficacy to make [them] known. Now although Aristotle says, in the beginning of the first book of *On the Soul*,[193] that there are two

esse rationes, quibus aliqua scientia alii scientiae praestare potest, vel enim praestat nobilitate subiecti vel τῇ ἀκριβείᾳ,[50] prior quidem ratio fortasse in ipsis contemplativis scientiis posteriori praeferenda est, ut maior sit illa praestantia, quae à subiecti nobilitate sumitur, quàm quae à certitudine cognitionis, ut videtur significare Aristoteles in ultimo capite primi libri de Partibus animalium. Sed in logica, in qua non res, sed instrumenta cognoscendi tractantur, contrario modo res se habet.

7   Cùm enim rerum nobilitates in logica considerandae et inter se conferendae non sint, sed solùm instrumenta, quae apta sint nos ad rerum absconditarum cognitionem ducere illud est dicendum nobilius instrumentum, quod sit ἀκριβέστερον et scientiam pariat certiorem. Ob id si duo logica instrumenta proponantur, quorum alterum eiusdem rei vel rei aequè ignotae scientiam maiorem et certiorem pariat, quàm alterum, illud certè nobilius dicendum erit. Quòd si non modò rei aequè ignotae, sed etiam rei ignotioris et cognitu difficilioris cognitionem certiorem nobis praestet,[51] quàm alterum rei minùs ignotae et facilioris cognitu, adhuc maior illius nobilitas et excellentia iudicanda erit. Talis autem est methodus demonstrativa respectu resolutivae, demonstrativa namque certiorem scientiam parit, cùm non solùm notificet rem esse, sed etiam cur sit, resolutiva verò solùm declaret rem esse, certior enim scientia est, qua rem esse per suam causam cognoscimus, quàm ea, qua sine causa rem cognoscimus, ut inquit Aristoteles. Praeterea illud, quod per demonstrationem potissimam notificatur, est naturaliter ignotius et difficilius cognitu, quàm id, quod per methodum resolutivam; siquidem accidens secundùm rerum naturam non est

ways by which any science can excel another science — it excels in either nobility of subject or *tēi akribeiai*[194] — the first way, of course, has to be preferred perhaps to the latter in contemplative sciences themselves, as the excellence that is taken from the nobility of subject is greater than that [taken] from certainty of knowledge, as Aristotle appears to indicate in the last chapter of the first book of *On the Parts of Animals*.[195] But in logic, in which not the things but the instruments for knowing [them] are treated, it happens in the contrary way.

Since in logic, what are to be considered and compared to each 7 other are not the nobilities of things, but only the instruments that are able to lead us to knowledge of hidden things, that instrument has to be said to be more noble that is *akribesteron* (more certain) and brings forth more certain scientific knowledge. On account of this, if two logical instruments are set out, of which one brings forth greater and more certain scientific knowledge than the other, regarding the same thing or something equally unknown, that one will certainly have to be said to be the more noble. And if it ensures for us not only more certain knowledge of something equally unknown but also of something more unknown and more difficult to know than the other [does] of something less unknown and easier to know, then the nobility and excellence of that one will have to be judged greater. Such is demonstrative method with regard to resolutive. For the demonstrative brings forth more certain scientific knowledge, since it makes known not only that something is, but also why it is; but resolutive makes clear only that something is. For scientific knowledge by which we know that something is by means of its cause, is more certain than that by which we know something without the cause, as Aristotle says.[196] Moreover, that which is made known by means of demonstration *potissima* is naturally more unknown and more difficult to know than that which [is made known] by means of resolutive method, since indeed an accident is, according to the nature of things, not

per se notum, quia pendet à causa externa; substantia verò cùm ab alia causa non pendeat, per se nota dicitur secundùm naturam, quòd si ignota sit, id non est secundùm ipsam rei naturam, sed propter nostram infirmitatem, ipsa enim rei natura per se cognoscibilis est, quoniam etiam per se est, accidens verò non est cognoscibile nisi per aliud, quia etiam per aliud est.

8   Illud igitur, quod per methodum demonstrativam declaratur, naturam habet ignotiorem quàm id, quod per methodum resolutivam notificatur, quare methodus demonstrativa multò efficacior est methodo resolutiva, cùm rei ignotissimae secundùm naturam scientiam pariat certissimam, in qua animus acquiescit, neque de illa re aliquid praeterea quaerit. Praestantius igitur instrumentum est, proinde primaria debuit esse ipsius consideratio in logica, methodi autem resolutivae secundaria.

9   Adde quòd logica est disciplina auxiliatrix humani ingenii, nobis enim instrumenta parat, quibus ad res cognoscendas iuvemur, praecipua igitur illa logicae pars iudicanda est, quae nobis in ea re auxiliatur, in qua maiore auxilio indigemus maiore autem egemus auxilio ad cognoscendas illas res, quae secundùm propriam naturam sunt ignotae, quàm ad illas, quae secundùm naturam sunt notae, illae namque, egent per se instrumento logico, quo innotescant, hae verò non per se. Demonstratio igitur potissima ordinem naturae servat et cognitionem earum rerum tradit, quae secundùm naturam debent cognosci per aliud, proinde per discursum et methodum logicam. Ob id primaria debuit esse eius consideratio in logica tanquàm instrumenti utilioris et nobilioris, secundaria verò methodi resolutivae, ut fusè in praecedentibus declaratum est.

known *per se*, because it depends on an external cause, but a substance, since it does not depend on another cause, is said to be, according to nature, known *per se*. Now if it is unknown, that is not according to the very nature of the thing, but on account of our inadequacy. For the very nature of the thing is knowable *per se*, since it also exists *per se*, but an accident is not knowable except by means of another, because it also exists by means of another.

That which is made clear by means of demonstrative method, 8 therefore, has a nature more unknown than that which is made known by means of resolutive method. And so demonstrative method is much more effectual than resolutive method, since it brings forth the most certain scientific knowledge of something most unknown according to nature; our soul acquiesces in this and does not inquire after anything else about that thing. It is, therefore, the more excellent instrument. And so consideration of it ought to be primary in logic, and [consideration] of the resolutive method secondary.

Add in that logic is a discipline that is an aid to the human wit. 9 For it brings forth the instruments by which we are helped in knowing something. That part of logic, therefore, has to be judged principal that is an aid to us in that in which we need more aid. Moreover, we need more aid in knowing those things that are unknown according to their proper nature, than those things that are known according to nature. For the former need *per se* a logical instrument by which they may become known, but the latter do not *per se*. Demonstration *potissima*, therefore, maintains the order of nature and conveys knowledge of those things that, according to nature, ought to be known by means of another, that is, by means of discursive movement and logical method. On account of this, consideration of it ought to be primary in logic, as of an instrument more useful and more noble, but [consideration] of resolutive method secondary, as was made clear at length in the preceding.

10   Ad secundum argumentum neganda est consequentia, licèt
enim logica sit facultas instrumentalis, tamen non est necessarium
ut quicquid tractatur in logica, instrumentum notificandi sit, sed
vel instrumentum vel saltem ad logica instrumenta respectum aliquem habens. Nomen enim et verbum non sunt sciendi instrumenta, tractantur tamen in logica tanquàm materia logicorum instrumentorum. Quatuor quaestiones declarantur ab Aristotele in
principio secundi libri Posteriorum, nec tamen sunt instrumenta
logica, sed scopi et fines instrumentorum. Definitio igitur in logica
tractatur non ut instrumentum logicum, sed ut principium vel ut
finis logicorum instrumentorum, ut antea declaravimus.

11   Ad tertium dicimus definitionem ut instrumentum sciendi neque à metaphysico neque à logico neque ab alio ullo considerari,
quoniam ipsa tale instrumentum non est.

12   Ad Averroem, qui saepe dixit primum librum Posteriorum esse
de demonstratione, secundum[52] verò de definitione, iam respondi-
mus ac mentem Averrois declaravimus, qui non putavit secundum
librum esse de definitione ut de altero instrumento re distincto à
demonstratione, sed in primo de demonstratione agi quatenus est
demonstratio, in secundo verò de eadem demonstratione quatenus
est definitio.

13   Unus relinquitur locus apud Averroem, qui aliquam videtur
difficultatem facere, sed ipsa re et Averrois mente benè intellecta
non est quòd nos eius verba perturbent, sed ad rectum sensum
trahenda sunt, ut facilè trahi possunt, verba autem Averrois, quae
in commentario 42 septimi Metaphysicorum leguntur, sunt haec,
'Considerare de definitionibus est commune logico et philosopho,
sed duobus diversis modis, logicus enim considerat de definitionibus secundùm quod[53] est instrumentum, quod inducit intellectum

To the second argument: The consequence has to be denied. 10
For, even granting that logic is an instrumental faculty, it is nevertheless not necessary that whatever is treated in logic is an instrument for making [something] known, but [only that it is] either
an instrument or at least [something] having some relationship to
logical instruments. For noun and verb are not instruments for
knowing scientifically; nevertheless they are treated in logic as the
matter of logical instruments. Four questions are made clear by
Aristotle in the beginning of the second book of the *Posterior* [*Analytics*];[197] they are nevertheless not logical instruments, but the
goals and ends of instruments. Definition, therefore, is treated in
logic not as a logical instrument, but as the beginning-principle or
as the end of logical instruments, as we made clear earlier.

To the third, we say that definition is not considered by the 11
metaphysician, by the logician, or by anyone else as an instrument
for knowing scientifically, since it is not such an instrument.

To Averroës, who often said that the first book of the *Posterior* 12
[*Analytics*] is about demonstration and the second about definition,[198] we have already responded and clarified what Averroës had
in mind. He did not hold that the second book is about definition
as about another instrument distinct in reality from demonstration. Instead, in the first, demonstration is dealt with insofar as
it is demonstration, and in the second, the same demonstration
insofar as it is a definition.

One passage in Averroës remains that appears to make some 13
difficulty [for us], but if the issue itself and what Averroës had in
mind are well understood, then his words would not disturb us;
but they have to be construed in the correct sense, as they easily
can be construed. For Averroës' words, which are read in commentary 42 to the seventh [book] of the *Metaphysics*,[199] are these: "To
consider definitions is common to the logician and the philosopher, but in two different ways. For the logician considers definitions according to what is an instrument that induces the intellect

ad intelligere quidditates rerum; philosophus autem secundùm quod[54] significat naturas rerum.' His verbis duos sensus tribuere possumus, primùm quidem possumus dicere Averroem nominare definitionem in numero plurali, cùm dicat, 'de definitionibus,' deinde instrumentum in numero singulari, quare non videtur definitionem appellare instrumentum, sed solùm dicere definitiones considerari à logico quatenus datur instrumentum aliquod logicum, quod ducit intellectum ad cognoscendas quidditates rerum, id est earum definitiones, quasi dicat definitionem à logico considerari quatenus per instrumentum logicum innotescit, instrumentum autem est demonstratio.

14   Possumus etiam concedere Averroem vocare definitionem instrumentum logicum, non tamen quòd sit re distinctum à demonstratione, sed quatenus est idem, quod demonstratio, ut nos quoque suprà declaravimus. Nam demonstratio est definitio et definitio est demonstratio, ideò definitio qua ratione potest vocari demonstratio, eadem ratione potest dici instrumentum, quo ducimur ad cognoscendum quid est, demonstratio enim ducit ad cognoscendum quid est. Quòd haec sit mens Averrois patet considerantibus alia eius verba in eodem loco, non potest enim dicere quòd definitio ut extra demonstrationem sumpta et ut ab ea distincta instrumentum sit declarans quidditatem rei, quia statim considerationem hanc primo philosopho attribuit dicens 'philosophus autem quatenus significat quidditates rerum' definitio enim respectu naturae et quidditatis rerum non est nisi eius significatrix, sed ad ipsius ignotae cognitionem nos ducere non potest. Hac igitur ratione asserit Averroes eam à primo philosopho considerari, non à logico, quare ut instrumentum significandi quidditatem est considerationis metaphysicae, quo fit ut eadem ratione à logico

to understand the quiddities of things, but the philosopher according to what signifies the natures of things." We can ascribe two senses to these words. First, of course, we can say that Averroës names definition in the plural (since he says "definitions") but then instrument in the singular. And so he does not appear to call definition an instrument, but only to say that definitions are considered by the logician insofar as there is some logical instrument that leads the understanding to knowing the quiddities of things, that is, their definitions, as if he were to say that definition is considered by the logician insofar as it makes [something] known by means of a logical instrument; and that instrument is demonstration.

Yet we can concede that Averroës calls definition a logical instrument, not however that it is distinct in reality from demonstration, but insofar as it is the same as demonstration, as we also made clear above. For demonstration is definition and definition is demonstration. And so, just as definition can be called demonstration, it can in the same way be said to be an instrument by which we are led to knowing what [something] is (*quid est*), for demonstration leads to knowing what [something] is (*quid est*). That this is what Averroës has in mind is patent to those [carefully] considering his other words in the same passage. For he cannot be saying that definition, as taken outside of demonstration and as distinct from it, is an instrument making the quiddity of something clear, because he at once attributes this consideration to the philosopher of metaphysics, saying, "but the philosopher insofar as it signifies the quiddities of things."[200] For definition is not with regard to the nature and quiddity of things, except as its signifier; it cannot lead us to knowledge of the unknown itself. For this reason, therefore, Averroës asserts that it is considered by the philosopher of metaphysics, not by the logician; wherefore, as an instrument for signifying the quiddity, it is a metaphysical consideration. Hence it cannot be considered by the logician in the same way but in some

14

considerari non possit, sed aliqua alia. Idque nos antea demonstravimus, definitionem ut instrumentum significandi quidditatem ad metaphysicum pertinere. Instrumentum autem notificandi ipsa per se definitio dici non potest, sed solùm quatenus in ipso demonstrationis discursu posita sumitur et quatenus est demonstratio, sic enim est instrumentum, quia demonstratio instrumentum est.

15   Haec absque dubio est Averrois mens in eo loco, non quam adversarii putant, ita enim manifestam[55] esse ostendimus Averrois de definitione sententiam in logicis libris, ut si eo in loco id diceret, quod illi volunt, sibi ipse adversaretur. Quòd igitur definitionem dicat esse instrumentum logicum, concedimus; quòd autem re distinctum à demonstratione, negamus, idque probare debent adversarii, quia nullo unquàm in loco Averroem hoc dixisse asseveramus.

: XXII :

Cur Aristoteles in logica de methodis tantùm,
non de ordinibus egerit.

1   Caeterum postquàm de ordinibus ac methodis verba fecimus et tractationem Aristotelis de methodis exposuimus, quaerere non sine ratione aliquis posset, cùm non minùs ordo, quàm methodus, logicum instrumentum sit, cur Aristoteles in logica de ordinibus nihil docuit, cùm de methodis diligentissimè ac fusissimè disseruerit? Quaestio haec illorum animos valdè torquere potest, qui nefas esse putant existimare Aristotelem in aliqua sua tractatione° mancum et diminutum fuisse, quasi necesse fuerit ipsum omnia novisse et omnia, quae novit, scripsisse.

other. And we demonstrated this earlier: As an instrument for signifying quiddity, definition pertains to the metaphysician. But definition itself cannot *per se* be called an instrument for making known, but only insofar as it is taken [as] placed in the very discursive movement of a demonstration and insofar as it is a demonstration. For thus is it an instrument, because demonstration is an instrument.

Without doubt, this, and not what our opponents hold, is what 15 Averroës has in mind in this passage. For we showed that in the logic books Averroës' position on definition is so manifest that if in this passage he said what they want, he would have been opposing himself. That he says that definition is a logical instrument, we therefore concede, but we deny that [it is] distinct in reality from demonstration. And our opponents ought to prove this, because we have averred that in no passage did Averroës ever say it.

: XXII :

Why in logic Aristotle dealt only with
methods and not with orders.

But now that we have offered up some words on orders and methods, and have laid out Aristotle's treatment on methods, someone could inquire, and not without reason, since order no less than method is a logical instrument, why did Aristotle in [his] logic teach nothing about orders, given that he discussed methods most carefully and at the greatest length? This question can greatly torment the soul of those who hold that it is a sin to judge Aristotle to have been deficient and wanting in any of his treatments,° as if it were necessary that he knew everything and that he wrote down everything he knew.

2   Ego verò, etsi in admiratione ingenii Aristotelis nulli mortalium secundus esse volo, tamen credo ipsum neque omnia scribere neque omnia cognoscere potuisse, neque etiam veritatem in omnibus, quae scripsit, ita esse assecutum, ut numquàm errare potuerit, quippe qui homo fuit, non Deus. Propterea dubium hoc apud me non magni momenti est, Aristoteles enim logicae inventor fuit, primusque eam redegit ad artem, ut ipse testatur in calce secundi libri Elenchorum sophisticorum, quare si totam perficere non potuisset vel etiam noluisset, mirandum profectò non esset.

3   Veruntamen ne fugere huiusce rei difficultatem, quam aliqui fortasse magnam putabunt, videamur, aliquam etiam rationem afferre possumus cur Aristoteles in libris logicis nullam tractationem de ordinibus facere voluerit, praesertim cùm manifestum sit ipsum ea, quae ad ordines attinent, non ignorasse, tum quia libros suos omnes ordinatissimè composuit tum etiam quia utriusque ordinis aliquando mentionem fecit,[56] et utriusque naturam egregiè expressit, ut compositivi in prooemio primi libri Physicorum, resolutivi autem in contextu 23 septimi Metaphysicorum et simul utriusque in quarto capite primi Ethicorum.

4   Duas igitur ob causas credo Aristotelem in logica nil de ordinibus dicere voluisse, unam quidem quòd logicam propter discentes, non propter docentes scribere constituit. Nam methodorum cognitio non solùm docentibus, sed etiam discentibus necessaria vel saltem maximè utilis est. Ordinum autem notitiam° iis quidem, qui scribere aliquam disciplinam volunt, non parum prodesse manifestum est; sed iis, qui discere volunt, vel nihil vel parùm utilitatis afferre potest, cùm satis esse videatur rudis illa et confusa ordinis notitia,° quam naturaliter insitam quisque rationis compos

Now although I want to be second to no mortal in admiration 2
of Aristotle's wit, I nevertheless believe that he could not have
known everything and written down everything, and also that he
did not so secure the truth in everything he did write that he
could never have erred. He was of course a man, not God. And so
this doubt is, to me, of no great moment. For Aristotle was the
discoverer of logic, and he was first to reduce it to an art, as he
himself attests at the end of the second book of the *Sophistical Refutations*,[201] and so if he could not perfect [it] as a whole, or even did
not want to, it is surely not to be wondered at.

Nevertheless, lest on this issue we appear to avoid a problem 3
that some perhaps will hold to be great, we can still bring forward
some reason why Aristotle in the logic books wanted to make no
treatment on orders, especially since it is manifest that he himself
was not ignorant of what is pertinent to orders, both because he
composed all his own books in a most orderly way and also because he sometimes made mention of both orders and expressed
the nature of both very well, as [for example] of the compositive in
the proem of the first book of the *Physics*,[202] of the resolutive in
text no. 23 of the seventh [book] of the *Metaphysics*,[203] and of both
together in the fourth chapter of the first book of the *Ethics*.[204]

I believe, therefore, that Aristotle wanted to say nothing in [his] 4
logic about orders, because of two things. [(a)] One, of course, is
that he decided to write logic for the sake of those learning, and
not for the sake of those teaching. For knowledge of methods is
necessary, or at least most useful, not only for those teaching but
also for those learning. And it is manifest that knowledge° of orders is of no small profit to those, of course, who want to write
down some discipline [i.e., some field of teaching], but to those
who want to learn, it can bring nothing or little that is useful,
since that crude and confused knowledge° of order that, implanted
naturally, anyone in possession of reason normally has, appears to

habere solet, nempè illa prius tractanda esse, sine quorum cognitione caetera non benè percipi possunt. Ideò videmus plurimos in libris Aristotelis interpretandis ne verbum quidem de ordine fecisse, reliqua tamen diligentissimè declarasse. Quo enim ordine libros Analyticos, libros de Moribus ac libros Metaphysicos scripserit, non considerarunt. De ordine autem in scientia naturali servato pauca quaedam dixerunt, quia ab ipso Aristotele in prooemio primi libri Physicorum occasionem huius considerationis sumpsere.

5   Cognitio autem methodorum discentibus omnibus perutilis est, sed quandoque etiam necessaria. Evenit enim saepenumerò ut ostensionem aliquam Aristotelis vel alius authoris intelligere tanti momenti sit, ut ea non intellecta locum illum intelligere nequeamus, quia non parùm refert qualisnam ea ostensio sit, an demonstratio ab effectu an à causa et an à causa proxima an à remota, qui enim proximam causam esse putat, causam aliam non quaerit, sed quaestioni plenè satisfactum esse arbitratur. Qui verò remotam esse existimat, aliam propinquiorem inquirit. Sed quisnam inficiari[57] potest ad rerum cognitionem acquirendam plurimùm pertinere hoc cognoscere? Nimirum an hoc illius causa sit, an effectus, et, si causa, qualisnam causa? Ob id interpretes Aristotelis saepe de methodo ab ipso servata controversantur et magnas faciunt disputationes.

6   Quoniam igitur Aristoteles logicam scripsit, ut nos iuvaret ad scientias capessendas, non ad tradendas, ideò de methodis tantùm, non de ordinibus scribere voluit, sola enim methodus scientiam rerum ignotarum tradit, ordo verò nihil notificat, sed solùm disponit.

7   Altera ratio esse potest, quia certum atque indubitatum Aristoteli fuit duos tantùm ordines dari, solos enim cognovit compositivum et resolutivum, ut ipse significavit in capite quarto primi

be enough; that is, that those things, without knowledge of which others cannot be grasped well, have to be treated first. And so we see that many, in commenting on Aristotle's books, offered not even a word about order, though they clarified the rest most carefully. They did not consider in what order he wrote the *Analytics* books, the *Ethics* books, and the *Metaphysics* books. About the order maintained in natural science, however, they said a few things, because they took the occasion for consideration of it from Aristotle himself, in the proem of the first book of the *Physics*.[205]

Moreover, to all those learning, knowledge of methods is very 5 useful, but sometimes it is also necessary. For it oftentimes happens that to understand a presentation by Aristotle or another author is of such moment, that, if it is not understood, we cannot understand the passage, because it matters not a little what sort of presentation it is, whether demonstration from effect or from cause and whether from proximate cause or from remote. For whoever holds that it is a proximate cause, inquires after no other cause, but thinks that the question has been fully answered. And whoever judges that it is remote, asks about another one closer. But who can deny that to know this pertains greatly to acquiring knowledge of things—whether this is cause or effect of that, and if cause, what sort of cause? On account of this, Aristotle's commentators often had disputes and great debates about the method he maintained.

Since, therefore, Aristotle wrote logic so as to help us grasp the 6 sciences, not to convey them, he wanted to write only about methods, not about orders. For only method conveys scientific knowledge of things unknown. Order makes nothing known; it only disposes.

[(b)] Another reason [that Aristotle did not write about order] 7 can be because it was certain and indubitable to Aristotle that there are only two orders. For he knew only compositive and resolutive, as he himself indicated in the fourth chapter of the first

libri Ethicorum. Et compositivum solis contemplativis, resolutivum solis operatricibus disciplinis aptari posse existimavit, ut eius verba declarant in prooemio primi libri Physicorum et in contextu 23 septimi Metaphysicorum.

8   Hoc autem cùm ita se habeat, ordinum declaratio paucis verbis absolvi potuit, non enim tanta in eis difficultas inerat, ut peculiarem tractationem, qua ad regulas artemque redigerentur, postularent. Propterea in prooemio primi Physicorum breviter Aristoteles ordinis compositivi naturam expressit et eum in solis scientiis contemplativis locum habere significavit, nullis aliis nixus fundamentis, quàm iis, quae in Posterioribus Analyticis iacta erant, ibi enim de methodo demonstrativa loquens dixerat scientiam cuiusque rei perfectam haberi ex cognitione proximae causae, at causa proxima non cognoscitur nisi prius causae remotae cognoscantur, sequitur itaque à causis remotis ad proximas esse progrediendum, si rerum scientia habenda sit, hic autem est ordo compositivus, qui in scientiarum traditione necessarius est.

9   Ex iis igitur, quae in libris Analyticis et in illo prooemio dicuntur, satis significatam ab Aristotele habemus ordinis compositivi naturam; quemadmodum resolutivi in contextu 23 septimi Metaphysicorum.

10   Quoniam igitur paucis verbis utriusque ordinis conditio declarari poterat, ut paucis ipsam Aristoteles declaravit, ideò cognovit eos tractatione logica non admodum indiguisse; et satis habuit de methodis accuratissimè scribere et in iis solis persistere; quam tractationem ita egregiè ac (si dicere convenit) ita divinè persecutus est, ut eorum librorum excellentiam ego admirari nunquàm desinam et in eorum recta intelligentia facillimam et optimam universae[58] philosophiae cognitionem constitutam esse non dubitem.

book of the *Ethics*.[206] And he judged that compositive could be applied only to contemplative disciplines and resolutive only to practical, as his words in the proem of the first book of the *Physics*[207] and in text no. 23 of the seventh [book] of the *Metaphysics*[208] make clear.

Moreover, since this is so, the clarification of orders could be 8 completed in a few words, since there was not so much of a problem with them that they demanded their own treatment, in which they might be reduced to rules and art. And so in the proem of the first book of the *Physics*,[209] Aristotle briefly expressed the nature of compositive order and indicated that it has a place only in contemplative sciences, having relied on no other foundations than those that were laid down in the *Posterior Analytics*. For there,[210] speaking about demonstrative method, he had said that perfect scientific knowledge of any thing is gotten from knowledge of the proximate cause. But a proximate cause is not known unless remote causes are known first. And so it follows that there has to be a progressing from remote causes to proximate ones if scientific knowledge of things is to be had. And this is compositive order, which is necessary in conveying the sciences.

From the things, therefore, that are said in the *Analytics* books 9 and in that proem, we have the nature of compositive order sufficiently indicated by Aristotle, just as of resolutive in text no. 23 of the seventh [book] of the *Metaphysics*.[211]

Since, therefore, what is characteristic of each order could be 10 made clear in a few words—just as Aristotle made it clear in a few—he thus knew that they were not at all in need of a logical treatment and it was enough to write about methods in the most precise way and to stick with them alone; he executed that treatment so well and (if it is appropriate to say) so divinely, that I will never cease to admire the excellence of these books and do not doubt that by correctly understanding them the easiest and optimal knowledge of philosophy as a universal whole is established.

11   Has igitur ob causas° credere possumus Aristotelem de ordinibus in logica non egisse. Eas quidem ego aliorum gratia afferre
volui, quod enim ad me attinet, satisfacere mihi ea una ratio potest, non scripsit, quia scribere noluit. Quòd si quis nil horum,
quae diximus, recipiat, aliud quidpiam probabilius, si potest, inveniat.

We can believe, therefore, that Aristotle did not deal with or- 11
ders in logic for these reasons.° I wanted to bring these forward for
the sake of others, of course, but as for me, this one reason can
satisfy me: He did not write [about orders], because he did not
want to. But if anyone is receptive to none of what I have said, let
him discover something else more probable, if he can.

DE REGRESSU

ON REGRESSUS

Quid sit regressus, et quid circulus.

1 Considerans Aristoteles in primo libro Posteriorum Analyticorum circularem demonstrationem, qua prisci quidam philosophi omnia sciri posse asseverabant, eam omnino refutavit et inutilem esse efficaciter ostendit. Fit autem circularis demonstratio quando demonstrata ex propositionibus conclusione ex ea vicissim eaedem propositiones demonstrantur, fitque è conclusione propositio et è propositione conclusio, quemadmodum longa oratione in secundo libro Priorum Analyticorum declaratur.

2 Verùm in illa confutatione ita locutus est Aristoteles, ut non omnem circularem ostensionem reiecisse videatur, sed aliquam admisisse, quam complures interpretes recipientes et à circulo illo seiungere volentes regressum appellarunt.

3 Alii tamen fuere, qui non modò circulum, sed regressum quoque ab Aristotele confutatum esse existimarunt, rationibus ita apparentibus° nixi, ut rei meo quidem iudicio[1] per se manifestae tenebras offuderint, eamque difficultatis ac dubitationis plenam reddiderint.

4 Ego igitur veritatis amore ductus, quam absque ulla offuscatione omnibus conspicuam esse vellem, de hac re aliqua dicere et tum ratione tum Aristotelis testimonio regressum negandum° non esse ostendere constitui. Quod equidem paucis verbis me praestiturum spero, quandoquidem non multorum argumentorum congerie, sed sola rei, de qua loquimur, intima scrutatione et exemplorum naturalium diligenti declaratione veritatem apertissimam futuram esse minimè dubito.

What a regressus is and what a circle is.[1]

Aristotle, in the first book of the *Posterior Analytics*, considering cir- 1
cular demonstration, by which some earlier philosophers averred
that everything could be known scientifically, altogether confuted
it and effectually showed it to be useless.[2] Circular demonstration
occurs when, the conclusion having been demonstrated from the
premises, the same premises are demonstrated in turn from that
[conclusion] and a premise is made out of the conclusion and a
conclusion out of a premise, as is made clear in a long speech in
the second book of the *Prior Analytics*.[3]

In that refutation, however, Aristotle spoke such that he ap- 2
pears not to have rejected all circular presentation but to have admitted one sort, which many commentators took up and, wanting
to separate it from that circle, called *regressus*.[4]

Nevertheless there were others who judged that not only the 3
circle but regressus too was confuted by Aristotle. They relied on
reasons so specious,° that they cast shadows over an issue that is
in my judgment *per se* manifest and rendered it full of problems
and doubts.

Led, therefore, by love of the truth, which I wanted to be 4
plainly apparent to everyone, without any obfuscation, I have decided to say something about this issue and to show, both by reason and by Aristotle's testimony, that regressus is not to be rejected.° I trust that I really will do this in few words, since I have
no doubt at all that the truth will become most plain, not by a
heap of many arguments, but just by a very deep scrutinizing of
the issue about which we speak and by a careful clarification of
natural examples.

5   Ante omnia intelligendum est quid per circulum, quid per regressum intelligamus. Circulus quidem, ut ex Aristotele colligimus, est si ex A, demonstremus B deinde revertentes ex B demonstremus A et utraque sit potissima demonstratio, quae per causam proximam declaret propter quid effectus sit. Hunc possibilem non esse ostendit in memorato loco Aristoteles, propterea quòd sequeretur idem respectu eiusdem causam esse et effectum, prius et posterius, notius et ignotius secundùm naturam. Sequeretur etiam, nil aliud per circulum demonstrari, quàm idem ex seipso, à primo namque ad ultimum demonstraremus A esse, quia A est, haec autem sunt absurda manifesta.

6   Regressus verò est inter causam et effectum, quando reciprocantur et effectus est nobis notior, quam causa, cùm enim semper à notioribus nobis progrediendum sit, prius ex effectu noto causam ignotam demonstramus, deinde causa cognita ab ea ad effectum demonstrandum regredimur, ut sciamus propter quid est.

7   Hoc igitur inter circulum et regressum interest, quòd in circulo uterque processus est demonstratio propter quid, quod ne excogitabile quidem est; in regressu autem prior processus est demonstratio quòd, posterior verò est demonstratio propter quid.

8   Quid igitur sit regressus quantum ad habendam nominis intelligentiam sufficit manifestum est; est enim reciprocata quaedam demonstratio, qua postquàm causam ignotam ex effectu noto demonstravimus, maiorem propositionem convertimus et eundem effectum per eandem causam demonstramus, ut sciamus propter

CHAPTER I

Before all [else], it has to be understood what we understand by　5
circle and what by regressus. The circle, of course, as we can gather
from Aristotle, is when we demonstrate B from A and then, turning around, demonstrate A from B, and each [demonstration] is a
demonstration *potissima* that, by means of the proximate cause,
clarifies what the effect is on account of (*propter quid*)[5]. In the passage referred to,[6] Aristotle shows that this is not possible, since it
would then follow that [one] selfsame thing would, with regard to
[another] selfsame thing, be [both] cause and effect, [both] prior
and posterior, and according to nature [both] more known and
more unknown. It would also follow that nothing would be demonstrated by means of the circle other than the same thing from
itself, for from first to last, we would have demonstrated that A is,
because A is. But all this is manifestly absurd.

Regressus, on the other hand, is between cause and effect, when　6
they reciprocate and the effect is more known to us than the cause.
For since progressing always has to be from what is more known
to us, we first demonstrate unknown cause from known effect;
then, the cause [now] known, we regress from it to demonstrate
the effect, with the result that we know scientifically what it is on
account of (*propter quid*).[7]

The difference, therefore, between the circle and regressus is　7
that in a circle each procedure is a demonstration *propter quid* —
this is not even imaginable — but in regressus the first procedure is
a demonstration *quòd* and the latter is a demonstration *propter
quid*.[8]

What regressus is, therefore, is manifest enough to have an un-　8
derstanding of the term: it is some sort of reciprocated demonstration, by which, after we have demonstrated unknown cause from
known effect, we convert the major premise[9] and demonstrate
the same effect by means of the same cause, so that we may know

quid sit. In sequentibus autem naturam et conditiones ipsius regressus et differentiam, qua à circulo discrepat, melius declarabimus.

: II :

In quo argumenta illorum, qui
regressum negant,° proponuntur.

1   Illi, qui regressum negant,° tribus potissimùm fundamentis nituntur, quorum duo sunt illa eadem, quibus Aristoteles contra circulum usus est, tertium ipsi addunt, quod efficacissimum et demonstrans esse arbitrantur.

2   Primum est, si daretur regressus, idem esset notius et ignotius eodem, cùm enim in omni demonstratione à notioribus ad ignotiora progrediendum sit et prior processus sit ab effectu ad causam, oportet igitur effectum causa notiorem esse. Atqui posterior processus est ab eadem causa ad eundem effectum, ergo in eo oportet causam effectu esse notiorem, eadem igitur causa eodem effectu erit simul notior et ignotior, quod quidem esse non potest. Quòd si quis dicat hoc non esse absurdum, si alio et alio modo sumatur notius et ignotius, nam causa est notior effectu secundùm naturam, effectus verò est notior nobis, hoc difficultatem non tollit, neque argumentum solvit, quoniam utraque demonstratio à nobis et propter nos ipsos fit, non propter naturam, ideò utraque fieri debet ex notioribus nobis. Quare argumentum viget ad

scientifically what it is on account of (*propter quid*). In what follows then, we will better clarify the nature and characteristics of regressus itself and the difference by which it is different from the circle.

: II :

In which the arguments of those who reject° regressus are set out.

Those who reject° regressus rely chiefly on three foundations, two of which are the same ones that Aristotle used against the circle. They add a third, which they think is demonstrative and most effectual.

The first is this:[10] If there were a regressus, [one] selfsame thing would be [both] more known and more unknown than [another] selfsame thing. For since in every demonstration there has to be a progressing from more known to more unknown and the first procedure is from effect to cause, it must be, therefore, that the effect is more known than the cause. But the latter procedure is from the selfsame cause to the selfsame effect; in this [procedure], therefore, it must be that the cause is more known than the effect. The selfsame cause, therefore, will at the same time be more known and more unknown than the selfsame effect, and this, of course, cannot be. Now if someone were to say that this is not absurd if "more known" is taken in one way and "more unknown" in another—for according to nature cause is more known than effect but to us effect is more known—this does not get rid of the problem or do away with the argument, since each demonstration is done by us and on account of us ourselves, not on account of nature. Therefore, both ought to be made from what is more known to us. And so the argument applies to showing that, if the

ostendendum quòd, si effectus est notior nobis, quàm causa et per ipsum invenimus causam, non possumus postea à causa regredi ad effectum, nam propter quod aliquid tale est, illud est magis tale, at propter effectum fit nobis nota causa, ergo semper magis ac firmius cognoscemus effectum, quàm causam, regredi igitur à causa ad effectum nunquàm licebit, nisi à minùs noto ad magis notum progressus fiat, quae est vana et inutilis demonstratio.

3   Secundum argumentum est, dato regressu nihil aliud per ipsum ostenditur, quàm idem per seipsum, qui est processus inutilis, priore namque processu ex A effectu demonstramus B causam, posteriore autem ex B causa demonstramus A effectum, quare à primo ad ultimum nil aliud demonstramus, quàm si A est, A est.

4   Haec sunt duo argumenta, quibus Aristoteles usus est ad circulum refellendum, quibus regressus quoque reprobari videtur.

5   Tertium argumentum aliqui addunt, quod validissimum esse putant et in eo tota huiusce rei difficultas consistit, quemadmodum in eius solutione tota veritatis declaratio constituta est. Argumentum est hoc, quando facimus priorem processum vel cognoscimus effectum pendere ab illa causa vel ignoramus. Si ignoramus, nunquàm illam causam inveniemus, ut si ignoremus fumi causam esse ignem, non magis dicere possumus, fumus est, ergo ignis, quàm fumus est, ergo asinus, quare non possumus uti demonstratione ab effectu, qui est prior progressus, quoniam maior illius demonstrationis ignota nobis erit, est enim maior, ubicumque est fumus, ibi est ignis, quam notam esse oportet, si extruenda sit demonstratio ab effectu ad causam.

6   Si verò praecognoscimus illam esse illius effectus causam, ergo posterior progressus supervacaneus est, id enim omne, quod ab eo

effect is more known to us than the cause, and we discover the cause by means of it [i.e., the effect], we cannot afterward regress from cause to effect. For that on account of which something is such and such, is [even] more such and such; but the cause becomes known to us on account of the effect; therefore we will always know the effect more and more firmly than the cause. To regress from cause to effect, therefore, is never permissible unless the progression occurs from less known to more known, and such a demonstration is vain and useless.

The second argument is this:[11] In a given regressus, nothing is shown by means of it other than the selfsame thing by means of that thing itself, and this is a useless procedure. For in the first procedure, we demonstrate cause B from effect A and in the latter we demonstrate effect A from cause B. Therefore from first to last, we demonstrate nothing other than that if A is, then A is.

These are the two arguments that Aristotle used to refute the circle; by them regressus too appears to be condemned.

Some add a third argument, which they hold to be the strongest; and in this [argument] the whole problem of this issue consists, just as the whole clarification of the truth is constituted by its solution. The argument is this: When we perform the first procedure, either we know that the effect depends on that cause or we do not know. If we do not know, we will never discover that cause, just as if we do not know that the cause of smoke is fire, we can no more say, "There is smoke, therefore there is fire," than "There is smoke, therefore there is an ass." And so we cannot use demonstration *ab effectu* (from effect), which is the first progression, since the major of that demonstration will be unknown to us. For the major is, "Wherever there is smoke, there is fire," which must be known if a demonstration from effect to cause is to be built up.

But if we know beforehand what is the cause of that effect, then the latter progression is superfluous, for everything we desire from

desideramus, à priore processu consequimur. Nam quod in secundo processu quaerimus, est cognoscere propter quid effectus
sit, hoc autem solo priore processu adipiscimur, cùm enim prae-
noscamus° eam illius effectus causam esse, cùm primùm invenimus
causam illam dari, simul invenimus propter quid ille effectus sit,
prior igitur processus sufficiens est ad demonstrandum et causam
esse et propter quid sit effectus, quare altero illo processu opus
non est.

: III :

In quo regressum dari ostenditur.

1 Contrarium sententiam ex Aristotele multis in locis sumimus,
nam in capite tertio primi libri Posteriorum argumentans contra
eos, qui circulum ponebant, regressum excipere videtur. Dicit
enim inconveniens esse quòd idem sit eodem notius et ignotius;
postea subiungit id non esse inconveniens, si diversis modis accipiatur notius et ignotius, ut altera quidem demonstratio sit simpliciter demonstratio, altera verò non simpliciter, sed demonstratio
quòd, quae sit à notioribus nobis. Concedit°² ergo regressum, in
quo prior processus est à notioribus nobis et demonstratio quòd;
posterior verò est à notioribus secundùm naturam et demonstratio
simpliciter dicta. Eandem sententiam legimus apud Aristotelem in
contextu 92 secundi libri.³

2   In capite autem decimo eiusdem libri declarans differentiam
inter demonstrationem propter quid et demonstrationem quòd,

it we gain by the first procedure. For what we inquire after in the second procedure is to know what the effect is on account of (*propter quid*), but we obtain this in the first procedure alone. For since we know° beforehand that this is the cause of that effect — since we first discover there is that cause — we discover at the same time what that effect is on account of (*propter quid*). The first procedure, therefore, is sufficient for demonstrating both that the cause is and what the effect is on account of (*propter quid*). Therefore there is no need for that other procedure.

: III :

In which it is shown that there is regressus.

We get the contrary position from Aristotle in many passages. In 1 the third chapter of the first book of the *Posterior* [*Analytics*],[12] arguing against those who posited the circle, he appears to exempt regressus. For he says that it is inconsistent for [one] selfsame thing to be [both] more known and more unknown than [another] selfsame thing; but he afterward adds that it is not inconsistent, if more known and more unknown are taken in different ways, so that one demonstration is a demonstration *simpliciter* and the other is not [a demonstration] *simpliciter*, but a demonstration *quòd*, which is from things more known to us. He therefore does allow° regressus, in which the first procedure is from what is more known to us and is a demonstration *quòd*, but the latter is from what is more known according to nature and is called a demonstration *simpliciter*. We read this same position in Aristotle in text no. 92 of the second book.[13]

And in the tenth chapter of that same book,[14] clarifying the dif- 2 ference between demonstration *propter quid* and demonstration

manifestè regressum facit, ex eisdem enim terminis utramque demonstrationem construit, prius quidem ab effectu demonstrat causam, deinde conversa propositione maiore ex causa illa demonstrat eundem effectum et dicit evenire ut hoc modo utrumque monstremus, scilicet effectum per causam et causam per effectum et modum etiam declarat dicens maiorem propositionem prioris demonstrationis fieri maiorem etiam posterioris demonstrationis, si simpliciter convertatur, minorem verò fieri conclusionem et conclusionem fieri minorem, ut si dicamus, omne non scintillans propè est, planetae non scintillant, ergo planetae propè sunt, quae est demonstratio ab effectu. Deinde dicamus, quod propè est, non scintillat, planetae propè sunt, ergo planetae non scintillant, quae est demonstratio propter quid. Si quis autem verba Aristotelis eo in loco benè consideret, videbit haec non dici exempli tantùm gratia, sed quia Aristoteles asserit evenire quandoque ut hac reciprocata demonstratione utamur et facta priore demonstratione alteram quoque posteriorem ex eisdem terminis construamus. Nam si exempla tantummodò utriusque demonstrationis afferre voluisset non opus fuisset ex eisdem terminis utramque constituere; modumque docere, quem servare debeamus.

3   Eandem sententiam ex Aristotele colligimus in prooemio primi libri Physicorum, nam in principio eius libri dicit in cognitione rerum naturalium procedendum esse à causis ad effecta, quae est demonstratio propter quid, postea animadvertens causas illas offerri nobis incognitas et ab ignotis non esse progrediendum, subiungit eas prius ex effectis tanquàm notioribus nobis indagandas et inveniendas esse, quae est demonstratio quòd. Manifestè igitur proponit se in scientia naturali usurum esse regressu, neque putat

quòd, he manifestly performs a regressus. For he constructs each demonstration from the same terms; first, of course, he demonstrates cause from effect, then, after converting the major premise, he demonstrates the same effect from that cause. He says it happens that in this way we demonstrate each: effect by means of cause, and cause by means of effect. And he also clarifies the way [we do so], saying that the major premise of the first demonstration also becomes the major of the latter demonstration, if it is converted *simpliciter*,[15] but the minor becomes the conclusion and the conclusion becomes the minor, as if [for example] we were to say, "Everything not twinkling is near; planets do not twinkle; therefore planets are near." This is a demonstration *ab effectu*. Then let us say, "What is near does not twinkle; planets are near; therefore planets do not twinkle." This is a demonstration *propter quid*. Moreover, if anyone considers Aristotle's words in that passage well, he will see that they are not said only for the sake of example, but because Aristotle asserts that it sometimes happens that we use this reciprocated demonstration, and the first demonstration having been made, we construct the latter too from the same terms. If he had wanted to bring forward merely examples of each demonstration, there would have been no need to constitute both from the same terms, and we ought to teach the way that we [ourselves] maintain.

We gather the same position from Aristotle in the proem of the first book of the *Physics*.[16] In the beginning of the book he says that in knowledge of natural things, there has to be a proceeding from causes to effects; this is demonstration *propter quid*. Afterward, noting that those causes are presented to us as unknown and the progressing should not be from unknown, he adds that they [i.e., the causes] have to be tracked down and discovered first from effects as from [things] more known to us; this is demonstration *quòd*. He manifestly sets out, therefore, that he himself will use regressus in natural science. He does not hold that either

3

alterum solum processum sufficere, sed utroque utendum esse censet. Sic Themistius locum illum interpretatur et regressum ibi manifestissimè ponit; et Averroes in commentariis 24 et 97 primi libri Posteriorum, ubi dicit regressum saepe in usu esse in scientia naturali, ut nos exemplis ab ea scientia petitis apertissimè demonstrabimus.

4   Rationem addere possumus efficacissimam ad hoc comprobandum, contingere potest ut effectus sit confusè notus, quia ipsius causa ignoretur, tunc igitur duo sunt, quae demonstratione indigent, causa quidem demonstratione eget, qua dari ostendatur; effectus verò demonstratione eget, qua distinctè per suam causam cognoscatur. Sed non potest una et eadem demonstratio utramque operam praestare, una enim demonstratio unius rei est, non plurium, quare si per hanc demonstrationem causa notificatur, non potest per eandem notificari effectus.

5   Praeterea si adversariorum sententia et ipsorum ratio admitteretur vel nulla unquàm daretur demonstratio ab effectu, vel si daretur eadem esset demonstratio propter quid. Qua enim demonstratione causa ignota notificatur, eadem etiam propter quid effectus sit notificaretur, omnis igitur demonstratio ab effectu esset dicenda tum demonstratio quòd tum demonstratio propter quid, quod risu dignum est.

procedure alone is sufficient; he deems it that both have to be used. This is how Themistius interprets that passage;[17] he posits regressus there most manifestly, as does Averroës in commentaries 24 and 97 to the first book of the *Posterior* [*Analytics*],[18] where he says that regressus is often in use in natural science, as we will demonstrate most plainly with examples taken from that science.

We can add very effectual reasoning to confirm this. It can happen that an effect is known confusedly because its cause is unknown. There are, therefore, then two things that are in need of demonstration. The cause needs a demonstration by which it may be shown that there is [such], and the effect needs a demonstration by which it may be known distinctly by means of its cause. But one and the same demonstration cannot really do the work of both, for one demonstration is of one thing, not of many. Therefore if the cause is made known by means of this demonstration, the effect cannot be made known by means of the same [demonstration].

Moreover, if the position of our opponents and their reasoning were admitted, either there would never be any demonstration *ab effectu*, or if there were, it would be the same as a demonstration *propter quid*. For by whatever demonstration an unknown cause is made known, by that same [demonstration], what the effect is on account of (*propter quid*) would be made known. Every demonstration *ab effectu*, therefore, would have to be said to be both a demonstration *quòd* and a demonstration *propter quid*, and that is ridiculous.

: IV :

*In quo declaratur, qualis sit in regressu primus
processus, exemplo sumpto demonstrationis factae
ab Aristotele in primo libro Physicorum.*

1 In hac difficultate res ipsa, si benè intelligatur, omnem dubitationem tollet. Idcirco declarandum est diligentissimè quomodo regressus fiat et quae in singulis eius partibus cognita nobis et quae incognita sint et quomodo. Hactenus enim satis levem huiusce rei cognitionem habemus, quae ut propositos nodos explicare valeamus[4] minimè sufficiens est.

2 In primis non est ignoranda distinctio illa satis apud philosophos trita et vulgata, cognitio nostra duplex est, alteram confusam vocant, alteram verò distinctam, et utraque tum in causa tum in effectu locum habet. Effectum confusè cognoscimus, quando absque causae cognitione novimus ipsum esse;° distinctè verò, quando per cognitionem causae. Illa quidem dicitur cognitio quòd est, haec verò propter quid est et simul etiam quid est, quoniam idem est cognoscere quid est et cognoscere propter quid est, ut docet Aristoteles in secundo libro Posteriorum. Loquitur autem de utraque cognitione Aristoteles in contextu 178 primi libri Posteriorum et in 39 secundi.

3 Causa verò quatenus causa est per causam sciri non potest, quia causam aliam non habet. Si namque causam habeat priorem, eam habet quatenus est effectus, non quatenus est causa. Datur tamen causae quoque cognitio tum confusa tum distincta. Confusa

: IV :

In which it is clarified of what sort is the first procedure in a regressus, using the example of a demonstration made by Aristotle in the first book of the Physics.

In this problem, the issue itself, if it is well understood, will get rid 1
of every doubt. And so it has to be clarified most carefully how
regressus occurs and what things in each of its parts are known to
us and what are unknown, and in what way. For up until now we
have had a knowledge of this issue so slight that it is not at all sufficient for us to untie the knots that have been set out.

In the first place, this distinction, familiar and commonplace 2
enough in the philosophers, should not be ignored: Our knowledge is twofold. They call the one confused and the other distinct,
and each has a place both in cause and in effect. We know an effect
confusedly when we know that it exists° [but] without knowledge
of the cause, and distinctly when [we know that it exists] by
means of knowledge of the cause. The former is said to be knowledge *quòd* [i.e., knowledge that something is the case], the latter
[knowledge] *propter quid* [i.e., knowledge of what something is on
account of] and also, at the same time, [knowledge] *quid est* [i.e.,
knowledge of what something is], since to know what something
is (*quid est*) and to know what it is on account of (*propter quid*) are
the same thing, as Aristotle teaches in the second book of the
Posterior [*Analytics*].[19] Aristotle also speaks about each knowledge
in text no. 178 of the first book of the *Posterior* [*Analytics*][20] and in
no. 39 of the second.[21]

But a cause insofar as it is a cause, cannot be known scien- 3
tifically by means of a cause, because it does not have another
cause. For if it has a prior cause, it has it insofar as it is an effect,
not insofar as it is a cause. Nevertheless, of cause too there is both

quidem quando ipsam[5] esse cognoscimus, sed quidnam sit ignoramus; distincta verò quando cognoscimus etiam quid sit, et ipsius
naturam penetramus.

4   Hoc declarato totam ipsius regressus seriem ordinatim consideremus et exemplum aliquod nobis proponamus, in quo ipsam regressus naturam melius inspiciamus, nempè effectum notiorem sua
causa, et cum ea reciprocabilem. Sumamus demonstrationem
Aristotelis in primo libro Physicorum, qua ex generatione, quae
substantiarum est, ostendit materiam primam dari, ex effectu noto
causam ignotam.[6] Generatio enim sensu nobis cognita est, subiecta verò materia maximè incognita.

5   Accepto igitur subiecto proprio, scilicet corpore naturali ca-
duco, cui primo utraque inest, in eo demonstratur causam inesse,
propterea quòd in eodem inest effectus et est demonstratio quòd,
quae ita formatur, ubi est generatio, ibi est subiecta materia, at in
corpore naturali est generatio, ergo in corpore naturali est materia.
In hac demonstratione minor propositio est nobis nota confusè;
quoniam generari quidem et interire naturalia corpora cernimus,
sed causam ignoramus. Maior verò propositio quanquàm sensu
non cognoscitur, aliqua tamen adhibita mentali consideratione facilè innotescit.

6   Qua ratione ipsam declarat Aristoteles in primo libro Physicorum, nam accidentium mutationem subiectum habere videmus,
substantiarum non videmus, attamen re aliquantum considerata
cognoscimus ita in omni mutatione esse oportere, ut in accidentium mutatione intuemur. Itaque inductionem illam facimus,
quam Averroes ad Dialecticae inductionis differentiam solet appellare demonstrativam. Dialectica namque inductio, cùm in materia

confused knowledge and distinct knowledge. It is, of course, confused when we know that it is, but we are ignorant of what it is, and distinct when we also know what it is and fathom its nature.

Now that this has been clarified, let us consider in order the 4 whole sequence of a regressus itself and set out for ourselves some example in which we may better observe the very nature of regressus — namely an effect more known than its cause and reciprocal with it. Let us take Aristotle's demonstration in the first book of the *Physics*,[22] by which he shows, from the generation that is of substances, that there is first matter — unknown cause from known effect. For generation is known to us by sense, but underlying subject matter is altogether unknown.

Given a proper underlying subject, therefore, namely, corrupt- 5 ible natural body, to which both [cause and effect] first belong, the cause is demonstrated to belong to it on account of the fact that the effect belongs to the same. And this is a demonstration *quòd* that is formed as follows: Where there is generation, there is underlying subject matter there; and in natural body there is generation; therefore in natural body, there is matter. In this demonstration, the minor premise is known to us confusedly, since we do discern that natural bodies are generated and pass away, of course, but do not know the cause. And although the major premise is not known by sense, it nevertheless easily becomes known by some applied mental consideration.

In the first book of the *Physics*, Aristotle clarifies it [i.e., this 6 applied mental consideration] this way: We see that change of accidents has an underlying subject; we do not see this of substances; but nevertheless, once this has been considered a little, we know it must be so in every change, just as we see in change of accidents. And so we perform that induction that Averroës, to differentiate it from dialectical induction, normally calls demonstrative. For dialectical induction, since it is performed in changeable

mutabili et contingente fiat, nil roboris habet nisi omnia singularia sumantur nullo praetermisso. Inductio autem demonstrativa fit in materia necessaria et in rebus, quae essentialem inter se connexionem habent, ideò in ea non omnia sumuntur particularia, quoniam mens nostra quibusdam inspectis statim essentialem connexum animadvertit, ideoque spretis reliquis singularibus statim colligit universale, cognoscit enim necessarium esse ut ita res se habeat in reliquis, ex eo enim quòd aliquod praedicatum alicui per se inest, licet inferre, ergo de omni praedicatur. Dum igitur mens nostra naturam mutationis seu generationis considerat, cognoscit essentiale ipsi ac necessarium esse hoc praedicatum, nempè subiectam aliquam materiam, quae mutationem ipsam recipiat, quemadmodum testimonio sensus in accidentium generatione rem se habere cognovit, ideò ex variis accidentium mutationibus infert hanc universalem, omnis mutatio habet subiectum, sub qua postea hanc maiorem sumit, ubi generatio, ibi subiecta materia.

7   Hunc discursum facit longo sermone Aristoteles in primo libro Physicorum, sed potissimùm in contextu sexagesimosecundo, quae est vera eius loci interpretatio, quam pauci cognoverunt.

8   Haec maioris propositionis cognitio non est nisi confusa, quia licèt praedicatum sit causa subiecti, attamen non cognoscitur ut causa, novimus enim mutationem omnem habere subiectam materiam, non tanquàm effectum ipsius materiae, sed tanquàm perpetuò coniunctam cum materia subiecta. Existimamus enim mutationem et materiam nexu ita necessariò iunctas esse, ut mutatio sine materia nunquàm inveniri[7] queat. Propterea cùm illam in corpore naturali manifestè intueamur; ex ea colligimus alteram quoque, quam non intuemur, in eodem inesse.

9   Hinc fit ut huius quoque conclusionis cognitio sit confusa, quia solùm quòd insit corpori naturali materia invenimus atque cognoscimus, ipsius autem conditiones et naturam et definitionem

and contingent matter, has no weight, unless all the singulars are taken account of, with none overlooked. But demonstrative induction is performed in necessary matter and in things that have an essential connection to each other. In it, not all particulars are taken account of, since once some have been inspected, our mind at once notes the essential connection,[23] and then, leaving aside the singulars, at once gathers the universal. For it knows that it is necessary that it be so in the rest, for from the fact that some predicate belongs to something *per se*, it may be inferred that it is, therefore, predicated of all (*de omni*). When our mind considers, therefore, the nature of change or of generation, it knows that this predicate is essential to it and necessary, that is, that there is some underlying subject matter that is receptive to the change itself, just as the testimony of sense knew this to be so in the generation of accidents. Therefore from the various changes in accidents it infers this universal, "Every change has an underlying subject," under which it afterward gets this major, "Where there is generation, there is underlying subject matter."

Aristotle makes this discursive movement in a long discourse in the first book of the *Physics* but chiefly in text no. 62.[24] This is that passage's true interpretation, which few knew.   7

This knowledge of the major premise is nothing but confused, because, granted that the predicate is the cause of the subject, nevertheless it is not known as cause. For we knew that every change has underlying subject matter — not as an effect of matter itself, but as perpetually conjoined to the underlying subject matter. For we judge that change and matter are joined by a bond so necessarily that change without matter can never be discovered.[25] Therefore, since we manifestly see the former in natural body, we gather from it that the other, which we do not see, belongs to it also.   8

And thus it happens that knowledge of this conclusion too is confused, because we discover and know only that matter belongs to natural body but do not know its characteristics, nature, and   9

ignoramus. Huius autem ratio in promptu est, quoniam nulla res dat alteri id, quod ipsa non habet, effectus est confusè tantùm nobis cognitus, ideò non potest nobis tradere distinctam causae cognitionem. Notus enim ipse quòd est° non declarat nisi quòd causa sit; non tamen ut ipsius causa, quia ipse quoque non est notus ut illius causae effectus; sed solùm ut res quaedam ab illo nunquàm separabilis.

10   Ita igitur dicere omnino necessarium est, si rem ipsam diligenter expendimus, effectus enim confusè cognitus non reddit causam cognitam nisi confusè, causa verò confusè cognita, dum ignoratur quid sit, non potest cognosci ut causa.

11   Hic itaque est primus processus in regressu, quo solam invenimus inhaerentiam causae in subiecto proposito, non tamen prout illius effectus causa est, sed ut praedicatum quoddam necessarium et inseparabile.

: V :

Quòd facto primo processu non statim regredi
ad effectum possimus, sed mediam quandam
considerationem interponi necesse sit.

1   Causa ita inventa videretur statim ab ea regrediendum esse ad effectum demonstrandum propter quid sit. Attamen hoc nondum facere possumus cùm enim, ut modò dicebamus, nil det id, quod non habet, nos autem per regressum quaeramus cognitionem effectus distinctam, hanc nobis causa confusè tantùm cognita tradere non potest, sed eam prius distinctè cognitam fieri oportet, quàm ab ea ad effectum regrediamur.

definition. The reason for this is right at hand. The effect is known to us only confusedly and so cannot convey to us distinct knowledge of the cause, since nothing gives to another that which it does not itself have. That the known thing itself exists,° makes nothing clear except that there is a cause — though not as its cause, because it is not known as an effect of that cause, but only as something never separable from it.

It is, therefore, altogether necessary to say this, if we weigh the   10
issue carefully: An effect known confusedly does not render a cause known except confusedly; and a cause known confusedly, when what it is, is unknown, cannot be known as a cause.

And so this is the first procedure in a regressus, by which we   11
discover only the inherence of the cause in a given subject, not in that it is the cause of that effect, but as some sort of necessary and inseparable predicate.

: V :

That we cannot regress straightaway to the effect once the
first procedure has been performed, but that it is necessary
for some sort of intermediate consideration to be interposed.

After the cause has been discovered it would appear that there   1
could at once be a regressing from it to demonstrating what the effect is on account of (*propter quid*). But we cannot yet do this, since, as we just said, nothing can give that which it does not have, and by means of a regressus we inquire after distinct knowledge of the effect. The cause, known only confusedly, cannot convey this to us. It must first be made known distinctly, before we may regress from it to the effect.

2   Facto itaque primo processu, qui est ab effectu ad causam, ante-quàm ab ea ad effectum retrocedamus, tertium quendam medium laborem intercedere necesse est, quo ducamur in cognitionem dis-tinctam illius causae, quae confusè tantùm cognita est, hunc aliqui necessarium esse cognoscentes vocarunt negotiationem intellectus, nos mentale ipsius causae examen appellare possumus seu menta-lem considerationem, postquàm enim causam illam invenimus, considerare eam incipimus, ut etiam quid ea sit cognoscamus, qualis autem sit haec mentalis consideratio et quomodo fiat, à ne-mine vidi esse declaratum, quamvis enim aliqui dicant mediam hanc intellectus negotiationem interponi, tamen quomodo per eam ducamur in cognitionem causae distinctam et quae sit vis huius negotiationis, non ostenderunt. Non parvum igitur operae pretium faciemus, si de hac re aliqua dixerimus.

3   Duo sunt, ut ego arbitror, quae nos iuvant ad causam distinctè cognoscendam, unum quidem cognitio quòd est, quae nos praepa-rat ad inveniendum quid sit, quando enim in re aliquid praenosci-mus,° in ea aliquid aliud indagare et invenire possumus; ubi verò nihil praenoscimus,° nil unquàm inveniemus, ut docet Aristoteles in primo capite primi libri Posteriorum et ut optimè declarat ibi Themistius in expositione secundi contextus illius capitis. Quando igitur causam illam dari invenimus[8] apti sumus ad quaerendum et inveniendum quid illa sit.

4   Alterum verò, sine quo illud non sufficeret, est comparatio cau-sae inventae cum effectu, per quem inventa fuit, non quidem cognoscendo hanc esse causam et illum esse effectum, sed solùm rem hanc cum illa conferendo. Sic enim fit ut ducamur paulatim ad cognitionem conditionum illius rei et una inventa conditione

And so, the first procedure, which is from effect to cause, having been performed, before we go back from it [i.e., the cause] to the effect, it is necessary that there intercede some third intermediate effort by which we are led into distinct knowledge of that cause, which was known only confusedly. Some, knowing that this is necessary, called it negotiation of [the] understanding.[26] We can call it a mental examination of the cause or a mental consideration.[27] For after we discover that cause, we start to consider it, so as also to know[28] what it is (*quid est*). What sort of thing this mental consideration is and how it is done, I have not seen made clear by anyone. For even though some say that this intermediate negotiation of [the] understanding is interposed, they nevertheless have not shown how we are led by means of it into distinct knowledge of the cause and what the power of this negotiation is. We will do something much worth the work, therefore, if we say something about this.

There are two things, I think, that help us in knowing the cause distinctly. The first, of course, is knowledge *quòd*, which prepares us for discovering what it is (*quid est*). For when we know° beforehand that something is in a thing, we can track down and discover something else in it, but where we know nothing beforehand,° we will never discover anything, as Aristotle teaches in the first chapter of the first book of the *Posterior [Analytics]*,[29] and as Themistius clarifies in the best way there in what he laid out about text no. 2 of that chapter.[30] When we discover, therefore, that there is that cause, we are able to inquire after and discover what that is.

The other thing, without which the first would not suffice, is a comparison of the discovered cause with the effect by means of which it was discovered—not, of course, by knowing that this is the cause and that is the effect but only by comparing the latter thing with the former. For thus it happens that we are led little by little to knowledge of the characteristics of the former thing, and, once one characteristic has been discovered, we are helped to

iuvemur ad aliam[9] inveniendam, donec tandem cognoscamus hanc esse illius effectus causam.

5   Quae omnia eodem sumpto exemplo melius intelligentur. Ex generatione invenimus esse in corpore naturali materiam substantiali formae subiectam. Nondum quidem cognoscimus materiam ipsius generationis causam esse, sed solùm cognoscimus non posse mutationem illam fieri absque aliqua subiecta materia, proinde materiam à mutatione inseparabilem esse. Quòd si quis dicat, nonne materia invenitur ab Aristotele ut principium, ut patet legentibus primum librum Physicorum? Ergo quam primùm est inventa, cognoscitur ut causa illius effectus; ad hoc dicimus materiam inveniri quidem ut principium respectu corporis naturalis, sed non respectu generationis, nempè ut Aristoteles ibi eam inveniat ut causam generationis, hoc enim nondum cognoscitur in eo libro. Cuius quidem rei testimonium clarum habemus in primis verbis primi libri de generatione et interitu, ubi Aristoteles agendum sibi proponens de generatione amplissimè sumpta dicit inquirendas esse eius causas, indicat[o] ergo esse adhuc nobis incognitas generationis causas. Postea verò in eo primo libro causam perpetuae generationis quaerit, eamque esse dicit naturam primae materiae. In primo igitur libro Physicorum causae generationis non cognoscuntur; sed causae internae corpus naturale constituentes, corporis enim naturalis principia quaerenda proposuerat Aristoteles in eo libro, sed non generationis vel mutationis, huius enim causae declarandae proponuntur in ipso initio libri de Ortu et interitu.

6   Materia igitur per generationem inventa nondum cognoscitur esse causa generationis, quia quid sit ipsa materia adhuc ignoratur. Propterea Aristoteles, qui principiorum cognitionem nobis tradere voluit non modò confusam, verùm etiam distinctam, quantam naturalis philosophus habere potest, coepit materiae inventae

discover another, until finally we know that this is the cause of that effect.

This will all be better understood once the same example is 5 taken up. From generation we discover that in natural body there is underlying subject matter of the substantial form. We do not yet know, of course, that matter is the cause of generation itself; we know only that this change cannot happen without some underlying subject matter, and accordingly that matter is inseparable from change. Now if someone were to say—"Is matter not discovered by Aristotle as a beginning-principle, as is patent to those reading the first book of the *Physics*? Therefore what is first discovered is known as the cause of that effect."—to this we say: Matter, of course, is discovered as a beginning-principle with regard to natural body, but not with regard to generation, that is, as Aristotle discovers it there as the cause of generation, for this is not yet known in the book. We have clear testimony for this, of course, in the first words of the first book on generation and passing away,[31] where Aristotle, setting out to deal with generation taken in the widest sense, says that its causes have to be asked about. He therefore indicates° that the causes of generation are not yet known to us. And then afterward, in the first book, he asks after the cause of perpetual generation and says that it is the nature of first matter.[32] In the first book of the *Physics*, therefore, causes of generation are not known, but internal causes constituting natural body [are]. For in that book Aristotle had set out to inquire after the beginning-principles of natural body, but not [those] of generation or of change; for the causes of the latter are set out to be clarified in the very start of the book *On Coming to Be and Passing Away*.[33]

Matter discovered by means of generation, therefore, is not yet 6 known to be the cause of generation, because what matter itself is is as yet unknown. And so Aristotle, who wanted to convey to us not just confused knowledge of beginning-principles, but also distinct—to the extent that a natural philosopher can have it—began

conditiones et naturam investigare, idque fecit à contextu 57 eius libri usque ad 70, in quo inquit se docuisse quid sit materia.

7   In ea autem parte ad inveniendum quid sit materia ita progressus est, docuit in primis quomodo à privatione differat, materiae namque officium est substare contrariis, et ea recipere; contrariorum verò officium est se ab eadem materia mutuò pellere. Ideò materia manet sub utroque còntrario et nunquàm interit; et, cùm omnium naturalium corporum principium et materia esse debeat, debet etiam esse talis, quae omnes formas, et omnes privationes recipere apta sit. Itaque nulli formae, nulli certae naturae, nulli affectioni addicta esse debet secundùm propriam eius naturam, sed ab omnibus libera et immunis, quia intus apparens prohibet extraneum. Debet igitur materia secundùm suam naturam carere omnibus formis et omnium recipiendarum potestatem habere. Haec absque dubio est natura materiae ut nihil sit actu, sed potestate omnia, in hunc enim sensum cadunt omnia, quae ibi de primae materiae natura dicuntur ab Aristotele. Et inde colligitur ipsa materiae definitio, materia est primum subiectum, ex quo permanente naturalia omnia corpora fiunt.

8   Hoc igitur est illud mentale examen, seu negotiatio illa intellectus, quam facit ibi Aristoteles ad materiam distinctè cognosendam. Ad eam autem cognitionem nos duxit tum cognitio illa praecedens confusa quòd materia detur tum eius comparatio cum generatione, ex qua fuit inventa. Sic enim paulatim discimus quem locum in generatione habeat ipsa materia et quodnam eius officium sit, idque ab aliorum principiorum muneribus distinguimus. Officio cognito conditiones singulae in lucem prodeunt, quae ut fungi tali officio possit[10] necessariae sunt.

to investigate the characteristics and nature of the discovered matter, and he did so from text no. 57 of the book up to no. 70,[34] in which he says that he himself has taught what matter is.

In that part, moreover, to discover what matter is,[35] he pro-   7
gressed as follows. He taught in the first place how it differs from privation, for the function of matter is to underlie (*substare*) contraries and take them on, and the function of contraries is, in return, to impress themselves upon that same matter. Therefore matter remains under each contrary and never passes away, and, since it ought to be the matter and beginning-principle of all natural bodies, it also ought to be such that it is able to take on all forms and all privations. And so, according to its own proper nature, it ought to be committed to no form, no definite nature, no affection, but be unconstrained by anything and not bound by anything, because what appears within [that is such and such] bars the external [that is such and such]. Matter, therefore, according to its own nature, ought to lack all forms and have the potential to take on all [forms]. Without doubt this is the nature of matter — that it is actually nothing but potentially everything — for under this sense falls everything that is said there by Aristotle about the nature of first matter. And from that, the very definition of matter is gathered:[36] Matter is the first underlying subject, from which, persisting,[37] all natural bodies come to be.

This, therefore, is that mental examination or that negotiation   8
of [the] understanding that Aristotle performs there for knowing matter distinctly. And both the preceding, confused knowledge that there is matter, and its comparison with the generation from which it was discovered, led us to this knowledge. And thus we learn, little by little, what place matter itself has in generation and what its function is, and we distinguish that from the jobs of other beginning-principles. Once the function is known, each of the characteristics that are necessary for it [i.e., matter] to be able to execute such a function come to light.

9   Haec autem omnia postquàm cognovimus, atque perpendimus, facilè nobis innotescit talem materiam generationis causam esse. Quoniam enim potestatem habet recipiendi omnes formas et nullam certam formam sibi praescribit, sed aequè apta est recipere formam et eius privationem, ideò facit ut nullum materiam habens possit esse perpetuum, sed ex necessitate aliquando intereat et ex eo aliud generetur. Hoc considerare est regressum facere, qui est postremus progressus tradens nobis ex distincta causae cognitione distinctam cognitionem effectus, quam antea non habebamus, quando enim cognoscimus illam esse causam illius effectus, facimus propositionem maiorem et iamiam, si volumus, in promptu habemus demonstrationem propter quid, proposito namque subiecto, in quo materiam inesse novimus, quae est propositio minor, in corpore naturali inest materia, colligimus in eodem inesse generationem; et est potissima demonstratio.

10   Sed hunc ultimum processum non facit in eo primo libro Physicorum Aristoteles, quoniam eius consilium in eo libro non est generationis notitiam° tradere per suas causas, sed solùm ex generatione confusè cognita nos in primorum principiorum cognitionem perducere. Ideò satis habuit in eo libro primum illum processum facere ad primam materiam inveniendam, quae est cognitio confusa, deinde etiam mentale illud examen, quo distinctam quoque materiae cognitionem adipisceremur.

11   At in primo libro de Generatione et interitu regressum facit à materia iam distinctè cognita ad generationem distinctè cognoscendam, eaque est demonstratio propter quid.

12   Ex tribus igitur partibus necessariò constat regressus, prima quidem est demonstratio quòd, qua ex effectus cognitione confusa ducimur in confusam cognitionem causae. Secunda est consideratio illa mentalis, qua ex confusa notitia° causae distinctam eiusdem cognitionem acquirimus. Tertia verò est demonstratio potissima,

Now after we have come to know and have examined all this, it easily becomes known to us that such matter is the cause of generation. For since it has the potential of taking on all forms, arrogates to itself no determinate form, and is equally able to take on a form and its privation, it therefore makes it that anything having matter cannot be perpetual, but out of necessity sometimes perishes, and from it something else is generated. To consider this is to perform the regressus, that is, the final progression, conveying to us, from distinct knowledge of the cause, distinct knowledge of the effect, which we did not have earlier. For when we know that that is the cause of that effect, we form the major premise, and then, if we want, we have right at hand a demonstration *propter quid*. For once the underlying subject to which we know matter belongs has been set out—this is the minor premise, "Matter belongs to natural body"—we gather that generation belongs to it also; and this is a demonstration *potissima*.

But Aristotle does not perform this last procedure in the first book of the *Physics*, because his intent in that book is not to convey knowledge° of generation by means of its causes, but only to lead us through to knowledge of first beginning-principles from generation known confusedly. And so he had enough in that book to perform the first procedure for discovering first matter—this is confused knowledge—and then also that mental examination by which we may also obtain distinct knowledge of matter.

But then in the first book of *On Generation and Passing Away*, he performs a regressus from matter now known distinctly to knowing generation distinctly, and this is a demonstration *propter quid*.[38]

Regressus, therefore, is necessarily composed out of three parts. The first, of course, is demonstration *quòd*, by which we are led from confused knowledge of the effect to confused knowledge of the cause. The second is that mental consideration, by which we acquire distinct knowledge of the cause from confused knowledge° of it. And the third is a demonstration *potissima*, by which we are

qua ex causa distinctè cognita ad distinctam effectus cognitionem tandem perducimur. Primam ac secundam partem in exemplo proposito habemus in primo libro Naturalis auscultationis, tertium verò in primo libro de Generatione et interitu.

13 Ex his autem, quae diximus, manifestum esse potest non posse benè cognosci quòd haec sit causa huius effectus, nisi natura et conditiones illius causae cognoscantur, per quas apta est talem effectum producere. Hoc enim ad propositae dubitationis solutionem magni momenti est.

: VI :

In quo ea, quae de regressu dicta sunt,
declarantur alio sumpto exemplo
ex octavo libro Physicorum.

1 Haec omnia, quae de regressu atque eius partibus diximus, possunt aliis pluribus exemplis è naturali philosophia sumptis declarari et comprobari. Nobis autem alterum solum accipere satis sit, in quo tota rei veritas apertissimè conspicitur. Idque sumitur ex octavo libro Naturalis auscultationis, quem locum diligenter considerare neque inutile neque à suscepta provincia alienum erit, quandoquidem non ea solùm, quae ab Aristotele aliisque probatis authoribus de rebus logicis scripta sunt, expendere decrevimus, sed etiam illa, quae in philosophiae et aliarum disciplinarum traditione ad logicae artis usum pertinentia observari possunt, notare et examinare; praesertim quando in iis locis ipsum artificium logicum à nullo alio declaratum ac fortasse etiam à nullo intellectum fuisse invenerimus. Quod equidem de his duobus, quibus uti constitui,

finally led on from the cause known distinctly to distinct knowledge of the effect. We have the first and second parts in the example set out in the first book of the *Nature Lectures* [i.e., the *Physics*][39] and the third in the first book of *On Generation and Passing Away*.[40]

So now, from all that we have said, it can be manifest that it 13 cannot be known well that this is the cause of this effect, unless the nature and characteristics of that cause, by means of which it is able to produce such an effect, are known. And this is of great moment for solution to the proposed doubt.

: VI :

In which the things that have been said about
regressus are clarified by another example,
taken from the eighth book of the Physics.

All these things that we have said about regressus and its parts can 1 be clarified and confirmed with many other examples taken from natural philosophy. But it is enough for us to take up just one other; in it the whole truth of the issue is seen most plainly. It is taken from the eighth book of the *Nature Lectures* [i.e., the *Physics*]. To consider the passage carefully will be neither useless nor foreign to the task undertaken. For we decided to weigh not only that which was written about logical issues by Aristotle and other proven authors, but also to note and examine those things that can be observed pertaining to the use of logical arts in conveying philosophy and other disciplines, especially when we discovered in those passages that the very craft of logic was not made clear by anyone else and perhaps not even understood by anyone. For my part I dare to claim this with respect to the two examples that I

exemplis audeo profiteri, nam tum ea, quae de regressu ex primo libro Physicorum diximus, tum illa, quae mox de eodem ex octavo libro dicturi sumus (seu vera seu falsa sint, quod quidem aliis iudicandum relinquo) ego primus inveni.[11]

2   In illo igitur octavo libro Aristoteles ex aeterno motu, quem in principio eius libri dari° ostenderat, demonstrat dari primum motorem aeternum et est demonstratio quòd confusam nobis tradens primi aeterni motoris cognitionem, scilicet quòd sit, non quid sit.

3   Sed statim adversus hoc dubitare non iniuria quispiam posset, videtur enim statim cognosci etiam quid sit, si vera illa sunt, quae paulò superius diximus, causam non posse cognosci ut causam, nisi prius quid ea sit cognoscatur, quo fit ut, si causa ut causa cognoscatur, eadem sit etiam cognita quid sit. Sed primus motor simulatque inventus est ex aeterno motu, est etiam cognitus ut motor, ergo ut causa, quia dicere motorem est dicere causam effectricem motus, hoc autem videtur adversariis maximè suffragari, cùm videatur eodem processu simul demonstrari et causam esse et propter quid effectus sit, quia causa statim invenitur prout causa est.

4   Attamen, si benè rem hanc expendamus, id minimè verum est, nam est quidem motor causa effectrix motus, sed aeterni motus, qui est species motus, propria causa non est motor latè sumptus, sed motor certis quibusdam conditionibus praeditus, per quas vim habet movendi perpetuò, his igitur conditionibus ignoratis non cognoscitur primus motor esse causa aeterni motus, nisi universaliter et confusè. Hae autem primi motoris conditiones per primum

have decided to use. I was the first to discover both the things we said about regressus using the first book of the *Physics* and also those that we will soon say about it using the eighth book. (Whether they are true or false, of course, I leave for others to judge.)

In that eighth book, therefore, from eternal motion, which in 2 the beginning of the book he had shown to exist,°[41] Aristotle demonstrates that there is an eternal first mover. This is a demonstration *quòd*, conveying to us confused knowledge of the first, eternal mover, namely that it is the case (*quòd*), not what it is (*quid est*).

But someone could at once and not unjustly raise a doubt 3 against this. For it appears that what it is (*quid est*), is also known at once, if what we said a little earlier is true, that a cause cannot be known as a cause unless what it is (*quid est*), is known first. Hence if a cause is known as a cause, what the same [cause] is (*quid est*), is also known. But as soon as the first mover is discovered from eternal motion, it is also known as a mover and therefore as a cause, because to say "mover" is to say efficient cause of motion. Now it appears that our opponents are fully in favor of this, since both that the cause is and what the effect is on account of (*propter quid*) appear to be demonstrated at the same time by the same procedure, because the cause is at once discovered in that it is a cause.

But nevertheless, if we weigh the issue well, this is not at all 4 true. For, of course, a mover is an efficient cause of motion, but the proper cause of eternal motion, which is a species of motion, is not a mover taken in the broad sense, but a mover endowed with some definite characteristics by means of which it has the power to move [other things[42]] perpetually. As these characteristics are unknown, therefore, the first mover is not known to be the cause of eternal motion, except universally and confusedly. And these characteristics of the first mover do not become known by means

illum, quem diximus, processum non innotescunt, ideò nec statim regredi possumus ad demonstrandum propter quid sit ipse aeternus motus, quia motor est adhuc confusè cognitus. Propterea necesse fuit Aristotelem facere mentalem illam considerationem seu negotiationem intellectus, qua conditiones essentiales primi motoris in lucem prodirent, si non omnes, saltem aliquae, scilicet illae, quas ad cognoscendam causam aeterni motus cognoscere necessarium erat. Facit hoc Aristoteles in particula 52 eius libri, dum ostendit primum motorem, qui aeterni motus causa sit, immobilem penitus esse debere, ut neque per se neque ex accidenti sit mobilis, quoniam videmus motores, qui per se sunt immobiles, ex accidenti autem mobiles, non posse movere perpetuò. Deinde in particula quinquagesimatertia regressum facit à motore primo aeterno immobili ad motum primum aeternum et est demonstratio propter quid.

5   Caeterùm ad plenam rei, de qua loquimur, intelligentiam non est silentio praetereundum artificium maximum Aristotelis in eo libro à nemine cognitum,° quod nos primi, dum eum librum publicè interpretaremur, decimo ab hinc anno patefecimus. Quaerunt multi rationem tractationis de impartibilitate primi motoris, quam in postremo eius libri capite legimus. In hac una plurimi consentiunt, scopus erat praecipuus Aristotelis in eo libro agere de primo aeterno motore, ideò caput illud ultimum est veluti totius libri coronis et apex, in quo Aristoteles declarat summas et praestantissimas primi motoris conditiones, ad quas naturalis philosophus poterat pervenire. Quare apud eos caput illud ultimum est finis et scopus Aristotelis in eo octavo libro.

6   Sed horum sententia longè abest à veritate et à consilio Aristotelis, cuius scopus non est agere de primo aeterno motore, sed solùm de aeterno motu, hic enim est res naturalis et accidens naturale, ille verò est res supernaturalis, cuius consideratio ad solum metaphysicum pertinet. De aeterno autem motore loquitur

of that first procedure that we spoke of. Therefore we cannot at once regress to demonstrating what eternal motion itself is on account of (*propter quid*),[43] because the mover is still known confusedly. It was, therefore, necessary for Aristotle to perform that mental consideration or negotiation of [the] understanding, by which the essential characteristics of the first mover came to light —if not all [of the characteristics], at least some, that is, those that it was necessary to know so as to know the cause of eternal motion. Aristotle does this in passage no. 52 of the book,[44] where he shows that the first mover, which is the cause of eternal motion, ought to be completely immobile, so that it is mobile neither *per se* nor by accident, for we see that movers that are immobile *per se* but mobile by accident cannot move [other things] perpetually. Then in passage no. 53,[45] he performs the *regressus* from immobile, eternal, first mover to eternal first motion, and this is a demonstration *propter quid*.

Now for a full understanding of the issue we are speaking about, Aristotle's great skill in that book, recognized° by no one, is not to be passed over in silence. We were the first to uncover this, when we made public comments on the book, ten years ago.[46] Many inquire after the reason for the treatment on indivisibility of the first mover that we read in the book's final chapter.[47] In this one thing most agree: Aristotle's principal goal in the book was to deal with the first, eternal mover. So, that last chapter is like the whole book's crown and apex, in which Aristotle clarifies the greatest and most excellent characteristics of the first mover that the natural philosopher could reach. Therefore, for them, that last chapter is Aristotle's end and goal in the eighth book.

But their position is far from the truth and from Aristotle's intent. His goal is not to deal with the first, eternal mover, but only with eternal motion. For the latter is a natural thing and a natural accident, while the former is something supernatural; consideration of it pertains only to the metaphysician.[48] Moreover, the

philosophus in eo octavo libro propter aeternum motum, non propter se, quia non poterat aeternus motus sciri, nisi per suam aequatam causam demonstraretur, causa autem est motor aeternus. Scopus igitur eius libri non est cognoscere primum motorem, sed perfectè cognoscere aeternum motum per suam causam, ut fuit Alexandri sententia à Simplicio relata; et ut ipse Aristoteles in ipso eius libri initio testatur.

7   Idem confirmat epilogus, quem ante illud ultimum caput in 77 contextu legimus, in eo enim Aristoteles colligit se id tractasse, quod proposuerat, scilicet ostendisse motum aeternum esse et quis ille sit, et quae sit causa perpetui motus et eam esse motorem immobilem. Caput itaque ultimum, quod post eum epilogum legitur, scriptum est ab Aristotele quasi extra primarium eius consilium in eo libro; et ratio, quae ipsum movit, fuit haec, cognovit Aristoteles illud mentis examen leviter et imperfectè factum esse in contextu 52 ideoque non benè cognitam esse illam primi motoris conditionem, per quam est aptus movere perpetuò, proinde debilem[12] et infirmum fuisse regressum à causa non benè cognita ad effectum. Conditio enim primi motoris, per quam potest facere aeternum motum, est haec, quòd est abiunctus à materia, ita enim fit ut in movendo non fatigetur, et perpetuò moveat. Conditionem hanc Aristoteles in contextu 52[13] valdè leviter attigit, solùm enim dixit primum motorem non esse ullo modo mobilem neque per se neque ex accidenti, quod significabat ipsum esse à materia separatum, nam si esset iunctus materiae, non posset saltem ex accidenti non moveri ad motum materiae. Hoc tamen Aristoteles exprimere non ausus est, quia vidit esse conditionem supernaturalem et

philosopher speaks about the eternal mover in the eighth book for the sake of eternal motion, not for its own sake, because eternal motion could not be known scientifically unless it was demonstrated by means of its own coextensive cause, and the cause is the eternal mover. The goal of the book, therefore, is not to know the first mover, but to know eternal motion perfectly by means of its own cause, as was Alexander's position, related by Simplicius,[49] and as Aristotle himself attests in the very start of the book.[50]

The summary, which we read before that last chapter in text no. 77,[51] confirms the same thing. For in it Aristotle gathers that he treated what he had set out [to treat], namely, that he showed that there is eternal motion, and which [motion] that is, and what the cause of perpetual motion is, and that this [cause] is an immobile mover. So the last chapter, which is read after this summary, was written by Aristotle as if outside his primary intent in the book, and the reason that moved him was this: Aristotle knew that in text no. 52[52] the examination by the mind was performed lightly and imperfectly, and therefore that the characteristic of the first mover, by means of which it is able to move [other things] perpetually, was not well known. Accordingly, the regressus from the cause not well known to the effect was inadequate and weak. For the characteristic of the first mover by means of which it can bring about eternal motion is this, that it is separated from matter. For thus it happens that in moving [other things] it does not wear out, and it moves [other things] perpetually. In text no. 52, Aristotle touched on this characteristic very lightly. For he said only that the first mover is not in any way mobile, either *per se* or by accident; this indicated that it is separated from matter. For if it were joined to matter, it could not, at least by accident, not be in motion with motion of the matter. But Aristotle did not dare to expressly say this, because he saw that this characteristic was supernatural and

7

transcendentem limites Physicos, ideò cum moderatione eam insinuare satis habuit, ut modus loquendi, quo ibi utitur, demonstrat, nam dicit credibile esse, quòd primus motor aeternus, cùm semper moveat, non sit ullo modo mobilis neque per se neque ex accidenti. His enim paucis verbis mentalem illam considerationem absolvit.

8   Postea verò facto regressu, et tota de aeterno motu tractatione completa Aristoteles, qui in illo brevi mentis examine non plenè sibi satisfecerat, constituit in calce eius libri, ac veluti extra eum librum et praeter ipsius consilium illam mentalem considerationem facere pleniorem et apertè docere primum motorem esse impartibilem et à materia penitus separatum, idque facit conferendo ipsum cum effectu, ex quo innotuerat, nempè cum aeterno motu, cùm enim semper moveat, non potest esse materialis; quia nullus motor materialis potest perpetuum motum efficere. Huius quidem tractationis proprius locus erat ille contextus 52 sed Aristoteles data opera ac tanquàm excusationis gratia usus est hac artificiosa inordinatione, ut significaret se non ignorasse tractationem illam excedere limites naturales et ut appareret eam neque praetermissam fuisse neque penitus immistam tractationi naturali.

9   Neque id reprehensione ulla dignum est, quandoquidem sicuti rerum omnium, quae in universo sunt, admirabilis est, colligatio et nexus et ordo; ita in scientiis contingere necessum fuit, ut colligatae essent et mutuum sibi auxilium praestarent. Divina quidem scientia quantum à naturali iuvetur in libro duodecimo Metaphysicorum legere possumus. Naturalis autem à divina accipit primam causam et eam reddit modo quodam naturalem, quatenus eam considerat ut causam accidentium naturalium, nempè in octavo libro Physicorum ut causam aeterni motus et in tertio libro de

transcended the limits of the *Physics*. He therefore took it as enough to introduce it with moderation, as the way of speaking that he uses there demonstrates. For he says that it is believable that the eternal first mover, although it always moves [other things], is not in any way mobile, neither *per se* nor by accident. And with these few words he completed that mental consideration.

But afterward, the regressus having been performed and the 8 whole treatment on eternal motion having been completed, Aristotle, who had not fully satisfied himself with that brief examination by the mind, decided, at the end of the book,[53] and as if outside the book and beyond his intent, to perform that fuller mental consideration and to teach plainly that the first mover is indivisible and completely separated from matter. He does this by comparing it with the effect by which it had become known, that is, with eternal motion. For since it always moves [other things], it cannot be material, because no material mover can effect perpetual motion. The proper place for this treatment was, of course, that text no. 52, but Aristotle used this artificial disorder deliberately and as an excuse so that he could indicate that he was not ignorant that this treatment exceeded the limits of the natural and so that it would not appear to have been either overlooked or completely mixed up with the treatment of natural things.

Nor is this deserving of any censure, since just as the connec- 9 tion, bond, and order of all things that are in the universe are admirable, so, it was necessary to touch [on these] in the sciences, so they might be gathered together and furnish each other with mutual aid. In the twelfth book of the *Metaphysics*, of course, we can read how much the divine science is helped by the natural. And moreover, the natural accepts first cause from the divine and renders it in some way natural, insofar as it [i.e., the natural science] considers it [i.e., the first cause] as the cause of natural accidents—in the eighth book of the *Physics* as the cause of eternal

Anima ut mentis humanae illuminatricem et causam nostrae intellectionis. Utroque autem in loco Aristoteles coactus est aliquas supernaturales eius conditiones attingere, leviter tamen id fecit et magna cum moderatione et absque diligenti earum declaratione, ut quisque iudiciosus atque eruditus vir utroque in loco inspicere potest. Nos enim satis superque de his locuti sumus. Haec autem omnia cum diligentia explicare voluimus, ut tres illae in regressu necessariae partes optimè cognoscerentur.

10   Alia quoque naturalia exempla perpendere possemus, quae aliis consideranda et cum his, quae declaravimus, conferenda relinquimus.

: VII :

Quòd tres memoratae partes in omni regressu si non semper
tempore,[14] *saltem natura distincta reperiantur.*

1   Quoniam autem ostensum est tres esse partes in regressu hoc ordine dispositas, ut prima sit demonstratio causae ex effectu; secunda sit causae inventae consideratio; tertia demum demonstratio eiusdem effectus ex illa eadem causa; sciendum° est ac summoperè annotandum, nihil esse absurdi si quandoque contingat has tres partes tempore distinctas non esse. Satis enim est si ratione et natura distinguantur. Eodem temporis momento incipit homo esse rationis particeps et esse risibilis, ratio tamen haec diversa esse cognoscit, etsi tempore non separantur.

2   Cùm igitur contingere possit ut aliqua causa, cùm primùm inventa est, absque ullo temporis intervallo perfectè nota fiat seu

motion and in the third book of *On the Soul*[54] as the illuminator of
the human mind and the cause of our understanding. In each passage, Aristotle was forced to touch on some supernatural characteristics of it, but he did so lightly, with great moderation,
and without a careful clarification of them, as any judicious and
learned man can observe in each passage. We have now spoken
more than enough about these things. But we wanted to explicate
all of them carefully so that those three necessary parts in a regressus might be known optimally.

We could also examine other natural examples, but we leave   10
them for others to consider and compare with these that we have
made clear.

: VII :

That the three parts referred to are found distinct in every
regressus, if not always in time, at least by nature.

Now since it has been shown that the three parts in a regressus are   1
disposed in this order — that is, the first is a demonstration of
cause from effect, the second is a consideration of the discovered
cause, and lastly the third is a demonstration of the same effect
from that same cause — it has to be understood° and most diligently noted that it is nothing absurd if it sometimes happens that
these three parts are not distinct in time. It is enough if they can
be distinguished by reason and by nature. In the same moment of
time, [a] man starts to participate in reason and to be risible. Nevertheless although these are not separated in time, reason knows
they are different.

Since, therefore, it can happen that some cause, when it is first   2
discovered, becomes perfectly known without any interval of time,

propter ipsius evidentiam seu propter ingenii nostri perspicacitatem, fortasse aliquis existimare posset unam atque eandem demonstrationem ea, quae diximus, omnia simul praestare, quae adversariorum sententia fuit dicentium eadem demonstratione ostendi et causam esse et propter quid sit effectus. Id tamen minimè verum est, quoniam unius rei una est demonstratio et una demonstratio unius rei est, non plurium.

3   Hoc autem ut melius intelligatur, tali exemplo utamur, particulari quidem ac sensili, qualibus Aristoteles quoque uti solet, sed tamen ad rem declarandam idoneo et accommodato, cùm idem in rebus universalibus postea quisque considerare possit. Ex inspectione fumi argumentatur aliquis et ostendit ignem ibi esse, quem non videt, ea una demonstratione videtur demonstrari et ignem ibi esse et cur fumus fiat. Quod tamen verum non est, illum enim decipit celeritas intellectus in horum cognitione, quoniam demonstratio ignis ex fumo suapte natura non declarat nisi ignem esse. Quòd si videmur etiam cognoscere propter quid sit ille fumus, id non est per eandem demonstrationem, sed per alteram, qua ex igne fumum demonstramus et rationem reddimus cur ibi sit fumus. Qui duo discursus vel ita parvo ac insensili tempore distinguuntur, ut eodem momento fieri videantur. Vel etiam si adversariis condonemus utrumque eodem tempore fieri, saltem ratione secernuntur, aliud enim est ex fumo ignem, aliud est ex igne fumum demonstrare.

4   Hoc igitur dicendum est quando eveniat ut causa ex effectu inventa statim cognoscatur illius effectus causa esse, per hoc enim ipsa regressus natura non labefactatur. Quoniam enim in illa

either on account of its evidentness or on account of the perspicacity of our wit, someone could perhaps judge that one and the same demonstration furnishes at the same time everything that we said. This was the position of our opponents, saying that both that the cause is and what the effect is on account of (*propter quid*) are shown by the same demonstration. But this is not at all true, since of one thing there is one demonstration, and one demonstration is of one thing, not of many.

So that this may be better understood, let us use an example, a 3
particular and sensible one, of course, of the sort that Aristotle normally uses also, but [one] nevertheless fitting and well accommodated to making this issue clear, since afterward anyone can consider the same in universal things. From the inspection of smoke, someone argues and shows that there is fire there that he does not see. By this one demonstration, it appears to be demonstrated both that there is fire there and why smoke comes to be. But this is not true. For in knowing these [two things], the speed of [his] understanding deceives him. For the demonstration of fire from smoke, by its very nature, makes clear nothing except that there is fire. And if we appear also to know what that smoke is on account of (*propter quid*), it is not by means of the same demonstration but by means of another, [one] by which we demonstrate smoke from fire and render the reason why there is smoke there. These two discursive movements are distinguished either by such a little and insensible time that they appear to happen at the same moment, or, even if we grant to our opponents that both happen at the same time, they are at least separated by reason. For to demonstrate fire from smoke is one thing, [to demonstrate] smoke from fire is another.

This has to be said, therefore, when it happens that a cause 4
discovered from an effect is at once known to be the cause of that effect; the very nature of regressus is not undermined by this.

media consideratione mentali, per quam innotescere diximus quid sit illa causa, ad hanc cognitionem ducimur per collationem° causae inventae cum effectu, ex quo inventa fuit; ideò contingere potest ut dum causam esse demonstramus ex effectu, simul etiam quid illa sit inveniamus. Attamen aliud est cognoscere causam esse,° aliud est cognoscere quid sit et demum aliud est scire propter quid sit effectus ille. Has enim omnes cognitiones si non tempore, saltem natura et ratione distinctas esse necesse est.

: VIII :

De differentiis, quibus regressus
à circulo discrepat.

1 Per ea, quae dicta sunt, si benè declarata à nobis est natura regressus, facile erit tum regressus et circuli ab Aristotele reprobati discrimen inspicere tum dubia omnia et argumenta contraria solvere.

2 Differt regressus à circulo et ratione formae et ratione materiae et ratione finis. Primùm quidem ratione formae, quoniam circulus (ut nomen ipsum significat) est transitus ab eodem ad idem, instar figurae Geometricae, à qua per translationem nominatus est. Quoniam igitur in primo processu non ab una propositione, sed à duabus ad conclusionem progredimur, oportet ex eadem conclusione vicissim utramque propositionem monstrari, non alteram solùm, si circulus esse debeat.° Oportet itaque prima conclusione monstrata

For since in that intermediate mental consideration, by means of which we said that what the cause is becomes known, we are led to this knowledge by means of a comparison° of the discovered cause with the effect from which it was discovered, it can, therefore, happen that when we demonstrate *that* a cause is from an effect, we may at the same time discover what that [i.e., the cause] is. But nevertheless it is one thing to know that the cause exists,° another to know what it is (*quid est*), and yet another to know scientifically what that effect is on account of (*propter quid*). For it is necessary that, if not in time, at least by nature and by reason, all these [types of] knowledge are distinct.

: VIII :

*On the differentiae by which regressus
is different from the circle.*

By means of what we have said, if the nature of regressus has been 1
well clarified by us, it will be easy both to observe the discriminating difference between regressus and the circle condemned by Aristotle and to do away with all doubts and contrary arguments.

Regressus differs from the circle by reason of form, by reason of 2
matter, and by reason of end. First, of course, [it differs] by reason of form, since the circle (as the name itself indicates) is a passage from one thing to the selfsame thing, just like the geometric figure after which, figuratively, it was named. And since, therefore, in the first procedure we progress to the conclusion not from one premise but from two, if there is° to be a circle, then each premise, not just one, must be demonstrated in turn from the same conclusion. And so, once the first conclusion has been demonstrated, one must take it as a [new] major premise, and then by joining to it

eam ut propositionem maiorem sumere et ei adiecta minore conversa demonstrare maiorem; deinde eandem ut minorem accipiendo, eique maiorem conversam apponendo demonstrare minorem. Sic enim ex conclusione propositio utraque demonstratur et is est verè circulus, quemadmodum declarat Aristoteles in tertio capite primi libri Posteriorum, ubi circularem demonstrationem refellit et fusius in secundo libro Priorum Analyticorum, ubi ex pluribus quoque demonstrationibus circulum perfectum constare asserit, sed nobis tres[15] considerasse satis sit.

3   In regressu autem non utramque propositionem ab initio acceptam demonstramus, sed solam minorem, ut docet Aristoteles in capite decimo primi libri Posteriorum, postquàm enim causam ex effectu ostendimus, maiorem propositionem convertimus, cui adiecta conclusione loco minoris concludimus minorem illius prioris demonstrationis.

4   At ex minore conversa et conclusione maiorem demonstrare nullo pacto possumus, tum quia in maiore praedicatur de effectu causa sua proxima et immediata, quare maior indemonstrabilis est; tum quia in minore praedicatur accidens de subiecto, ideò si minor convertatur, fit praedicatio contra naturam, quod de propositione maiore non evenit, ea enim sine ulla absurditate converti potest ob eam rationem, quam aliàs in libro nostro de Propositionibus necessariis fusè declaravimus.

5   Formae igitur ratione[16] regressus non est circulus, quia non ad idem penitus revertitur, cùm non utramque propositionem demonstret, sed alteram solùm.

6   Sed neque ratione[17] materiae, materia namque demonstrationis est medius terminus, ex quo utraque propositio conflatur, ut ait Aristoteles in 48 contextu secundi libri Posteriorum et ut ibi declarat optimè Themistius. Materia quidem circuli tota similis est

the minor converted, demonstrate [what was] the major, and then, by accepting the same [i.e., the first conclusion] as a [new] minor and adjoining to it the major converted, demonstrate [what was] the minor. For thus each premise is demonstrated from the conclusion; this truly is a circle, just as Aristotle makes clear in the third chapter of the first book of the *Posterior* [*Analytics*],[55] where he refutes circular demonstration, and at greater length in the second book of the *Prior Analytics*,[56] where he asserts that a perfect circle is composed out of even more demonstrations. But let it be enough for us to have considered three.

Now in regressus, we do not demonstrate each premise given at the start, but only the minor, as Aristotle teaches in the tenth chapter of the first book of the *Posterior* [*Analytics*].[57] For after we show the cause from the effect, we convert the major premise; and once the conclusion is joined to it [i.e., the major premise] in place of the minor, we conclude the minor of that prior demonstration.

But in no way can we demonstrate the major from the converted minor and the conclusion, both because its own proximate and immediate cause is predicated of the effect in the major — and so the major is indemonstrable — and because in the minor the accident is predicated of the subject — and so if the minor were converted, the predication would be contrary to nature. This does not happen with the major premise; it can be converted without any absurdity for the reason that we made clear at length elsewhere, in our book *On Necessary Premises*.[58]

By reason of form, therefore, regressus is not a circle, because it does not completely return to the same [point]; it does not demonstrate each premise, but only one.

But it is not [a circle] by reason of matter, also. For the matter of a demonstration is the middle term; from it each premise is assembled, as Aristotle says in text no. 48 of the second book of the *Posterior* [*Analytics*][59] and as Themistius makes optimally clear there.[60] The whole matter of a circle, of course, is similar and of

et eiusdem generis, quoniam in utroque progressu medium est proxima causa maioris extremi, uterque enim est potissima demonstratio. Tota igitur circuli progressio est à causa proxima ad effectum, proinde est per viam totam similem, qualis est circumferentia figurae circularis, notato enim in ea puncto si per eam aliquid ab illo puncto moveatur, transit per viam, quae tota eiusdem generis est donec ad idem punctum revertatur.

7   At in regressu via non tota similis est, prior enim processus ab effectu fit, posterior verò à causa. Non est igitur tota eiusdem generis ea materia, ex qua fit regressus, cùm medium in primo processu sit effectus; in secundo autem sit causa, quare non est circulus, quia in circulo medium in utroque processu est causa.

8   Demum ex fine distinguuntur, quia finis circuli est scientia eadem, à qua sumptum fuit primum demonstrationis initium, ut si propositiones sint A conclusio verò B tam processus prior ab A ad B quàm posterior ab B ad A quemadmodum diximus, est potissima demonstratio, in qua nulla cognitio locum habet nisi perfecta et distincta, nam à causa distinctè cognita progreditur ad effectum distinctè cognoscendum. Quare prior processus est ab A distinctè cognito ad B distinctè cognoscendum et posterior similiter à B distinctè cognito ad A distinctè cognoscendum. Quocirca idem prorsus in circulo est finis, atque principium, nam à primo ad ultimum transitus est ab A distinctè cognito ad A distinctè cognoscendum, quae verè est circuli natura.

9   Non sic regressus, sed finem habet diversum à principio, ut optimè notavit Alexander referente Ioanne Grammatico in contextu 11 primi libri de Anima,[18] finis enim est scientia distincta effectus, quae dicitur scientia propter quid est, à qua non fuit

the same genus, since in each progression, the middle [term] is the proximate cause of the major extreme, for each is a demonstration *potissima*. The whole progression of the circle, therefore, is from the proximate cause to the effect. Accordingly, its complete path is just like the circumference of a circular figure. For once a point is marked on it, if something is moved along it from that point, that thing passes along a way that is wholly of the same kind until it returns to the same point.

But in a regressus, the whole way is not the same, for the first 7 procedure occurs from the effect, but the latter from the cause. The whole matter, therefore, from which the regressus occurs, is not of the same genus, since the middle [term] in the first procedure is the effect, but in the second it is the cause. Accordingly it is not a circle, because in a circle the middle [term] in each procedure is the cause.

Lastly, they [i.e., regressus and circle] are distinguished by end, 8 because the end of a circle is the same scientific knowledge from which the first start of the demonstration was taken. If the premises are A and the conclusion B, just as the first procedure is from A to B, so the latter is from B to A, as we said. This is a demonstration *potissima*; in it no knowledge except perfect and distinct has a place. For there is progression from the cause known distinctly to knowing the effect distinctly. And so the first procedure is from A known distinctly to knowing B distinctly, and, similarly, the latter is from B known distinctly to knowing A distinctly. And so in a circle, the end and the beginning are utterly the same. For from first to last the passage is from A known distinctly to knowing A distinctly. This truly is the nature of a circle.

Regressus is otherwise, for it has an end different from the be- 9 ginning, as Alexander (via a reference in John the Grammarian) noted so very well in text no. 11 of the first book of *On the Soul*.[61] For the end is distinct scientific knowledge of the effect, which is said to be scientific knowledge *propter quid*.[62] The beginning of the

sumptum primae[19] demonstrationis principium, primus enim processus fuit ab effectu confusè cognito ad causam confusè cognoscendam; postremus verò à causa distinctè cognita ad effectum distinctè cognoscendum, quare à primo ad ultimum non est ab eodem ad idem, quamvis enim sit ab effectu ad eundem effectum, tamen finis demonstrationis non est effectus ipse, sed eius scientia, haec autem non eadem est in fine regressus et in principio, quandoquidem à notitia° ipsius effectus confusa exordium sumitur, totus autem regressus in cognitione eiusdem desinit perfecta et distincta, est igitur à primo ad ultimum progressus à cognitione confusa ad distinctam eiusdem effectus scientiam, qui nihil in se absurditatis habet, quia naturalis est nobis haec progrediendi ratio à confusa ad distinctam eiusdem rei cognitionem. Igitur ratione[20] finis non potest regressus appellari circulus.

: IX :

In quo adversariorum argumenta solvuntur.

1   Ad argumenta adversariorum responsio ex iis, quae dicta sunt, facilè sumitur, ad primum enim, quod erat, dato regressu sequeretur idem esse eodem notius et ignotius, dicimus totum argumentum esse concedendum, id enim, si alio et alio modo sit, non est inconveniens, nam in primo processu effectus est notior causa, in postremo autem causa est notior effectu, cum hoc tamen discrimine,

first demonstration was not taken from this. For the first procedure was from the effect known confusedly to knowing the cause
confusedly, but the final [procedure was] from the cause known
distinctly to knowing the effect distinctly. Therefore, from first to
last, it is not from one thing to the selfsame thing. For even
though it is from an effect to the same effect, nevertheless the end
of the demonstration is not the effect itself, but scientific knowledge of it. And this [i.e., scientific knowledge] is not the same at
the end of the regressus and at the beginning, since the beginning
is taken from confused knowledge° of the effect itself while the
regressus as a whole ends in perfect and distinct knowledge of the
same [effect]. From first to last, therefore, the progression is from
confused knowledge to distinct scientific knowledge of the same
effect. This has no absurdity in it, because this way of progressing
from confused to distinct knowledge of the same thing is natural
to us. By reason of the end, therefore, regressus cannot be called a
circle.

: IX :

In which the arguments of our opponents are done away with.

From the things that have been said, a response to the arguments   1
of our opponents[63] is easily had. To the first — which was that, in
a given regressus, it would follow that [one] selfsame thing would
be [both] more known and more unknown than [another] selfsame thing — we say that the whole argument has to be conceded,
but it is not inconsistent if it is [more known] in one way and
[more unknown] in another. For in the first procedure, the effect
is more known than the cause, and in the final the cause is more
known than the effect, but with this discriminating difference:

quòd effectus est notior causa cognitione confusa, at causa est notior effectu cognitione distincta. Nec dicimus illum esse nobis notiorem, hanc verò non nobis, sed natura. Sed asserimus ambo nobis notiora esse, diversis tamen modis, effectum quidem confusè, causam verò distinctè. Et sicuti per effectum confusam causae cognitionem adipiscimur, ita per causam ad distinctam effectus cognitionem pervenimus.

2   Ad secundum negamus regressum esse processum ab eodem ad idem, est enim à confusa ad distinctam eiusdem rei cognitionem, à cognitione quòd sit ad cognitionem propter quid sit. Circulus autem est à distincta cognitione ad distinctam, quare est prorsus ab eodem ad idem. Patet igitur duo haec argumenta ab Aristotele contra circulum adducta regressui non officere.

3   Postremum argumentum erat, in priore processu vel notum nobis est causam esse illius effectus causam, vel ignotum. Si notum, ergo prior processus non modò demonstrat causam esse,[21] verùm etiam propter quid sit effectus, quare posterior processus supervacaneus est. Si ignotum, ne priore quidem processu uti possumus, cùm maior propositio sit ignota. Nec magis dicere licet, fumus est, ergo ignis, quàm fumus est, ergo asinus, cùm ignoremus causam fumi ignem esse.

4   Ad hoc dicimus ignorari à nobis causam esse illius effectus causam, dum priore processu utimur. Cùm autem obiiciunt, ergo priore processu uti non possumus, quia maiorem propositionem ignoramus, negandum est hoc, fallax enim est ratio à secundùm quid (ut vocant) ad simpliciter. Nam est quidem illa maior aliqua ratione ignota, non tamen simpliciter et omnino ignota, sed aliquo

The effect is more known than the cause by confused knowledge, but the cause is more known than the effect by distinct knowledge. We do not say that the former is more known to us, and the latter not to us, but by nature. Instead we assert that both are more known to us, but in different ways — the effect confusedly, the cause distinctly. And just as we obtain confused knowledge of the cause by means of the effect, so by means of the cause we arrive at distinct knowledge of the effect.

In response to the second [argument], we deny that regressus is 2 a proceeding from one thing to the selfsame thing. For it is from confused to distinct knowledge of the same thing, from knowledge *quòd* to knowledge *propter quid*. But the circle is from distinct knowledge to distinct; therefore it is utterly from one thing to the selfsame thing. It is, therefore, patent that these two arguments adduced by Aristotle against the circle are not detrimental to regressus.

The final argument was that in the first procedure either it is 3 known to us that the cause is the cause of that effect, or it is unknown. If it is known, then the first procedure not only demonstrates that the cause is, but also what the effect is on account of (*propter quid*) — whereby the latter procedure is superfluous. If it is unknown, we cannot even use the first procedure, since the major premise is unknown. We may no more say, "There is smoke, therefore there is fire," than, "There is smoke, therefore there is an ass," since we do not know that the cause of smoke is fire.

To this we say, when we use the first procedure, it is unknown 4 to us that the cause is the cause of that effect. But then when they object, "Then we cannot use the prior procedure, because we do not know the major premise," this has to be denied. For reasoning *a secundum quid ad simpliciter* ("from a certain respect to absolutely") is fallacious.[64] For, of course, that major [premise] is in some way unknown, but not absolutely and altogether unknown;

etiam modo cognita quantum satis est ad demonstrationem ab effectu construendam.

5   Similem errorem obiecit Aristoteles Platoni et quibusdam sophistis in primo capite primi libri Posteriorum, illi namque fallacibus argumentis nixi scientiam non dari ostendebant, quia non cognoscebant medium dari quoddam inter perfectam rei cognitionem et perfectam ignorantiam, media quippe inter utramque est cognitio confusa, qua posita illorum cavillationes corruunt.

6   Hunc eundem errorem adversariis obiicere possumus, dicunt enim, vel notum nobis est ignem esse causam fumi, dum maiorem illam sumimus, ubicumque est fumus, ibi est ignis, vel ignotum. Sic enim perfectam notitiam° et perfectam ignorationem solas accipiunt; unde sequitur, si perfectè sit cognitum, iam esse cognitum propter quid est fumus absque ope secundi processus. Si verò prorsus incognitum, nil ex ea maiore sic ignorata posse demonstrari. Quae nos utraque concedimus, quandoquidem neque omnino ignoramus coniunctionem et nexum harum duarum rerum, fumi et ignis, sed eam antea novimus, neque plenam eius scientiam habemus, quia adhuc ignoramus ignem esse causam fumi. Confusam igitur habemus fumi cognitionem, non distinctam, quare per priorem illum processum non demonstramus nisi inhaerentiam causae in subiecto, non dum novimus eam esse illius effectus causam.

7   Neque nos hoc exemplum perturbet, quòd videatur ante illam demonstrationem notum nobis esse, ignem esse causam fumi. Quoniam ea non est demonstratio, sed syllogismus quidam particularis exempli gratia ad rei declarationem adductus, quo ostenditur

it is at least known in some way, as much as is enough for constructing a demonstration *ab effectu*.

Aristotle objected to a similar error in Plato and some sophists, 5 in the first chapter of the first book of the *Posterior [Analytics]*.[65] For they, relying on fallacious arguments, showed that there is no scientific knowledge, because they did not know that there is something in the middle between perfect knowledge of something and perfect ignorance. The middle between them is, of course, confused knowledge. Once this is posited, their sophistry collapses.

We can object to this same error in our opponents. For they 6 say, either it is known to us that fire is the cause of smoke when we take up that major [premise], "Wherever there is smoke there is fire there," or it is unknown. They thus accept only perfect knowledge° and perfect ignorance, and from this it follows that if it is known perfectly, then what smoke is on account of (*propter quid*) is known without the work of the second procedure. And if it is utterly unknown, then from the major thus unknown nothing can be demonstrated. Both of these we concede, since neither are we altogether ignorant of the conjunction and bond between these two things, smoke and fire — we did know them earlier — nor do we have full scientific knowledge of them, because we as yet do not know that fire is the cause of smoke. We have, therefore, confused and not distinct knowledge of smoke. And thus by means of that first procedure we do not demonstrate [anything] except the inherence of the cause in the subject, when we did not yet know that it is the cause of that effect.

Let this example not disturb us. It could appear that before that 7 demonstration it is known to us that fire is the cause of smoke. This, however, is not a deonstration, but some sort of particular syllogism, adduced for making something clear for the sake of an example, in which it is shown that there is this fire that we do not

hunc ignem dari, quem non videmus, non ostenditur ignem simpliciter dari, id enim notum est et ignem esse causam fumi.

8   Propterea si exemplo veriore utamur, nullus remanebit adversariis cavillandi locus. Sumamus igitur demonstrationem Aristotelis
in primo libro Physicorum, quae talis est, ubi est generatio, ibi est
subiecta materia, at in corpore naturali est generatio, ergo in eodem est materia. Ante hanc demonstrationem non est nobis notum quòd materia sit causa generationis, nec tamen prorsus ignota
est illa propositio maior, ubi generatio, ibi materia, sed confusè
cognita, praenoscimus° enim esse cum omni mutatione coniunc-
tum ex necessitate subiectum aliquod, sed nondum ut ipsius mutationis causam. Idque satis est ad demonstrandum quòd in corpore
naturali materia insit. Quando enim duarum rerum nexum cognoscimus et earum alteram in subiecto aliquo existere inspicimus,
colligimus alteram quoque in eodem inesse. Demonstratio namque
ab effectu non simpliciter causam esse ostendit, ut dicamus, effectus est, ergo causa est, sed horum duorum connexu in maiore
propositione accepto ostendimus in subiecto aliquo tertio causam
inesse, propterea quòd in eodem inest effectus, quemadmodum in
libro nostro de speciebus demonstrationis copiosè declaravimus.

see. It is not shown that there is fire absolutely (*simpliciter*) — for that is known — nor that fire is the cause of smoke.

And so if we use a truer example, no place for sophistry will   8 remain for our opponents. Let us take, therefore, Aristotle's demonstration in the first book of the *Physics*,[66] which is like this: Where there is generation, there is underlying subject matter; and in natural body there is generation; therefore, in it also there is matter. Before this demonstration it is not known to us that matter is the cause of generation. Nevertheless, that major premise, "Where there is generation, there is matter there," is not utterly unknown; it is known confusedly. For we know° beforehand that some underlying subject is out of necessity conjoined with every change but not yet as the cause of the change itself. And this is enough for demonstrating that matter belongs to natural body. For when we know the bond between two things, and we observe that one of them exists in some subject, we gather that the other belongs to that same thing also. For demonstration *ab effectu* (from effect) does not show that there is a cause absolutely (*simpliciter*), as [when] we say, "There is an effect, therefore there is a cause." But the connection of these two having been accepted in the major premise, we show that the cause belongs to some third subject and therefore that the effect belongs to the same thing, just as we made plentifully clear in our book on the species of demonstrations.[67]

Note on the Text and Translation

※§?※

The Latin text is based on a full collation of the two editions published in Venice during Zabarella's lifetime,

A   [Editio Prima]. Venice: Paulo Meietti, 1578.
B   Editio Secunda. Venice: Paulo Meietti, 1586.

and two posthumous editions published in Germany,

C   Editio Tertia. Cologne: Lazarus Zetzner, 1597.
E   Editio Postrema. Frankfurt: Lazarus Zetzner, 1608.

A third posthumous German edition was consulted wherever a difference in the other four was discovered:

d   Editio Quarta. Cologne: Lazarus Zetzner, 1603.

These posthumous editions are not linear descendants of either of the Venice editions. They descend presumably from the edition published by Jean Mareschal in 1586–87, which has not been collated.*

    The text printed here generally follows B. The differences, though small, between it and A suggest the changes were made by

---

* Maclean, "Mediations of Zabarella," 47, states: "It can be deduced also from the same preface that Pace did not consult the author himself in advance about the publication of his works" [including the *Opera logica*]. The list of editions in the Bibliography is heavily indebted to Maclean's study. Copies of the Mareschal 1587 edition can be found at the Huntington Library and at the Bayerische Staatsbibliothek; the latter has made the edition available online in its digital collection.

the author himself. Punctuation here also follows the second edition, although some discretion has been exercised. None of the consulted editions had paragraph breaks; these have been added in keeping with the editorial practice of the I Tatti Renaissance Library. Generally, these breaks follow what look typographically in the source like boundaries between very long sentences. Semicolons in the Latin have been changed to periods where the English translation warrants it. Serial commas were regularly used in lists of only two items, and these have been silently removed.

In the sources, *et* was always rendered as an ampersand; it has here been expanded. The periods after Arabic numerals and after capital letters when used as symbols have been removed. The way numbers were presented in the source texts varied and these have been silently rationalized.

In the sources, the conjunction *quum* was always distinguished from the preposition *cum*. The first has been rendered here as *cùm*. Grave accents, such as those distinguishing *quòd* (adverb, conjunction) from *quod* (pronoun), *quàm* (adverb, conjunction) from *quam* (pronoun), and *secundùm* (preposition) from *secundum* (adjective) have been retained and silently rationalized in the few cases where it was deemed necessary. These distinguishing marks are often helpful in determining the sense and are occasionally disambiguating. An exception has been made for *-iùs/-ius*. Later editions attempted, however inconsistently, to use the accent to distinguish adjectives and adverbs of the comparative. The first edition did not, and that edition's practice has been adopted here.

Some editions used a circumflex to mark the genitive plural in the phrase *nostrûm cognoscentium*. These endings have silently been expanded to *-orum* and the phrase translated as "our knowing." The circumflex was sometimes used to distinguish the adverb *hîc* (here) from the demonstrative *hic* (this) and has here been retained.

Acute accents, used with enclitics, as *-ne*, *-am*, and *-que*, have been removed.

Spellings silently changed include *secu-* for *sequu-*, *-sid-* for *-syd-*, and *-m-* for *-n-* in compounds such as *quanvis* and *nanque*. Alternate spelling, such as *qua propter/quapropter* and *eapropter/propterea* have been silently rationalized. Banal typographical errors have been silently corrected.

Other variations are noted in the apparatus.

THE TRANSLATION

Zabarella's works were part of a tradition of carefully worded philosophical commentary and were part of an international dialogue conducted in multiple languages. To preserve the precision of the commentary and properly situate the texts in that dialogue, the translation here is rather literal, even wooden.

Attempts have been made to retain ambiguities. When insertion of an article could alter meaning and might not actually be warranted, the insertion is marked (e.g., "habit of [the] understanding"). Pronouns whose resolution is apparent in the Latin because of, say, gender, are similarly marked as insertions. Others are left unresolved.

Great care has been taken with vocabulary, with two goals in mind. The first goal is to employ terminology that would not be far from what Zabarella's English readers and discussants used in their own writings on the same subjects. Consequently, the reader of this translation may find old-fashioned uses of some words. For example, "to dispose" here means to arrange, not to discard; a disposition is an arrangement; "vain" means empty. "Speculative sciences" are theoretical ones, as opposed to practical; there is no suggestion that the science is unfounded or merely hypothetical, as the term might nowadays suggest. In the early seventeenth cen-

tury, the translation of *prima materia* was "first matter," it was not yet "prime matter," and so the former is here used. Other words of later origin, especially technical ones such as "deduction," have been avoided.

The second goal is a highly consistent correspondence between Latin and English terms. For example, *cognoscere* is translated "to know" (as in Zabarella's time), even when "to understand" or "recognize" might seem better to our ear. "To understand" is reserved for *intelligere*. *Ratio* has a few regular translations, chosen based on context: "reason" or "reasoning" when standing alone, "rational way of . . ." in phrases such as "rational way of ordering." Cognates are common, such as "manifest" for *manifestus*, "apparent" for *apparens*. Throughout, exceptions and irregular choices are marked with a superscripted ring on both Latin and English; e.g., *rationibus ita apparentibus° nixi* (they relied on reasons so specious°).

It is hoped that reading the translation here will feel no more unfamiliar than reading a comparable text written by an English logician of the early seventeenth century.

Terms of particular importance include the following.

via	way
methodos	method
inferre	infer
tradere	convey (more literally: deliver, pass on)
ducere/duci	lead/be led
inducere	induce
deducere	deduce

Some associations here that would be plain to Zabarella are inevitably lost in English. Greek *hodos* is a path or way; the compound *meth-odos* a pursuit along a path. Zabarella draws attention to the connection between Greek *methodos* and Latin *via* at *Meth. l. 1 c. 2. par. 1, l. 1 c. 7 par. 4, l. 3 c. 2 par. 2.* For Zabarella, "to infer"

(*inferre*, to carry from point to point) and "to be led" (*duci*) have strong literal associations with "way" and "method," while "deduce" (used very seldom), "induce," and "lead" (used often in the passive) have stronger affinities than the English would suggest. The phrase "method by which we are led to knowledge" (*Meth.* l. 4 c. 10 *par.* 3) has for Zabarella a naturalness, even an inescapability, that is easy for us to overlook. A similar association is echoed in his frequent statement that knowledge is conveyed by method.

deducere	deduce
inferre/illatio	infer/inference
concludere/conclusio	conclude, is conclusive, be a conclusion/ conclusion
colligere/collectio	gather/gathering

Of these four, *colligere* is by far the most common and has the widest range of meanings. It is always translated here as "gather" but occasionally — and especially in *Meth.* l. 3 — seems to refer specifically to deductive reasoning. For Zabarella, the conclusion of a syllogism is "gathered" from its premises. And it may be that a *collectio* (gathering) sometimes means what we would now call a "deduction," a much later term. Note also use of the term in *Meth.* l. 4 c. 13.

declarare	make clear, clarify
declaratio	clarification

The senses are very close to "explain" and "explanation," but the alternatives used here keep close the association with *clarus* (clear), a property of some mental content and a property central to Zabarella's epistemology (and later Descartes').

ratio	reason, reasoning, rational way (as in *ratio ordinandi*, rational way of ordering)

Zabarella frequently uses *ratio* with a gerund in the genitive, as *ratio ordinandi*. The bare meaning is "the way" or "the manner" of doing something, but "right, proper, and rational" is generally implied. The usage places Zabarella squarely within humanistic discussions of *ratio dicendi*, *ratio docendi*, etc., the proper manner of speaking, the right way of teaching, etc. Even *ratio* by itself has a sense of "the correct way of reasoning."

disponere   dispose
dispositio   disposition

The sense is "to arrange," not "to discard." "Disposition" is synonymous with "order."

tractare   treat
tractatus   treatise
tractatio   treatment, tract

A *tractatio* is both an examination of some topic and the body of text containing that examination. So it can be a passage, a section of a book, or even a complete book on a specialized topic.

cognoscere   know
noscere°   know°

Holyoake explains that *noscere* means "to know" and *cognoscere* means "to come to know." Zabarella uses the first very seldom and the second very often. Though *cognoscere* is here translated simply as "to know" (exceptions are noted), it should be remembered that the term does carry a sense of "coming to know," rather than just knowing, and indeed the meaning always hovers close to our "recognize." Remembering this is vital for understanding Zabarella's epistemology generally.

cognitio	knowledge
notitia°	knowledge°

In Zabarella's day, "knowledge," not "cognition," was the standard translation for *cognitio*, and the practice is maintained here. It seems to accord well with his meaning (though *cognitio* always carries some sense of "recognition" as well; see above, s.v. "cognoscere"). Zabarella uses *notitia* much less frequently. English "knowledge" is used for both, though instances of *notitia* are marked as exceptions. It is left to the reader to determine whether Zabarella intends a difference in meaning. *Regr. l.* 9 *c.* 5–6 suggest he does not; *Meth. l.* 3 *c.* 1–2 may suggest he does. Compare also his commentary on the *Posterior Analytics l.* 1 *c.* 18 *t.* 134. Schicker used *erkenntnis* for both.

scire	know scientifically, have scientific knowledge
scientia	a science, the science, scientific knowledge
sciendum° *est*	it has to be understood° that
intelligere/intellectus/ intellectualis	understand/understanding/ of understanding

Zabarella almost always uses *scire* in the narrow sense corresponding to Aristotle's *epistasthai* and *scientia* in the sense of *epistēmē* (or the *epistēmē* of a particular subject). See Zabarella's commentary on *Posterior Analytics l.* 1 *c.* 2 71b9–15, *Comm. Post. An. l.* 1 *t.* 7. "Understand" is reserved for *intelligere* (with exceptions noted), though occasionally the sense is "to understand the meaning to be, to mean."

prior/notior nobis	prior/more known to us
prior/notior natura	prior/more known by nature

prior/notior secundùm naturam	prior/more known according to nature

In the dichotomy *notius natura* vs. *notius nobis*, which translates Aristotle's γνωριμώτερον τῇ φύσει (*gnōrimōteron tēi phusei*) vs. γνωριμώτερον πρὸς ἡμᾶς (*gnōrimōteron pros hemas*), it can appear that both *natura* and *nobis* are ablative. But at least for Zabarella, only the first is; the second is dative. In English the terms are "more known *by* nature" and "more known *to* us."

Moreover, Zabarella criticizes those who took "more known by nature" to indicate that nature was the knower. "Nature," he says, "knows nothing." The phrase should instead be understood to mean known by us in a way "that maintains the order of nature." Seeking, presumably, to distance himself from the common misconception, he generally uses *notior secundùm naturam* (more known according to nature) when speaking in his own voice and *notior natura* (more known by nature) when citing or paraphrasing his predecessors, even though he indicates that the two are synonyms.

For explications of Zabarella's position, see especially his commentary on *Posterior Analytics l.* 1 *c.* 2 71b33–72a7, *Comm. Post. An. l.* 1 *t.* 12, but also *Meth. l.* 4 *c.* 10 and *De Speciebus Demonstrationis c.* 17.

demonstratio quòd	demonstration *quòd*	*quòd*: that something
cognitio quòd	knowledge *quòd*	is the case
notitia° quòd	knowledge° *quòd*	

demonstratio quid est	demonstration *quid est*	*quid est*: what something is
cognitio quid est	knowledge *quid est*	thing is
notitia° quid est	knowledge° *quid est*	

demonstratio propter quid (est)	demonstration *propter quid*	*propter quid:* what something is on account of
cognitio propter quid (est)	knowledge *propter quid*	
notitia° propter quid (est)	knowledge° *propter quid*	
demonstratio à signo	demonstration *a signo*	*a signo:* from an indicator or sign
demonstratio ab effectu	demonstration *ab effectu:*	*ab effectu:* from an effect
demonstratio quia	demonstration *quia:*	*quia:* because
demonstratio potissima	demonstration *potissima*	*potissima:* of the strongest sort
demonstratio simpliciter	demonstration *simpliciter*	*simpliciter:* absolute

As early as Zabarella's own time, *potissima demonstratio* has been translated "perfect demonstration" or "most perfect demonstration," but this is misleading. As Zabarella explains in *Meth. l.* 3 *c.* 4 and *De speciebus demonstrationis* 12, *potissima demonstratio* translates the Greek *kurios apodeixin* and means the same as demonstration *propter quid.* Greek *kurios* means "powerful or strong."

A demonstration *propter quid,* an "on account of which" demonstration, is not a demonstration *of* a cause. The cause is in the premises of the syllogism as a middle term; it is not in the conclusion.

In *Meth. l.* 3 *c.* 4, Zabarella says that demonstration *a signo* and demonstration *quia* are the same and in *l.* 3 *c.* 16 uses *a signo* and *ab effectu* interchangeably. In *Regr.*, he explains that a demonstration *quod* is a demonstration *ab effectu.*

Zabarella uses *demonstratio simpliciter* only in *Regr. c.* 3, where he says it is demonstration from what is better known in accord with nature.

Note, especially for understanding *Meth. l.* 4, that "demonstration" and "demonstrative method" are not synonyms. Only some demonstrations qualify as demonstrative method.

certè	certainly
certus	certain, definite

Care should be taken not to read into the term too much difference between instances translated as "certain" and those translated as "definite." Literally, the word refers to what is seen and discerned. Holyoake says it has extended meanings such as "true," "necessary," and "manifest," "because what we discern is clear and indubitable." Zabarella has no reservation about using the term in the comparative and the superlative. He uses it both for "certain knowledge" and "a definite order," but the meanings are not as different for him as they are for us.

perfectus	perfect

The sense is nearly always "perfect and complete."

axioma	axiom
dignitas	axiom

In *Comm. Post. An. l.* 1 *t.* 14 and the index to *Opera Logica*, Zabarella says these two mean the same thing; the first term is the Greek, the second the Latin.

principium   beginning (as of a chapter)
          beginning-principle (as a premise in a syllogism)

Translators conventionally choose "beginning" or "principle" as the context suggests. But this masks the role in Zabarella of a principle as an ontological beginning. If this is not constantly borne in mind (especially in *l.* 2), it is often difficult to appreciate Zabarella's point or even just the meaning of some sentences. So, for *principium,* "beginning" for, say, the beginning of a chapter, is here used, but "beginning-principle" when "principle" would be a more traditional translation. The strategy parallels the emerging preference for the hyphenated "starting-point" among translators of *principium*'s Greek equivalent, αρχη (*archē*).

in (eo) quod quid est   because of what (it) is

This translates the Greek ἐν τῷ τί ἐστι.

proportione   in comparative relation

The term should probably be considered against the background of medieval theories of the analogy of proportion and proportionality.

temperies   temper
temperatura   balance

These are synonymous technical terms in medieval and Renaissance medicine referring to a balance of elements, humors, or properties. To maintain the difference, two English words are used, but Zabarella may or may not have thought there was any substantive difference.

species   species
genus   genus, kind

differentia   differentia, difference

In Zabarella's time the English phrase was "genus and difference" not "genus and differentia." John Locke used "differentia" in 1691; he did not need to explain the term, but he did ask his reader's indulgence for using it. Here, "genus" and "differentia" are used when the context involves the relationship between genus and species; "kind" and "difference" otherwise. But the distinction is not always sharp and would not have been a strong one for Zabarella and his readers.

discrimen   discriminating difference
distinctio   distinction

For Zabarella, generally, *discrimen* is ontological, *distinctio* is epistemological. A *discrimen* exists in the world; a *distinctio* is made by us. A *discrimen* is the difference by which one genus differs from another. A *distinctio* is the product of a cognitive act, the act of distinguishing, the act of recognizing a *discrimen*. When we distinguish, we make a distinction; it is we who distinguish A and B. It is not the difference that distinguishes them.

accidentia   accident
affectio   affection

In his commentary on *Posterior Analytics* l. 1 t. 57, Zabarella says that for Aristotle, "accident" and "affection" mean the same thing and says that if there is any useful distinction (as Averroës held) it would be that affections are accidents more known. But whether formally there is any difference, Zabarella consistently uses *affectio* when presenting an accident as a term in the conclusion of a syllogism in which a beginning-principle and an underlying subject are the other two terms. In the works here, Zabarella uses *accidentia* very seldom.

prima materia	first matter
prima motor	first mover

The conventional sixteenth- and early seventeenth-century English renderings are used. None of the terms were typically capitalized, and none are here. Only later did "prime matter" and "prime mover" gain currency.

immobilis motor	immobile mover

The Latin *immobilis* translates Aristotle's *akinēton*. Both the Greek and Latin can mean "immovable," "motionless," or "unmoved." Dictionaries in Zabarella's time always gave the first and sometimes the second, but many commentators then and now have preferred the third. The ambiguity is here retained.

philosophicus	philosopher
divinus philosophicus	philosopher of divinity
primus philosophicus	philosopher of metaphysics
metaphysicus	metaphysician, metaphysical
Metaphysicos	the *Metaphysics*

A *metaphysicus*, "metaphysician," is presumably the same as a *primus philosophicus*, "first philosopher," "philosopher of first philosophy," or here "philosopher of metaphysics," but a distinction is retained nonetheless. Zabarella does not use *theologia* or *theologicus*.

scriptor	writer
author	author

The meanings of *auctor* and *author*, what in earlier times were simply alternate spellings of one word, were distinguished in Renaissance dictionaries. The first suggested an originator, the second an authority, and Zabarella consistently uses the second. This is unsurprising given that he believes scientific learning should be

done largely by reading and studying established texts, and he is usually discussing such texts. The sense of authority should always be read into his term.

habitus   habit

A habit (*habitus*) is, generally, something one has (*habet*) in mind, and only specifically a mental disposition or customary practice. For Zabarella, knowledge one has of a certain subject matter is a habit. The ambiguity between the general and the specific meanings is central to Zabarella's argument in *l.* 1 *c.* 2–3.

anima   soul
animus   man's soul, our soul, my soul, someone's soul, etc.
mens   mind

Feminine *anima* translates Greek *psyche*, as in the title of Aristotle's book *De Anima* (*On the Soul*). The word refers to the soul in any animate being, the bundle of vital capacities that allow it to function as a living being. Masculine *animus* is the rational soul, the part or function of *anima* in man that thinks. In some phrases, Zabarella uses *animus* and *mens* interchangeably, but in others he sticks strictly to one or the other. For example, he frequently explains that, after it has obtained the conception it seeks, our soul (*animus*) rests (*acquiescit, conquiescit,* or *quiescit*). In Locke and Hobbes, it is the mind, not the soul, that rests. This suggests "mind" as a translation of *animus*. But there is surely an indication here of a substantive difference in psychological ontology between the late Renaissance and the early modern, so "mind" is here reserved for *mens*.

Notes to the Text

ॐᏚᏫᏚॐ

BOOK III

1. *A lacks* in qua significatione
   . . . de Anima
2. cognoscendi] cognoscendis
   d E
3. differt] differre *A*
4. *B, C, d, and E lack* Aristoteles
5. docturus] dicturus *d E*
6. libro secundo] lib. 2 *C d*]
   librum secundum *A B. This*
   suggests E derives from C, d, or
   a mutual source.
7. declarationem] declaratione *B*
8. alias] aliàs *B*
9. eam] etiam *d E*
10. *B, C, d, and E lack* quidem
11. *A lacks* penu
12. quot sint] quod sint *d E*
13. sunt] sint *A*
14. eius] ei *A B*
15. sumamus genus] sumamus
    igitur genus *A*
16. medium] medicum *A*
17. negavit] negat *B C d E*
18. hominis] homini *all editions*
19. *B, C, d, and E lack* ut

20. ignoratur] ignorantur *A*
21. erroris causa] erroris in
    causa *A*
22. potest] valet *A*
23. *A lacks* omnes
24. viam] etiam *A*
25. ea] id certè in *A*
26. ordo] ordo est *A*
27. *A lacks* aliisque sequentibus
28. *B, C, d, and E lack* esse
29. ignotum] notum *C d E*
30. cognito] cognitio *B*
31. definitionem] definitiones
    E d
32. non] nota *A*
33. *A lacks* Cùm enim logica
    . . . calidum sine calore. *C,*
    d, and E lack dicta.
34. igitur] enim *A*
35. idque] id *d E*
36. vellemus] velimus *A*
37. *C, d, and E lack* primo
38. loquimur] loquitur *A*
39. sumenda] servanda *A*
40. *A lacks* quoniam ipsum quid
    est . . . ut dictum est
41. ea] et *A*

42. *A lacks* Demonstratio enim
    est . . . quoad significatio-
    nem.
43. sufficit] sufficiat *B*
44. *B lacks* ea
45. *C, d, and E lack* est. *It is pre-
    sent in* AOAC.
46. tum manifestum est] mani-
    festum est tum *A*

47. ex] in *d E*
48. sint] sunt *A*
49. demonstrentur] demonstre-
    tur *A*
50. qua] quia *E*
51. *B, C, d, and E lack* omnibus
52. causa] in causa *A*
53. quia] quin *C d E*

BOOK IV

1. *corrected:* tractari *edd.*
2. ego] ergo *A*
3. *A lacks* in
4. causa] in causa *A*
5. enunciationes, verum vel fal-
   sum] enunciationes verum,
   vel falsum *A*] enunciationis
   verùm, vel falsum *B*
6. *B, C, d, and E lack* ut est medium
7. ipsum] ipsam *A*
8. possimus] valeamus *A*
9. ipsae] ipse *A*
10. colligunt] colligit *d E*
11. considerarent] conside-
    rent *A*
12. illi dicunt] illi tractari di-
    cunt *C d E. The addition ap-
    peared in neither A nor B. In
    A,* tractari *in the subsequent
    clause appeared just below* illi.
13. tonitrum] tonitru *all editions.
    The variant was not uncom-
    mon at the time, though this is*

*Zabarella's only use of it in
these works.*
14. aliquam aliam] aliam ali-
    quam *C d E*
15. *B lacks* de
16. quod] quòd *C d*
17. demonstrativam] definiti-
    vam *A*
18. effectus] affectus *E*
19. causae] causa *E*
20. possimus] valeamus *A*
21. naturam nota] non *B*
22. *B lacks* in
23. respectu] habito respectu *A*
24. insensilia] insensibilia *C d E*
25. *B, C, d, and E lack* est
26. significandi] significans
    C d E
27. sine] sive *A*
28. *A lacks* rectè
29. *B, C, d, and E lack* Aristote-
    les
30. Primaria] Primum *A*
31. *C, d, and E lack* de definito

32. adjectam] adjectum *A*
33. *B, C, d, and E lack* ideò
34. *A lacks* ut . . . Posteriorum
35. traditionem] definitionem *C d E*
36. *B, C, d, and E lack* quidem
37. causa est] in causa est *A*
38. cuiusnam] cuiusdam *d, E*
39. *C, d, and E lack* tum
40. Averrois] Aristotelis *C d E*
41. plurima] plura *C d E*
42. *C, d, and E lack* cognitionem
43. *d and E lack* re
44. sive] sine *A*
45. *A lacks* Averroes in . . . Posteriorum
46. *C, d, and E lack* ea

47. consideratur] consideratum *d E*
48. volentes] credentes *A*
49. et omni] et in omni *C d E*
50. τῇ ἀκριβείᾳ] τῇ ἀκριβεία *A B*
51. praestet] praestat *E*
52. secundum] secundùm *C*
53. secundùm quod] secundùm quòd *A B*
54. secundùm quod] secundùm quòd *A B*
55. manifestam] manifestum *C d E*
56. fecit] facit *C d E*
57. inficiari] inficiaris *A*
58. universae] universe *C*

DE REGRESSU

1. iudicio] indicio *E*
2. concedit] concedis *d E*
3. *A lacks* eandem sententiam . . . libri
4. *One of only two instances of* valere *in* De Methodis *and* De Regressu *not changed to something else after the first edition.*
5. ipsam] ipsum *d E*
6. ex effectu noto causam ignotam] proinde ex effectu noto causam ignotam declarat *A*
7. inveniri] evenire *E*
8. invenimus] inveniemus *d E*
9. aliam] alium *A*

10. possit] valeat *A*
11. inveni] inveniri *A*
12. debilem] mobilem *E*
13. 52] 70 *A*
14. *A lacks* tempore
15. tres] res *C d E*
16. ratione] habita ratione *A*
17. ratione] habita ratione *A*
18. *A lacks* ut optimè . . . de Anima
19. primae] primum *E*
20. igitur ratione] habita igitur ratione *A*
21. causam esse] causam non esse *E*

Note on References

For everything in *Opera Logica*, Zabarella's point of reference is the corpus (or what was thought to be the corpus) of Aristotle. Zabarella's *Commentary on the Two Books of the Posterior Analytics* itself is three-quarters as big as everything else in the *Opera* combined, and the other parts, including *On Methods* and *On Regressus*, are largely commentaries on specialized topics in scholastic Aristotelian philosophy.

In several ways, the Aristotelian corpus as Zabarella knew it differs from the corpus as we know it. One difference is that in Zabarella's community of colleagues and readers, the corpus circulated with Averroës' commentaries interleaved with blocks of Aristotle's writing. In many works, each block, called a *contextus*, had a number, here translated as "text no." and abbreviated *t*. For the works that had these numbers, they were the standard mechanism by which passages were cited.

To make it easier to correlate Zabarella's comments with other parts of his writings, with the writings of other commentators, and with published Renaissance printings, citations to Aristotle and to Averroës are to the divisions as Zabarella knew them. The modern Bekker, book, and chapter numbers that correspond to those sections are then given in parentheses. The citation does not attempt to identify, narrowly, the passage Zabarella is referring to, but rather identifies the full extent of the Renaissance unit in which the cited passage lies. Hence, "*Posterior Analytics l*. 2 *t*. 36–47 (93a1–19 = *l*. 2 *c*. 8–10)" indicates that the passage Zabarella cites and knew as *liber* 2, *contextus* 36 through 47, corresponds to Bekker 93a1–19 and that this precisely matches Bekker Book 2, chapters 8–10. "*On Interpretation c*. 1–4 (16a1–17a37 = *c*. 1–6)" indicates that what Zabarella knew as chapters 1 through 4, we know as 1

through 6. "*On the Soul l.* 1 *t.* 3 (402a7–10 in 1.1)" indicates that *liber* 1, *contextus* 3 corresponds to a passage, Bekker 402a7–10, that lies within what we know as Book 1, chapter 1. For the *Posterior Analytics*, boundaries in Zabarella's own edition (*Comm. Post. An.*) are presumed authoritative; for other titles, *AOAC* is used. For citations to Plato, the parenthetical reference is to Stephanus pagination. For citations to Galen, it is to Kühn pagination; for Avicenna, the numbering in Bakhtiar.

Volume 1 of *AOAC* has three parts. Though *AOAC*'s table of contents does not indicate this, the numbering of folios restarts partway through each of the three. The first numbered sequence in part 1 is here cited as *pt.* 1a, the second as *pt.* 1b; the first sequence in part 2 as *pt.* 2a and the second as *pt.* 2b. The second pair matches the labels used in the binding of the 1962 reprint by Minerva.

Though citations here are to *AOAC*, Zabarella's quotations of Aristotle and Averroës are generally not taken from the translations in *AOAC* or from those in other common Renaissance editions. The passages of Galen are not those by Niccolò Leoniceno. Thus, translations from Greek are possibly Zabarella's own. His Latin translations of Averroës would have had to come from a Hebrew edition. Some clues to what editions Zabarella was using are given in the notes.

Some writings that Zabarella and his contemporaries presumed were Aristotle's own we now believe to be otherwise. Zabarella presumes that titles and the order in the printed corpora were Aristotle's choice, and some of Zabarella's arguments rely on this.

The reader should be cautious about presuming that what Zabarella claims is stated in a cited passage is as unambiguously stated as Zabarella suggests.

Works with no author listed are Aristotle's or Zabarella's. Thus, "*Topics*" is Aristotle's; "Cicero, *Topics*" is Cicero's.

References for the Greek commentators are to *CAG*, taken from Schicker. These references can be used to find the relevant passages in the English translations in the series edited by Richard Sorabji, *Ancient Commentators on Aristotle* (London, 1987–).

ABBREVIATIONS

AOAC	Aristotle, Averröes, et al. *Aristotelis opera cum Averrois commentariis*. Venice: Iunctae, 1562. Reprint, Frankfurt am Main: 1962.
Bakhtiar	Laleh Bakhtiar. *The Canon of Medicine*. Great Books of The Islamic World, 1999.
CAG	*Commentaria in Aristotelem Graeca*. 23 vols. Berlin: Reimer, 1882–1909.
Comm. Post. An.	Jacopo Zabarella. *Commentarii in libros duos Posteriorum Analyticorum*. Venice: Meietus, 1582.
Galen	*Opera omnia*. Edited by Karl Gottlob Kühn. 22 vols. Leipzig: Knobloch, 1821–33.
Holyoake	Francis Holyoake. *Dictionarium Etymologicum Latinum*. Oxford: William Turner, 1627.
Meth.	Jacopo Zabarella. *On Methods*.
Regr.	Jacopo Zabarella. *On Regressus*.
Schicker	Jacopo Zabarella. *Über die Methoden = De methodis; Über den Rückgang = De regressu*. Translated with an introduction and commentary by Rudolf Schicker. Munich: Wilhelm Fink, 1995.

c.	*caput, capitulum*, chapter
d.	*doctrina*
fn.	*fen* (a subdivision in Avicenna, *Liber Canonis*)
fol.	folio
l.	*liber*, book
ln.	line
lo.	*locus*, locus
p.	page

· NOTE ON REFERENCES ·

par.	paragraph
pt.	part
q.	*quaesitum*, question
s.	*sectio*, section
sg.	segment of a page in *AOAC*
su.	*summa*
t.	*contextus*, text no.
tr.	*tractatus*, tract
v.	*volumen*, volume

Notes to the Translation

❧❧❧

ON METHODS

BOOK III

1. For the meaning of "gather" here and throughout Book 3, see Note on the Text and Translation, s.v. *colligere*.

2. Averroës, long commentary on *Posterior Analytics* l. 1 t. 1 (71a1–11 in l. 1 c. 1), AOAC v. 1 pt. 2a fol. 12 sg. E–fol. 17 sg. C.

3. See Note on the Text and Translation, s.v. *colligere*.

4. *On the Soul* l. 1 t. 4–5 (402a10–22 in l. 1 c. 1).

5. Reading "esse ignotum" as the object of "negat" and not of "necesse est."

6. See Notes to the Translation, s.v. *cognitio* and *notitia*.

7. *Prior Analytics* l. 1 s. 1 c. 1 (24a10–b30 = l. 1 c. 1).

8. See beginning of *Prior Analytics* l. 1 c. 5 (25b26–33 in l. 1 c. 4).

9. *Prior Analytics*. Enthymeme: l. 2 c. 34 (70a10–b6 in l. 2 c. 27); induction: l. 2 c. 29 (68b8–37 = l. 2 c. 23); example: l. 2 c. 30 (68b38–69a19 = l. 2 c. 24).

10. *Prior Analytics* l. 2 c. 30 (68b38–a19 = l. 2 c. 24).

11. Ibid. l. 2 c. 29 (68b8–37 = l. 2 c. 23).

12. Ibid. Zabarella's understanding of the chapter was common from late antiquity into the Renaissance, but John P. McCaskey, "Freeing Aristotelian Epagōgē from *Prior Analytics* II 23," *Apeiron* (Dec., 2007): 345–74, argues that the reading is incorrect.

13. *Prior Analytics* l. 1 s. 2 c. 3 (46a31–b40 = l. 1 c. 31).

14. Alexander, CAG v. 2.1 p. 333.

15. *Prior Analytics* l. 1 s. 2 c. 3 (46a31–b40 = l. 1 c. 31).

16. Alexander, CAG v. 2.2 p. 1 f.

17. Philoponus, CAG v. 13.3 p. 295 ln. 21 ff.

18. *Posterior Analytics l.* 1 *t.* 176 (87a1–20 in *l.* 1 *c.* 26).

19. Alexander, *CAG v.* 2.1 *p.* 333 *ln.* 21 f.

20. *Sophistical Refutations l.* 2 *c.* 8 (183a27–184b8 = *c.* 34).

21. Alexander, *CAG v.* 2.1 *p.* 18 *ln.* 20.

22. Ibid., *CAG v.* 2.2 *p.* 9 *ln.* 22 ff.

23. See Note on the Text and Translation, s.v. *certus*.

24. Eustratius, *CAG v.* 21.1 *p.* 4 *ln.* 34 ff.

25. Zabarella, *De propositionibus necessariis* (the fifth work in *Opera Logica*) *l.* 2 *c.* 16–17.

26. The term is not used by Aristotle (though see *Posterior Analytics* 72b14 *l.* 1 *c.* 3) but is by Greek commentators such as Alexander, Themistius, Philoponus (e.g., *CAG v.* 13.3 *p.* 31 *ln.* 12, translated in Richard Sorabji, ed., *The Philosophy of the Commentators 200–600 AD* (Duckworth, 2004), 9(d)3) and Eustratius.

27. See Note on the Text and Translation, s.v. *demonstratio quòd*.

28. *Meth. l.* 3 *c.* 19 *par.* 2 below.

29. *Meth. l.* 4.

30. Eustratius, *CAG v.* 21.1 *p.* 4 *ln.* 34 ff.

31. Ibid., *CAG v.* 21.1 *p.* 6 *ln.* 5 ff.

32. Ammonius, *CAG v.* 4.3 *p.* 34 f.

33. *Prior Analytics l.* 1 *s.* 2 *c.* 3 (46a31–b40 = *l.* 1 *c.* 31).

34. Ibid.

35. Averroës, long commentary on *Posterior Analytics l.* 2 *t.* 25 (91b32–92a5 in *l.* 2 *c.* 5), *AOAC v.* 1 *pt.* 2a *fol.* 436 *sg.* E–fol. 437 sg. F.

36. *Posterior Analytics l.* 1 *t.* 14 (72a14–24 in *l.* 1 *c.* 2).

37. Ibid. *l.* 1 *t.* 18–21 (72b5–25 in *l.* 1 *c.* 3).

38. Some are known *per se* actually and some potentially.

39. *Posterior Analytics l.* 1 *t.* 1 (89b23–31 in *l.* 2 *c.* 1).

40. *Physics l.* 1 *t.* 68 (191a2–7 in *l.* 1 *c.* 7).

41. Ibid. *l.* 1 *t.* 6 (184b15–22 in *l.* 1 *c.* 2).

42. See note to *Regr. c.* 4 *par.* 4.

43. *Posterior Analytics l.* 2 *t.* 42 (93b15–25 in *l.* 2 *c.* 8). See *Meth. l.* 1 *c.* 2 above and *Regr. c.* 4 *par.* 14.

44. "Res ipsa loquitur" is a well-known Ciceronian phrase (see *Pro Milone* 53).

45. *Prior Analytics l.* 1 *s.* 2 *c.* 3 (46a31–b40 = *l.* 1 *c.* 31).

46. *On the Heavens l.* 1 *t.* 7 (268b26–269a2 in *l.* 1 *c.* 2).

47. *On Meteorology l.* 1 *su.* 1 *c.* 1 (338a20–339a9 ≈ *l.* 1 *c.* 1).

48. A kind of infinitely self-referential babbling. See Aristotle, *Sophistical Refutations l.* 1. *c.* 3 and commentary thereon.

49. *Prior Analytics l.* 1 *s.* 2 *c.* 3 (46a31–b40 = *l.* 1 *c.* 31).

50. *Posterior Analytics l.* 2 *t.* 23 (91b12–27 in *l.* 2 *c.* 6); *t.* 73 (96b25–30 in *l.* 2 *c.* 13).

51. *Physics l.* 1 *t.* 5 (184b10–b14 in *l.* 1 *c.* 1).

52. *On Interpretation l.* 1 *c.* 4 (16b26–17a37 = *c.* 4–6).

53. *Physics l.* 1 *t.* 5 (184b10–b14 in *l.* 1 *c.* 1).

54. *Meth. l.* 3 *c.* 1 *par.* 2.

55. See *Posterior Analytics l.* 2 *t.* 6 (90a9–23 in *l.* 1 *c.* 2).

56. *On Interpretation l.* 1 *c.* 1 (16a1–18 = *c.* 1).

57. Aristotle does not exactly say this. That a concept is an image or similitude is a scholastic expansion.

58. *Physics l.* 1 *t.* 1–5 (184a10–b14 = *l.* 1 *c.* 1).

59. *Posterior Analytics l.* 2 *t.* 45 (93b35–94a7 in *l.* 2 *c.* 10).

60. Zabarella, *De natura logicae* (the first work in *Opera Logica*) *l.* 2 *c.* 2.

61. *Posterior Analytics l.* 1 *t.* 14 (72a14–24 in *l.* 1 *c.* 2); *t.* 79 in *Comm. Post. An.* but *t.* 80 in *AOAC* (76b35–39 in *l.* 1. *c.* 10).

62. See *Posterior Analytics l.* 1 *t.* 14 (72a14–24 in *l.* 1 *c.* 2).

63. *On the Soul l.* 1 *t.* 4–5 (402a10–402a22 in *l.* 1. *c.* 1).

64. Averroës, long commentary on *Posterior Analytics l.* 1 *t.* 1 (71a1–11 in *l.* 1 *c.* 1), *AOAC v.* 1 *pt.* 2a *fol.* 12 *sg.* E–*fol.* 17 *sg.* C.

65. *Posterior Analytics* l. 2 t. 19–25 (91a12–92a5 ≈ l. 2 c. 4–5).

66. *Topics* l. 6 c. 1 Zabarella probably means what we know as the first four chapters (139a24–143a28 = l. 6 c. 1–4). See note to *Meth.* l. 3 c. 15 par. 11 below.

67. *Posterior Analytics* l. 2 t. 37 (93a3–14 in l. 2 c. 8).

68. Ibid. l. 2 t. 38 (93a14–15 in l. 2 c. 8).

69. Ibid. l. 1 t. 1–6 (71a1–b8 = l. 1 c. 1).

70. *Topics* l. 6 c. 1. If Zabarella is referring to what we know as c. 4, 141a23–142b19, then he is providing a clue what edition he was using. In *AOAC*, this passage is indeed in what is called c. 1, but in some sixteenth-century editions, it is c. 3.

71. *Metaphysics* l. 1 t. 48 (992b30–33 in 1.9).

72. Ammonius, *CAG* v. 4.3 p. 36 *ln.* 21 ff.

73. Eustratius, *CAG* v. 21.1 p. 3 *ln.* 11 ff.

74. Aristotle does not say this. Maybe Zabarella is reading it into the end of *Prior Analytics* l. 2 c. 29 (68b8–37 = l. 2 c. 23).

75. Averroës, long commentary on *Physics* l. 1 t. 3 (184a21–23 in l. 1 c. 1), *AOAC* v. 4 fol. 7 *sg.* C–D.

76. Ibid. l. 1 t. 5 (184b10–14 in l. 1 c. 1), *AOAC* v. 4 fol. 7 *sg.* M–fol. 8 *sg.* H.

77. Zabarella, *De medio demonstrationis* (the ninth work in *Opera Logica*) l. 1 c. 17.

78. *patibilis*, what other writers called *possibilis* or *patiens*: "passive" or "patient," as opposed to "active" or "acting."

79. Aristotle does not say this exactly. Zabarella may be putting it together from his understanding of *Posterior Analytics* l. 2 t. 100–107 (99b15 –100b17 = l. 2 c. 19).

80. *Posterior Analytics* l. 2 t. 42 (93b15–25 in l. 2 c. 8–9).

81. Uncharacteristically, Zabarella is here using "demonstration" in the narrow sense, equivalent to "demonstrative method."

82. *Posterior Analytics* l. 1 t. 73 (76a32–36 in l. 1 c. 10).

83. Ibid. *l.* 1 *t.* 57 (75a39–44 in *l.* 1 *c.* 7). In the passage, Aristotle does not say "beginning-principles" (*archai; principia*) but "axioms" (*axiōmata; dignitates*). What Zabarella here calls *subjectum* he calls *subjectum genus* in his own edition of the *Posterior Analytics*, and *subjectum scientiae* in his commentary. Aristotle's phrase is *to genus to hupokeimenon*.

84. *Posterior Analytics l.* 1 *t.* 101–3 (79a2–32 in *l.* 1 *c.* 13–14); *t.* 178 (87a31–38 = *l.* 1 *c.* 27).

85. *Physics l.* 1 *t.* 1–5 (184a10–b14 = *l.* 1 *c.* 1).

86. Averroës, *Epitome in Libros Logicae Aristotelis (Short Commentary on Aristotle's Organon) c.* 1, AOAC *v.* 1 *pt.* 2b *fol.* 52 *sg.* C–*fol.* 53 *sg.* G.

87. Themistius, commentary on *Posterior Analytics l.* 1. *t.* 1 (89b23–31 in *l.* 1 *c.* 1), CAG *v.* 5.1 *p.* 44 *ln.* 2 f.

88. Ibid. *l.* 1 *t.* 93 (78a6–10 in *l.* 1 *c.* 12), CAG *v.* 5.1 *p.* 26 *ln.* 22 f.

89. Eustratius, CAG *v.* 21.1 *p.* 3 *ln.* 24 ff.

90. *Posterior Analytics l.* 1 *t.* 8 (71b15–19 in *l.* 1 *c.* 2).

91. Ibid. *l.* 1 *t.* 9 (71b19–25 in *l.* 1 *c.* 2). Zabarella is paraphrasing.

92. Zabarella, *De speciebus demonstrationis* (the sixth work in *Opera Logica*), and *De propositionibus necessariis* (the fifth work in *Opera Logica*).

93. Zabarella, *De medio demonstrationis* (the ninth work in *Opera Logica*) *l.* 2 *c.* 5.

94. *Prior Analytics l.* 2 *c.* 29 (68b8–37 = *l.* 2 *c.* 23).

95. *Posterior Analytics l.* 1 *t.* 134 (81a38–b9 = *l.* 1 *c.* 18).

96. Ibid. *l.* 2 *t.* 101–7 (99b17–100b17 in *l.* 2 *c.* 19). "And in the last chapter of the second" would seem to refer to *On Ethics* and not the *Posterior Analytics*, but the contents of the respective chapters suggests otherwise. Schicker, too, assumed that the *Posterior Analytics* was intended.

97. *Nicomachean Ethics l.* 6 *c.* 3 (1139b14–b36 = *l.* 6 *c.* 3).

98. *Prior Analytics l.* 2 *c.* 29 (68b8–37 = *l.* 2 *c.* 23). Though Zabarella's explanation of induction here is common, going back to the Neoplatonic commentators of late antiquity, it is not an accurate rendering of what Aristotle actually says in the chapter. See note to *Meth. l.* 3 *c.* 3 *par.* 5 above.

99. Zabarella's contrast here between "known according to nature" and "unknown according to nature" should not be presumed to be the same as his contrast between "prior and more known according to nature" and "prior and more known to us."

100. Averroës, long commentary on *Posterior Analytics* l. 2 t. 25, 29 (91b32–a5 in l. 2 c. 5, 92a23–27 in l. 2 c. 6), *AOAC v.* 1 pt. 2a *fol.* 436 *sg.* E–*fol.* 437 *sg.* F., *fol.* 444 *sg.* C–*fol.* 445 *sg.* E.

101. *Prior Analytics* l. 2 c. 21 (64b28–65a37 = l. 2 c. 16).

102. The Latin would support either "that are said to be indemonstrable" or "which are said to be indemonstrable." The difference here is substantive. What Zabarella says in the rest of this chapter seems to warrant "that."

103. *Physics* l. 1 t. 11 (185a11–17 in l. 1 c. 2).

104. *Posterior Analytics* l. 2 t. 101–7 (99b17–100b17 in l. 2 c. 19).

105. *Physics* l. 8 t. 52–53 (259b20–260a19 in l. 8 c. 6). See *Regr. c.* 6.

106. *On the Soul* l. 1 t. 11 (402b16–403a2 in l. 1 c. 1).

107. *Posterior Analytics* l. 2 t. 51 (94b8–21 in l. 2 c. 11).

108. *Nicomachean Ethics* l. 7 c. 8 (1150b29–1151a28 = l. 7 c. 8).

109. See *Physics* (l. 2 c. 9), *On Coming to Be and Passing Away* (l. 2 c. 11), and *On the Parts of Animals* (l. 1 c. 1).

110. Alexander, *CAG v.* 2.1 *p.* 1 *ln.* 3 ff.

111. *Nicomachean Ethics* l. 7 c. 8 (1150b29–1151a28 = l. 7 c. 8).

BOOK IV

1. See *Meth.* l. 3 c. 4 *par.* 6

2. Robert Grosseteste (ca. 1168–1253) was bishop of Lincoln from 1235 until his death; thus the Latin appellation.

3. Zabarella is referring to sections that begin at what we know as 25b26, the beginning of l. 1 c. 4, and 43a20, the beginning of l. 1 c. 27. In *AOAC* these points mark the beginnings of l. 1 s. 1 c. 5 and l. 1 s. 2 c. 1, respectively. It appears that in the edition Zabarella used, the first four chapters

in *AOAC* (the first three in our numbering) formed a first section, followed then by a second and third at the noted boundaries.

4. E.g., *Metaphysics l.* 7 *t.* 11–19 (1029b13–1031a14 in *l.* 7 *c.* 4–6); *t.* 37–43 (1036a26–1038a35 in *l.* 7 *c.* 11–12); *t.* 53–55 (1039b20–1040b4 = *l.* 7 *c.* 15); *l.* 8 *t.* 2 (1042a12–25 in *l.* 8 *c.* 1); *l.* 8 *t.* 15–16 (1045a7–b23 = *l.* 8 *c.* 6).

5. *Posterior Analytics l.* 2 *t.* 46 (94a7–14 in *l.* 2 *c.* 10).

6. Eustratius, *CAG v.* 21.1 *p.* 4 ff.

7. A presumption here is that it was Aristotle who gave the works their titles.

8. See Note on the Text and Translation, s.v. *demonstratio quòd*, etc.

9. *Posterior Analytics l.* 2 *t.* 6 (90a9–23 in *l.* 2 *c.* 2).

10. Ibid. *l.* 2 *t.* 36–47 (93a1–94a19 = *l.* 8 *c.* 10).

11. Ibid. *l.* 1 *t.* 95 (78a22–26 in *l.* 1 *c.* 13). In the translations of Aristotle through here, the Greek, *to prôton aition*, justifies the definite article.

12. *Posterior Analytics l.* 1 *t.* 7 (71b9–15 in *l.* 1 *c.* 2).

13. Averroës, long commentary on *Posterior Analytics l.* 1 *t.* 11 (71b29–33 in *l.* 1 *c.* 2), *AOAC v.* 1 *c.* 2a *fol.* 33 *sg.* F–*fol.* 35 *sg.* F.

14. *Posterior Analytics l.* 1 *t.* 11 (71b29–33 in *l.* 1 *c.* 2).

15. Ibid. *l.* 1 *t.* 95 (78a22–26 in *l.* 1 *c.* 13).

16. See Note on the Text and Translation, s.v. *demonstratio quòd*, etc.

17. *Posterior Analytics l.* 1 *t.* 95 (78a22–26 in *l.* 1 *c.* 13).

18. Ibid. *l.* 1 *t.* 99 (78b13–34 in *l.* 1 *c.* 13).

19. Ibid. *l.* 1 *t.* 48 (74b26–32 in *l.* 1 *c.* 6); *t.* 49 (74b32–75a7 in *l.* 1 *c.* 6).

20. Ibid. *l.* 1 *t.* 51 (75a12–17 in *l.* 1 *c.* 6).

21. Ibid. *l.* 1 *t.* 54 (75a28–31 in *l.* 1 *c.* 6).

22. Ibid. *l.* 1 *t.* 56 (75a35–39 in *l.* 1 *c.* 6).

23. Ibid. *l.* 1 *t.* 73–75 (76a32–76b11 in *l.* 1 *c.* 10).

24. Ibid. *l.* 2 *t.* 1–7 (89b23–90a34 in *l.* 2 *c.* 1–2). Zabarella's ordering of the four questions is not Aristotle's, and Zabarella's third and fourth are uncommon Latin renderings. In the more common ordering and render-

ing, these are: the *quòd* (for *an insit*), that something is the case, τὸ ὅτι, *to hoti*; the *propter quid* (for *cur*), what something is on account of, τὸ διότι, *to dioti*; the *an*, whether something exists, εἰ ἔστι, *ei esti*; and the *quid*, what something is, τί ἐστιν, *ti estiv*. The *quòd* and the *propter quid* are complex questions. The *an* and *quid* are simple questions.

25. *Posterior Analytics l.* 2 *t.* 47 (94a14–19 in *l.* 2 *c.* 10).

26. Ibid. *l.* 1 *t.* 1–7 (89b23–90a34 in *l.* 2 *c.* 1–2).

27. Ibid. *l.* 2 *t.* 46–47 (94a7–19 in *l.* 2 *c.* 10).

28. Ibid. *l.* 2 *t.* 8 (90a35–38 in *l.* 2 *c.* 3).

29. Ibid. *l.* 2 *t.* 47 (94a14–19 in *l.* 2 *c.* 10).

30. Ibid. *l.* 2 *t.* 46 (94a7–14 in *l.* 2 *c.* 10).

31. See *Meth. l.* 4. *c.* 14–15 below.

32. *Posterior Analytics l.* 2 *t.* 42 (93b15–25 in *l.* 2 *c.* 8–9).

33. Ibid. *l.* 2 *t.* 1 (89b23–31 in *l.* 2 *c.* 1).

34. Albert, *Alberti Magni Opera Omnia*, ed. A. Borgnet (Paris: 1890), *v.* 1 *tr.* 1 *c.* 5.

35. Robert Grosseteste, *Commentarius in Posteriorum Analyticorum Libros*, ed. P. Rossi (Florence, 1980), p. 287 ff.

36. *On the Soul l.* 1 *t.* 3 (402a7–10 in *l.* 1 *c.* 1).

37. Averroës, proem to long commentary on *Posterior Analytics l.* 2, AOAC *v.* 1 *pt.* 2a *fol.* 1 *sg.* B–*fol.* 9 *sg.* E.

38. Averroës, long commentary to *Posterior Analytics l.* 1 *c.* 1 (184a10–16 in *l.* 1 *c.* 1), AOAC *v.* 1 *pt.* 2a *fol.* 401 *sg.* E–*fol.* 403 *sg.* A.

39. Averroës, long commentary on *Metaphysics l.* 7 *t.* 42 (1037b8–27 in *l.* 7 *c.* 12), AOAC *v.* 8 *fol.* 194 *sg.* A–G.

40. A reference to Aristotle's list of ten in the *Categories*.

41. *Posterior Analytics l.* 1 *t.* 30 (73a34–37 in *l.* 1 *c.* 4).

42. The first of Aristotle's ten categories.

43. Averroës, commentary on *Metaphysics l.* 6 *t.* 1 (1025b3–28 in *l.* 1 *c.* 1), AOAC *v.* 8 *fol.* 144 *sg.* D–*fol.* 145 *sg.* C.

44. Themistius, *CAG v.* 5.1 *p.* 52 *ln.* 1 f., commenting on *Posterior Analytics l.* 2 *t.* 48 (94a20–24 in *l.* 2 *c.* 11).

45. *Prior Analytics l.* 1 *s.* 1 *c.* 5 (25b26–26b33 = *l.* 1 *c.* 4).

46. *Posterior Analytics l.* 1 *t.* 8 (71b15–19 in *l.* 1 *c.* 2).

47. Ibid. *l.* 2 *t.* 45 (93b35–94a7 in *l.* 2 *c.* 10).

48. Ibid. *l.* 2 *t.* 100 (99b15–99b17 in *l.* 2 *c.* 19). See note to *Meth. l.* 4 *c.* 7 *par.* 18.

49. *Posterior Analytics l.* 2 *t.* 45 (93b35–94a7 in *l.* 2 *c.* 10).

50. *On Interpretation l.* 1 *c.* 4 (16b26–17a37 = *c.* 4–6).

51. *Posterior Analytics l.* 2 *t.* 48 (94a20–24 in *l.* 2 *c.* 11).

52. Ibid. *l.* 2 *t.* 47 (94a14–19 in *l.* 2 *c.* 10).

53. Ibid. *l.* 2 *t.* 42 (93b15–25 in *l.* 2 *c.* 8–9).

54. Ibid. *l.* 2 *t.* 47 (94a14–19 in *l.* 2 *c.* 10).

55. Averroës, long commentary on *Metaphysics l.* 6 *t.* 1 (1025b3–28 in *l.* 1 *c.* 1), *AOAC v.* 8 *fol.* 144 *sg.* D–*fol.* 145 *sg.* C.

56. *Posterior Analytics l.* 2 *t.* 47 (94a14–19 in *l.* 2 *c.* 10).

57. Ibid. *l.* 2 *t.* 48 (94a20–24 in *l.* 2 *c.* 11).

58. Ibid. *l.* 2 *t.* 68 (96a11–19 in *l.* 2 *c.* 12).

59. Ibid. *l.* 2 *t.* 69 (96a20–32 in *l.* 2 *c.* 13).

60. The nominal, imperfect definition.

61. *Posterior Analytics l.* 2 *t.* 42 (93b15–25 in *l.* 2 *c.* 8–9).

62. Zabarella is referring to *Posterior Analytics* 96a36–97a6. In *Meth. l.* 4 *c.* 13 *par.* 7 below, Zabarella explains that this would be better called resolutive than compositive.

63. This echoes the first phrase in this paragraph.

64. *Posterior Analytics l.* 2 *t.* 69–84 (96a20–97b39 = *l.* 2 *c.* 13).

65. Ibid. *l.* 2 *t.* 85 (98a1–12) begins *c.* 14 in our editions, *c.* 13 in Zabarella's edition, *c.* 9 in *AOAC*. Text no. 100 (99b15–17) is the first sentence of our *c.* 19 but in Renaissance editions the final sentence of the penultimate chapter. Zabarella considers that sentence the end and summary of the

work proper and then treats the final chapter (*c.* 15 for him, *c.* 11 in *AOAC*, the rest of our *c.* 19) as something tacked on, outside the main body of the work. See *Meth. l.* 4 *c.* 10 *par.* 9 below.

66. See note immediately preceding.

67. Averroës, long commentary on *Posterior Analytics l.* 1 *t.* 11 (71b29–33 in *l.* 1 *c.* 2), *AOAC v.* 1 *pt.* 2a *fol.* 33 *sg.* F–*fol.* 35 *sg.* F.

68. Ibid. *l.* 2 *t.* 38 (93a14–15 in *l.* 2 *c.* 8), *AOAC v.* 1 *pt.* 2a *fol.* 457 *sg.* F–*fol.* 460 *sg.* A.

69. Averroës, *Epitome in Libros Logicae Aristotelis* (*Short Commentary on Aristotle's Organon*), *AOAC v.* 1 *pt.* 2b *fol.* 52 *sg.* C–*fol.* 53 *sg.* G. In *AOAC,* the first four chapters of the *Epitome,* which treat the first book of the *Posterior Analytics,* are all labeled *De Demonstratione.* Zabarella's statement suggests that in the edition he was using either all four were grouped as one chapter or only one of the four was labeled *De Demonstratione.*

70. Averroës, *Quaesita in Libros Logicae Aristotelis q.* 8 in "IX. Quaesita Demonstrativa in Libros Posteriorum Analyticorum Aristotelis," *AOAC v.* 1 *pt.* 2b *fol.* 118 *sg.* H–*fol.* 119 *sg.* G. Averroës' questions are numbered differently in different editions. The question Zabarella here cites is not numbered 10 in *AOAC* (1562), in the 1552 or 1575 editions of *AOAC,* or in the 1560 Venice edition by Cominus de Tridino.

71. Averroës, proem to long commentary on *Posterior Analytics l.* 1, *AOAC v.* 1 *pt.* 2a *fol.* 1 *sg.* B–*fol.* 9 *sg.* E.

72. Ibid.

73. On verification and formation, see Averroës, long commentary on *Posterior Analytics l.* 2 *t.* 1 (71a1–11 in *l.* 2 *c.* 1) and *Epitome in Libros Logicae Aristotelis* (*Short Commentary on Aristotle's Organon*), *AOAC v.* 1 *pt.* 2b *fol.* 36 *sg.* E.

74. Averroës, long commentary on *Posterior Analytics l.* 1 *t.* 11 (71b29–33 in *l.* 1 *c.* 2), *AOAC v.* 1 *c.* 2a *fol.* 33 *sg.* F–*fol.* 35 *sg.* F.

75. Averroës, *Quaesita in Libros Logicae Aristotelis q.* 2 in "IX. Quaesita Demonstrativa in Libros Posteriorum Analyticorum Aristotelis," *AOAC v.* 1 *pt.* 2b *fol.* 102 *sg.* G–*fol.* 103 *sg.* H.

76. *Posterior Analytics l.* 2 *t.* 45 (93b35–94a7 in *l.* 2 *c.* 10).

77. Normally for Zabarella, including here, "demonstration" and "demonstrative method" are not synonyms. Demonstrative method and resolutive method are both demonstrations. Demonstration *potissima* is demonstrative method. Demonstration *ab effectu* is a resolutive method. See also Note on the Text and Translation, s.v. *demonstratio quòd*.

78. *Prior Analytics l.* 1 *s.* 1 *c.* 1 (24a10–b30 = *l.* 1 *c.* 1)

79. *Posterior Analytics l.* 2 *t.* 100 (99b15–17 in *l.* 2 *c.* 19). See note to *Meth. l.* 4 *c.* 7 *par.* 18.

80. Alexander, commentary on *Prior Analytics*, CAG *v.* 2.1, commentary on the *Topics*, CAG *v.* 2.2; Ammonius, commentary on the *Categories*, CAG *v.* 4.4; Simplicius, commentary on the *Categories*, CAG *v.* 8. It is not fully clear what passages Zabarella is referring to in these commentaries. But in none of the possibilities does the commentator stress demonstration *potissima* (*kurios*) as distinct from any other sort of demonstration. Zabarella is reading in *potissima* based on the context.

81. In this chapter, "of things" (*rerum*) is probably meant to have a sense close to "of reality" or "of all things in the world." Cf., the *De Natura Rerum* by Lucretius and comparable tracts by subsequent authors.

82. It should not be simply taken for granted here that the distinction between naturally known and naturally unknown is the same as the distinction between prior and more known by nature and prior and more known to us. See Note on the Text and Translation, s.v. *prior/notior nobis*, *Meth. l.* 3 *c.* 19 *par.* 7, and the following chapter here, *c.* 11, "Solution to a doubt regarding known and unknown according to nature."

83. In this paragraph, in phrases translated with "existence," Zabarella uses the verb *esse* and not the potentially more technical *existere* (verb) or *existens* (noun).

84. Themistius, CAG *v.* 5.1 *p.* 6 *ln.* 1 ff.

85. *Posterior Analytics l.* 1 *t.* 12 (71b33–72a5 in *l.* 1 *c.* 2).

86. See Note on the Text and Translation, s.v. *prior/notius nobis*.

87. See Note on the Text and Translation, s.v. *prior/notius nobis*, and the references cited there.

88. *Posterior Analytics l.* 2 *t.* 42 (93b15–25 in *l.* 2 *c.* 8–9).

89. E.g., Ibid. *l*. 2 *t*. 95–102 (78a22–79a16 = *l*. 2 *c*. 13).

90. Ibid. *l*. 2 *t*. 106 (100a15–b5 in *l*. 2 *c*. 19).

91. Ibid. *l*. 2 *t*. 100 (99b15–17 in *l*. 2 *c*. 19). See note to *Meth*. *l*. 4 *c*. 7 *par*. 18.

92. As at *Posterior Analytics l*. 2 *t*. 80 (97b7–15 in *l*. 2 *c*. 13).

93. Demonstration *ab effectu* and, secondarily, induction.

94. *Meth*. *l*. 3 *c*. 19 *par*. 2.

95. *Posterior Analytics l*. 1 *t*. 12 (71b33–72a5 in *l*. 1 *c*. 2).

96. *Physics l*. 1 *t*. 2 (184a16–20 in *l*. 1 *c*. 1).

97. *haplos* in Aristotle's Greek

98. Averroës, long commentary on *Metaphysics l*. 7 *t*. 42 (1037b8–27 in *l*. 7 *c*. 12), *AOAC v*. 8 *fol*. 194 *sg*. A–G.

99. *Posterior Analytics l*. 2 *t*. 44 (93b29–35 in *l*. 2 *c*. 10).

100. Ibid. *l*. 2 *t*. 46 (94a7–14 in *l*. 2 *c*. 10).

101. Ibid. *l*. 2 *t*. 8 (90a35–38 in *l*. 2 *c*. 3).

102. Ibid. *l*. 2 *t*. 47 (94a14–19 in *l*. 2 *c*. 10).

103. The distinction here is not the same as the one between a definition that is the beginning-principle (*principium*) of a demonstration and one that is the conclusion (*conclusio*) of a demonstration.

104. *Posterior Analytics l*. 1 *t*. 178 (87a31–38 = *l*. 1 *c*. 27).

105. Ibid. *l*. 2 *t*. 42 (93b15–25 in *l*. 2 *c*. 8–9).

106. Ibid.

107. *Metaphysics l*. 12 *t*. 1 (1069a18–19 in *l*. 12 *c*. 1).

108. *Posterior Analytics l*. 2 *t*. 69–84 (96a20–97b39 = *l*. 2 *c*. 13).

109. See Note on the Text and Translation, s.v. *colligere*.

110. "This compositive way" is the one Zabarella believes would better be called "resolutive." "The power of induction" is the "syllogistic power" Zabarella refers to in the previous paragraph and later in this one. The induction provides the inference that what is true of all individuals is true of the species. It does not provide an inference from some individu-

als to all individuals nor is it used to identify which attributes are essential to the individuals.

111. Zabarella does not here say how we know that the parts of the definition apply to all the particulars, only that once we do, induction allows us to prove that they belong to the species.

112. For Zabarella, induction and demonstration *a signo* (elsewhere called *ab effectu*) are the two kinds of resolutive method.

113. *On the Soul* l. 1 t. 11 (402b16–403a2 in *l.* 1 c. 1).

114. *Posterior Analytics* l. 2 t. 42 (93b15–25 in *l.* 2 c. 8–9).

115. Presumably *l.* 2 t. 69–84 (94a20–97b39 = *l.* 2 c. 13). In AOAC this part is labeled *De compositione definitionis ex suis partibus, et quando sunt ignota quo pacto venentur.* In Zabarella's own later edition, the part is untitled.

116. Some Aristotelian commentators distinguish essential definitions from nominal ones. But through here, Zabarella uses "nominal," "quidditative," and "essential" as synonyms.

117. Averroës, long commentary on *Posterior Analytics* l. 2 t. 6 (90a9 23 in *l.* 2 c. 2), AOAC v. 1 pt. 2a *fol.* 406 *sg.* D–*fol.* 408 *sg.* F.

118. *Posterior Analytics* l. 2 t. 41 (93a33–b14 in *l.* 2 c. 8). See *On Meteorology* l. 2 *su.* 4 (369a10–370a33 = *l.* 2 c. 9). Doubt about Aristotle's account of causation in meteorology was common in the late sixteenth century; see Craig Martin, *Renaissance Meteorology: Pomponazzi to Descartes* (Baltimore: Johns Hopkins, 2011).

119. *Posterior Analytics* l. 2 t. 42–43 (93b15–28 in *l.* 2 c. 8–9).

120. The "quidditative definition" is not the perfect definition Zabarella was just discussing, but the imperfect, merely nominal one. See the next two paragraphs.

121. See the note to *Meth.* l. 4 c. 15 par. 1 below.

122. Averroës, long commentary on *Posterior Analytics* l. 1 t. 64 (75b30–32 in *l.* 1 c. 8), AOAC v. 1 pt. 2a *fol.* 145 *sg.* C–*fol.* 146 *sg.* A.

123. Themistius, CAG v. 5.1 p. 51 *ln.* 19 ff., commentary on *Posterior Analytics* l. 2 t. 46 (94a7–14 in *l.* 2 c. 10).

124. Zabarella now adds "formal" to his list of synonyms for the imperfect definition.

125. *Posterior Analytics* l. 2 t. 44 (93b29–35 in l. 2 c. 10)

126. In t. 44, Aristotle refers to a name "or other nominal speech." He does not call a type of definition "nominal."

127. *Posterior Analytics* l. 2 t. 46 (94a7–14 in l. 2 c. 10).

128. Ibid. l. 2 t. 44 (93b29–35 in l. 2 c. 10).

129. *De Anima* l. 1 t. 16 (403a29–b9 in l. 1 c. 1). About the position Zabarella here rejects, Sten Ebbeson writes, "[In examinations of the thirteenth century] one of the questions relating to the *Posterior Analytics* is why Aristotle said that a definition is a demonstrative proof, only 'differing in position.' The answer consists in a classification of definitions into formal, material and combined definitions. A formal definition of anger might be: 'Anger is a desire to reciprocate vexation.' The material definition might, for instance, be: 'Anger is a boiling of the blood around the heart.' The two combine in the combi-definition 'Anger is a boiling of the blood around the heart because of a desire to reciprocate vexation.' Now, with a little rearrangement of the elements the combi-definition emerges as a syllogism, namely: 'Every desire to reciprocate vexation is a boiling of the blood around the heart; anger is a desire to reciprocate vexation; therefore anger is a boiling of the blood around the heart.' So, some definitions, namely combi-definitions, may be considered demonstrations whose parts have been shuffled." ("Greek and Latin Medieval Logic," in *Greek-Latin Philosophical Interaction: Collected Essays of Sten Ebbesen* [Aldershot — Burlington: Ashgate, 2008], vol. 1, chap. 11.)

130. *Posterior Analytics* l. 2 t. 36–37 (93a1–14 in l. 2 c. 8).

131. Ibid. l. 2 t. 36–47 (93a1–94a19 = l. 2 c. 8–10).

132. Averroës, long commentary on *Posterior Analytics* l. 2 t. 45 (93b35–94a7 in l. 2 c. 10), AOAC v. 1 pt. 2a fol. 471 *sg.* E–*fol.* 473 *sg.* B.

133. Ibid. l. 2 t. 42 (93b15–25 in l. 2 c. 8–9), AOAC v. 1 pt. 2a *fol.* 467 *sg.* D–*fol.* 469 *sg.* C.

134. *Posterior Analytics* l. 1 t. 2 (71a11–17 in 1. 1 c. 1).

135. Ibid. l. 2 t. 1–47 (89b23–94a19 = l. 2 c. 1–10).

136. See Note on the Text and Translation, s.v. *demonstratio quòd*.

137. *Posterior Analytics l.* 2 *t.* 1–7 (89b23–90a34 = *l.* 2 *c.* 1–2).

138. Ibid. *l.* 2 *t.* 42–43 (93b15–28 in *l.* 2 *c.* 8–9).

139. Ibid. *l.* 2 *t.* 45 (93b35–94a7 in *l.* 2 *c.* 10).

140. Ibid.

141. Averroës, long commentary on *Posterior Analytics l.* 1 *t.* 11 (71b29–33 in *l.* 1 *c.* 2), AOAC *v.* 1 *pt.* 2a *fol.* 33 *sg.* F–*fol.* 35 *sg.* F.

142. Ibid. *l.* 2 *t.* 38 (93a14–15 in *l.* 2 *c.* 8), AOAC *v.* 1 *pt.* 2a *fol.* 457 *sg.* F–*fol.* 460 *sg.* A.

143. Averroës, *Epitome in Libros Logicae Aristotelis (Short Commentary on Aristotle's Organon)*, AOAC *v.* 1 *pt.* 2b *fol.* 52 *sg.* C–*fol.* 53 *sg.* G. See note to this same passage above, *Meth. l.* 4 *c.* 8 *par.* 1.

144. Averroës, *Quaesita in Libros Logicae Aristotelis q.* 8 in "IX. Quaesita Demonstrativa in Libros Posteriorum Analyticorum Aristotelis," AOAC *v.* 1 *pt.* 2b *fol.* 118 *sg.* H–*fol.* 119 *sg.* G. See note to this same passage above, *Meth. l.* 4 *c.* 8 *par.* 1.

145. Averroës, proem to long commentary on *Posterior Analytics l.* 1, AOAC *v.* 1 *pt.* 2a *fol.* 1 *sg.* B–*fol.* 9 *sg.* E.

146. See *Posterior Analytics l.* 1 *c.* 1 (89b23–31 in *l.* 1 *c.* 1).

147. *On Meteorology l.* 4 *su.* 1 *c.* 2 (378b26–379b9 in *l.* 4 *c.* 1).

148. Regarding the genus and subject, recall *Meth. l.* 4 *c.* 14 *par.* 2.

149. *Prior Analytics l.* 1 *s.* 1 *c.* 1 (24a10–b30 = *l.* 1 *c.* 1). This may have been labeled a proem in the edition Zabarella was using.

150. *Posterior Analytics l.* 2 *t.* 100 (99b15–17 in *l.* 2 *c.* 19). See note to *Meth. l.* 4 *c.* 7 *par.* 18.

151. Averroës, middle commentary on *Prior Analytics* (24a10–b30 = *l.* 1 *c.* 1), AOAC *v.* 1 *pt.* 1b *fol.* 1 *sg.* L–M. The Latin Zabarella gives is not Burana's or William of Luna's.

152. *Posterior Analytics l.* 2 *t.* 1–7 (89b23–90a34 = *l.* 2 *c.* 1–2).

153. Ibid. *l.* 2 *t.* 42 (93b15–25 = *l.* 2 *c.* 8–9).

154. Ibid. *l.* 2 *t.* 100 (99b15–17 in *l.* 2 *c.* 19), and Averroës, long commentary thereon, *AOAC v.* 1 *pt.* 2a *fol.* 557 *sg.* B–F. See note to *Meth. l.* 4 *c.* 7 *par.* 18.

155. *Posterior Analytics l.* 2 *t.* 100 (99b15–17 in *l.* 2 *c.* 19).

156. Ammonius, *CAG v.* 4.3 *p.* 1 *ln.* 5, commentary on Porphyry's *Isogoge.*

157. Averroës, long commentary on *Posterior Analytics l.* 2 *t.* 97 (99a16–29 in *l.* 2 *c.* 17), *AOAC v.* 1 *pt.* 2a *fol.* 548 *sg.* A–*fol.* 551 *sg.* C.

158. *Meth. l.* 4 *c.* 18 *par.* 10. Recall that Zabarella considers the bulk of our *l.* 2 *c.* 19, where induction is discussed, to be outside the body of the *Posterior Analytics* proper.

159. Zabarella, *De natura logicae* (the first work in *Opera Logica*) *l.* 2 *c.* 7.

160. *Prior Analytics l.* 1 *s.* 1 *c.* 1 (24a10–b30 in *l.* 1 *c.* 1).

161. *Posterior Analytics l.* 1 *t.* 1–6 (71a1–b8 = *l.* 1 *c.* 1).

162. Ibid. *l.* 1 *t.* 56 (75a35–39 in *l.* 1 *c.* 6–7).

163. Averroës, long commentary on *Posterior Analytics l.* 2 *t.* 100 (99b15–17 in *l.* 2 *c.* 19), *AOAC v.* 1 *pt.* 2a *fol.* 557 *sg.* B–E.

164. *Posterior Analytics l.* 2 *t.* 1 (89b23–31 in *l.* 2 *c.* 1).

165. Ibid. *l.* 2 *t.* 42 (93b15–25 in *l.* 2 *c.* 8–9).

166. Ibid. *l.* 2 *t.* 47 (94a14–19 in *l.* 2 *c.* 10).

167. Ibid.

168. Recall from *Meth. l.* 4 *c.* 14 *par.* 7 above that the definition that is the beginning-principle of demonstration is not one of the premises. It is the causal middle term.

169. *Posterior Analytics l.* 2 *t.* 47 (94a14–19 in *l.* 2 *c.* 10).

170. In the sixteenth and seventeenth centuries, "deduce" and "deduction" did not have the narrow technical meanings that they do now. They were not the opposites of "induce" and "induction."

171. Zabarella, *De medio demonstrationis* (the ninth work in *Opera Logica*) *l.* 1 *c.* 1.

172. *Posterior Analytics l.* 1 *t.* 7–17 (71b9–72b4 = *l.* 1 *c.* 2).

173. Ibid. *l.* 1 *t.* 28–37 (73a21–74a4 = *l.* 1 *c.* 4).

174. Ibid. *l.* 1 *t.* 56 (75a35–39 in *l.* 1 *c.* 6–7).

175. In "since those cannot exist without these," which depends on which is not stated unambiguously.

176. *Prior Analytics l.* 1 *s.* 2 (43a20–46b40 = *l.* 1 *c.* 27–31). In AOAC the section is titled *De abundantia propositionum, inventioneque medii syllogistici* (On furnishing of premises and discovery of the syllogistic middle).

177. Aristotle's word for place or locus is *topos* (pl. *topoi*), hence the book's title, the *Topics*.

178. Cicero, *Topics* 7 (OCT).

179. *Topics l.* 1 *c.* 3 (101b11–36 = *l.* 1 *c.* 4).

180. *Prior Analytics l.* 1 *s.* 2 (43a20–46b40 = *l.* 1 *c.* 27–31).

181. *Posterior Analytics l.* 1 *t.* 28–37 (73a21–74a4 = *l.* 1 *c.* 4).

182. Presumably commentary on *Posterior Analytics l.* 1 *t.* 30–32 (73a34–b4 in *l.* 1 *c.* 4). But Zabarella's *Commentarii in libros duos posteriorum Analyticorum* was not published until 1582. See note to *Meth. l.* 2 *c.* 16 *par.* 6.

183. The phrase "of the locus passage" translates as "loci." Generally, *locus* can be either a passage of text or a *topos*. But here the word means both. In Renaissance humanism, the *Topics* was read more as an ordered catalog than it commonly is now. Each topic, each *topos* or *locus*, had a number, usually placed in the margin, that corresponded to the text in which the topic was described.

184. *Posterior Analytics l.* 1 *t.* 28 (73a21–25 in *l.* 1 *c.* 4).

185. *Topics l.* 1 *c.* 3 (101b11–36 = *l.* 1 *c.* 4).

186. Themistius, CAG *v.* 5.1 *p.* 10 *ln.* 6 f.

187. Averröes, long commentary on *Posterior Analytics l.* 1 *t.* 28 (73a21–25 in *l.* 1 *c.* 4), AOAC *v.* 1 *pt.* 2a *fol.* 64 *sg.* D–*fol.* 65 *sg.* E.

188. *Posterior Analytics l.* 2 *t.* 3–7 (89b34–90a34 in *l.* 2 *c.* 1–2).

189. Ibid. *l.* 2 *t.* 46 (94a7–14 in *l.* 2 *c.* 10).

190. Eustratius, CAG *v.* 21.1 *p.* 1 *ln.* 4 ff.

191. Zabarella does not consider the possibility that the title was not assigned by Aristotle.

192. *Posterior Analytics l.* 2 *t.* 100 (99b15–17 in *l.* 2 *c.* 19).

193. *On the Soul l.* 1 *t.* 1 (401a1–4 in *l.* 1 *c.* 1).

194. The Greek noun means exactness, accuracy, precision. Latin translations of the passage in Zabarella's day included *secundum certitudinem, exquisitior,* and *in subtilitate.* From what follows, it appears Zabarella understood the Greek noun to mean *certitudo* (certainty).

195. *On the Parts of Animals l.* 1 *c.* 5 (644b22–646a4 = *l.* 1 *c.* 5).

196. *Metaphysics l.* 1 *su.* 1 *c.* 2 (982a4–983a23 = *l.* 1 *c.* 2).

197. *Posterior Analytics l.* 2 *t.* 1 (89b23–31 in *l.* 2 *c.* 1).

198. E.g., Averroës, proem to long commentary on *Posterior Analytics, AOAC v.* 1 pt. 2a *fol.* 1 *sg.* B–*fol.* 9 *sg.* E.

199. Averroës, long commentary on *Metaphysics l.* 7 *t.* 42 (1037b8–27 in *l.* 7 *c.* 12), *AOAC v.* 8 *fol.* 194 *sg.* A–G. Zabarella's Latin here is verbatim the vulgate translation, as in *AOAC.* See the following note.

200. Ibid. Zabarella here quotes the ending part of the passage that he just quoted, marking it as he did there, with quotation marks. But the words are different. Such liberties may explain why Zabarella's other quotations of Averroës do not match the common published translations. See note to *Meth. l.* 1 *c.* 2 *par.* 4.

201. *Sophistical Refutations l.* 2 *c.* 8 (183a27–184b8 = *c.* 34).

202. *Physics l.* 1 *t.* 1–5 (184a10–b14 = *l.* 1 *c.* 1).

203. *Metaphysics l.* 7 *t.* 23 (1032a28–b31 in *l.* 7 *c.* 7). See note to *Meth. l.* 1 *c.* 6 *par.* 3.

204. *Nicomachean Ethics l.* 1 *c.* 4 (1095a14–b13 = *l.* 1 *c.* 4).

205. *Physics l.* 1 *t.* 1–5 (184a10–b14 = *l.* 1 *c.* 1).

206. *Nicomachean Ethics l.* 1 *c.* 4 (1095a14–b13 = *l.* 1 *c.* 4).

207. *Physics l.* 1 *t.* 1–5 (184a10–b14 = *l.* 1 *c.* 1).

208. *Metaphysics l.* 7 *t.* 23 (1032a28–b31 in *l.* 7 *c.* 7). See note to *Meth. l.* 1 *c.* 6 *par.* 3.

209. *Physics l.* 1 *t.* 1–5 (184a10–b14 = *l.* 1 *c.* 1).

210. *Posterior Analytics l.* 1 *t.* 95–99 (78a22–b34 in *l.* 1 *c.* 13).

211. *Metaphysics l. 7 t. 23 (1032a28–b31 in l. 7 c. 7).* See note to *Meth. l. 1 c. 6 par. 3.*

ON REGRESSUS

1. For a seventeenth-century presentation of regressus in English, see John Newton, *An Introduction to the Art of Logicke* (London, 1671), *l. 2 c. 9.* Newton's material on regressus is virtually a translation (without citation) of Robert Sanderson, *Logicae Artis Compendium* (Oxford, 1615), "De demonstratione potissima," *pt. 3 c. 16.* Throughout his textbook, Newton cites Zabarella as authoritative.

2. *Posterior Analytics l. 1 t. 18–27 (72b5–73a20 = l. 1 c. 3).*

3. *Prior Analytics l. 2 c. 5–7 (57b18–59a41 = l. 2 c. 5–7).*

4. Though Zabarella defends the procedure with references to Aristotle, there is no term in Aristotle corresponding to "regressus." For the possible beginning of regressus theory in Aristotelian commentary, see Donald Morrison, "Philoponus and Simplicius on Tekmeriodic Proof," in *Method and Order in Renaissance Philosophy of Nature*, ed. Eckhardt Kessler (Ashgate, 1998), 1–22. For a history of regressus at Padua, see William A. Wallace, "Circularity and the Paduan Regressus."

5. See Note on the Text and Translation, s.v. *demonstratio quòd.*

6. *Posterior Analytics l. 1 t. 22–27 (72b25–73a20 in l. 1 c. 3).*

7. Ibid. *l. 1 t. 95–98 (78a22–b11 in l. 1 c. 13).* One of Aristotle's examples involves the fact that, because they are near, planets do not twinkle. The first demonstration: What does not twinkle is near; planets do not twinkle; therefore, planets are near. This demonstrates cause (nearness) from effect (not twinkling). The second demonstration: What is near does not twinkle; planets are near; therefore, planets do not twinkle. This demonstrates effect from cause. Some commentators have taken a regressus to be a combination of an induction and a deduction, but this is incorrect. Both halves of the regressus are deductive syllogisms.

8. See Note on the Text and Translation, s.v. *demonstratio quòd.*

9. In the technical sense of "convert." To convert a premise, exchange subject and predicate.

10. *Posterior Analytics* l. 1 t. 22–23 (72b25–32 in l. 1 c. 3).

11. Ibid. l. 1 t. 24–25 (72b32–73a5 in l. 1 c. 3).

12. Ibid. l. 1 t. 22–23 (72b25–32 in l. 1 c. 3).

13. Ibid. l. 2 t. 92 (98b16–24 in l. 2 c. 16).

14. In the first edition, "that same book" referred to Book 1. In the second edition, the immediately preceding sentence, referring to a passage in Book 2, got added, but this relative reference was not then changed, as it should have been. The example of twinkling bodies and associated discussion are at *Posterior Analytics* l. 1 t. 95–98 (78a22–b11 in l. 1). In our editions, this is in c. 13, as it is in Zabarella's own edition (with commentary), his *Comm. Post. An.*, first published in 1582 and then printed with later editions of *Opera Logica*. In AOAC, the passage appears in what is labeled c. 12 (though some chapter numbering nearby is misprinted). So presumably, Zabarella was using an edition in which this discussion and example were in a chapter labeled 10, but then he used a different numbering in his own edition, and did not retroactively change the reference here. Supporting evidence for this is provided by Zabarella's citation of the same passage again below as "the tenth chapter of the first book" (*Regr. c. 8 par. 3*).

15. "Conversion *simpliciter*" is swapping of subject and predicate without changing the quantity. Zabarella discusses it in his *Tabulae Logicae*, in the tables for the *Prior Analytics*.

16. *Physics* l. 1 t. 1–5 (184a10–b14 = l. 1 c. 1).

17. Themistius, CAG v. 5.2 p. 1 ff.

18. Averroës, long commentary on *Posterior Analytics* l. 1 t. 24 (72b32–73a1 in l. 1 c. 3), AOAC v. 1 pt. 2a fol. 58 sg. A–fol. 60 sg. C; long commentary on *Posterior Analytics* l. 1 t. 97 (78a39–b4 in l. 1 c. 13) v. 1 pt. 2a fol. 215 sg. F–fol. 216 sg. D.

19. *Posterior Analytics* l. 1 t. 6–7 (90a9–34 in l. 2 c. 2). See Note on the Text and Translation, s.v. *demonstratio quòd*.

20. *Posterior Analytics* l. 1 t. 178 (87a31–38 = l. 1 c. 27).

21. Ibid. l. 2 t. 39 (93a16–29 in l. 2 c. 8).

22. Throughout the book, but especially *Physics* l. 1 t. 57–70 (189b30–191a22 ≈ l. 1 c. 7), and within that, t. 62 (190a31–b5). The argument that because there is generation (an effect), there must be a first matter (the cause), is an expansion beyond what Aristotle actually says in these passages, but the interpretation goes back at least to Aquinas and Averroës. See Aquinas' commentary on *Physics* 1.7, lectures 12 and 13, and Averroës' commentary on the chapter, especially t. 68 and 69 (191a2–14 in l. 1 c. 7). In *Regr.* l. 4 c. 7 below, Zabarella acknowledges there is disagreement over the interpretation. Whether Aristotle believed in the existence of a first matter (or what is now called "prime matter") remains a subject of debate.

23. Possible translations of *quibusdam inspectis* include "in the ones inspected," "by means of the ones inspected," "the ones [taken account of] having been inspected," "once some have been inspected." "At once" translates as *statim*.

24. *Physics* l. 1 t. 62 (190a31–b5 in l. 1 c. 7). Zabarella's explanation here of *how* the inductive step occurs and *how* it is justified are not in the cited passage. As Zabarella says in the following sentence, few others interpreted the passage as he does.

25. be discovered] occur C.

26. The knowledge of cause by a "negotiation of the intellect" is cited by an earlier Paduan philosopher, Agostino Nifo (1470–1538), as one of four types of knowledge according to Aristotle. Nifo's work was evidently known to Zabarella. See Wallace, "Circularity and the Paduan Regressus," 84, 92.

27. John Newton offers no direct translation of *negotiatio*. He says that once "consideration" has been had, "distinct knowledg[e] of the cause" becomes "habituated and radicated in the understanding." *Art of Logicke,* l. 2 c. 9. A fruitful comparison can also be made to Everard Digby, *Theoria analytica* (London: Henrici Binneman, 1579).

28. The sense of "recognize" or "come to know" that *cognoscere* can carry should be kept in mind through this passage. See Note on the Text and Translation, s.v. *cognoscere*.

29. *Posterior Analytics* l. 1 t. 1–6 (71a1–b8 = l. 1 c. 1).

30. Themistius, paraphrasing *Posterior Analytics* l. 1 t. 2 (71a11–17 in l. 1 c. 1), CAG v. 5.1 p. 2 ln. 26 ff.

31. *On Coming to Be and Passing Away* l. 1 su. 1 t. 1 (314a1–315a3 in l. 1 c. 1).

32. Ibid. l. 1 su. 1 t. 13–20 (317a33–319a22 in l. 1 c. 3). Aristotle does not identify some "first matter" as unambiguously as Zabarella suggests.

33. Ibid. l. 1 su. 1 t. 1 (314a1–315a3 in l. 1 c. 1). As Zabarella knows the corpus, *On Coming to Be and Passing Away* comes after the *Physics*, with only *On the Heavens* in between. Zabarella presumes the sequence was by Aristotle's design and that Aristotle intentionally addressed, first, the beginning-principles of natural body in the *Physics* and then, afterward, the causes of generation in *On Coming to Be and Passing Away*.

34. *Physics* l. 1 t. 57–70 (189b30–191a22 in l. 1 c. 6–7).

35. We might expect Zabarella to now review the content of text no. 70, which in *AOAC* is the last text in the chapter, su. 4 c. 3, but he seems instead to recount the argument in texts no. 79 through 82 (191b35–192a34 in l. 1 c. 9), which in *AOAC* form the end of su. 4 c. 4 and the beginning of su. 4 c. 5.

36. *Physics* l. 1 t. 82 (192a26–34 in l. 1 c. 9).

37. Zabarella's use of *permanente*, "persisting," in *ex quo permanente*, follows Aristotle's Greek (192a32 in *Physics* l. 1 c. 9) closer than does the vulgate translation (as in *AOAC*) or Averröes' commentaries, so Zabarella was relying on a different translation, a different commentary, or Aristotle's Greek.

38. Zabarella uses the term *regressus* both for the third step (as here) and for the overall three-step process that the third step completes (as in the following sentence).

39. *Physics* l. 1 t. 57–70 (189b30–191a22 in l. 1 c. 6–7).

40. *On Coming to Be and Passing Away* l. 1 su. 1 t. 13–20 (317a33–319a22 in l. 1 c. 3).

41. *Physics* l. 1 t. 5–15 (251a16–252b6 in l. 8 c. 1).

42. Latin *movere* (without *se*) is normally transitive. When Zabarella says the first mover moves, he means that the first mover moves other things, not that the first mover is itself in motion.

43. See Note on the Text and Translation, s.v. *demonstratio*, re demonstration *of* a cause.

44. *Physics l.* 8 *t.* 52 (259b20–31 in *l.* 8 *c.* 6).

45. Ibid. *l.* 8 *t.* 53 (259b32–260a19 in *l.* 8 *c.* 6).

46. Presumably in his preliminary discourse to a course on Aristotle, *Physics* 8. See Dominique Bouillon, "Un discours inédit de Jacopo Zabarella préliminarie à l'exposition de la 'phisique' d'Aristote (Padue 1568)," *Atti e memorie dell'Accademia Galileiana di Scienze, Lettere, ed Arti in Padova,* cx/1 (1998–1999), 119–27. Zabarella's *In libros Aristotelis Physicorum commentarii* was not published until 1601, twelve years after he died. See note to *Meth. l.* 2 *c.* 16 *par.* 6.

47. *Physics l.* 8 *t.* 78–86 (266a10–267b26 = *l.* 8 *c.* 10).

48. See *Metaphysics l.* 12.

49. Simplicius, *CAG v,* 10 *p.* 1117 *ln.* 8 ff.

50. That is, the very start of Book 8, *Physics l.* 1 *t.* 1 (250b11–23 in *l.* 8 *c.* 1).

51. *Physics l.* 8 *t.* 77 (265b17–266a9 in *l.* 8 *c.* 9).

52. Ibid. *l.* 8 *t.* 52 (259b20–31 in *l.* 8 *c.* 6).

53. Ibid. *l.* 8 *t.* 78–86 (266a10–267b26 = *l.* 8 *c.* 10).

54. *On the Soul l.* 3 *t.* 18 (430a14–17 in *l.* 3 *c.* 5).

55. *Posterior Analytics l.* 1 *t.* 24–27 (72b32–73a20 in *l.* 1 *c.* 3).

56. *Prior Analytics l.* 2 *c.* 5–7 (57b18–59a41 = *l.* 2 *c.* 5–7).

57. *Posterior Analytics l.* 1 *t.* 95–98 (78a22–b11 in *l.* 1 *c.* 13). See note to *Regr. c.* 3 *par.* 2 above.

58. Zabarella, *De propositionibus necessariis* (the fifth work in *Opera Logica*) *l.* 2 *c.* 13.

59. *Posterior Analytics l.* 2 *t.* 48 (94a20–24 in *l.* 2 *c.* 11).

60. Themistius, *CAG v.* 5.1 *p.* 52 *ln.* 6 ff.

61. Philoponus, commentary on *On the Soul l.* 1 *t.* 11 (402b16–403a2), *CAG v.* 15 *p.* 43 *ln.* 1 ff.

62. It is not that scientific knowledge of an effect is the same as scientific knowledge of the cause, i.e., of what the effect is on account of, but that

to have *distinct* scientific knowledge of the effect is to have knowledge of the effect's cause.

63. *Regr. c.* 2.

64. A fallacy Aristotle describes in *Sophistical Refutations, c.* 5 166b38–167a21. Zabarella cites it also in *Meth. l.* 1 *c.* 9 *par.* 13.

65. *Posterior Analytics l.* 1 *t.* 4–6 (71a29–b8 in *l.* 1 *c.* 1).

66. *Physics l.* 1 *c.* 7 especially *t.* 62 (190a31–b5 in *l.* 1 *c.* 7). See *Regr. c.* 4 above.

67. Zabarella, *De speciebus demonstrationis* (the sixth work in *Opera Logica*) *c.* 4.

Bibliography

EDITIONS OF *ON METHODS* AND *ON REGRESSUS**

Opera logica. Venice: Paulo Meietti, 1578.

Opera logica. Editio Secunda. Venice: Giorgio Angelieri for Paulo Meietti, 1586.

Opera logica. In *Opera quae in hunc diem edidit.* Preface by Giulio Pace. [Frankfurt]: Jean Wechel for Jean Mareschal, 1586/87.

Opera logica. Preface by Johann Ludwig Hawenreuter. Basel: Konrad Waldkirch for Lazarus Zetzner and Pierre [and/or Jean] Mareschal, 1594.

Opera logica. Editio tertia. Preface by Johann Ludwig Hawenreuter. Cologne [but probably Neustadt, Basel, or Frankfurt]: Lazarus Zetzner, 1597. Photoreprint. Hildesheim: Olms, 1966.

Opera logica. Quarta editio. 3 vols. Venice: Paulo Meietti (vols. 1, 3) and Giovanni Antonio, Giacomo Franceschi, and Francesco Bolzetta (vol. 2), 1599–1601.

Opera logica. Editio Quarta. Preface by Johann Ludwig Hawenreuter. Cologne [but probably Neustadt, Basel, or Frankfurt]: Lazarus Zetzner 1602–3.

Opera logica. Venice: Paulo Meietti, and Treviso: Roberto Meietti, 1604.

Opera logica. Editio postrema. Preface by Johann Ludwig Hawenreuter. Frankfurt: Lazarus Zetzner, 1608.

Opera logica. Editio postrema. Preface by Johann Ludwig Hawenreuter. Frankfurt: Heirs of Lazarus Zetzner, 1622–23.

De methodis libri quatuor — Liber De regressu. With an introduction by Cesare Vasoli. Bologna: CLUEB, 1985. Contains a photoreprint of the 1578 edition.

---

* For an edition of the *Opera logica* printed in Venice in 1580, see Maclean, "Mediations of Zabarella," 56. A surviving copy of this imprint could not be located in a modern collection; it may be a ghost.

461

TRANSLATION

Schicker, Rudolf, trans. *Jacopo Zabarella: Über die Methoden (De methodis).* *Über den Rückgang (De regressu).* With introduction and notes by Schicker. Munich: Wilhelm Fink, 1995. Translation based on the 1578 edition, compared with the Frankfurt edition of 1608.

SECONDARY LITERATURE

Di Liscia, Daniel A., Eckhard Kessler, and Charlotte Methuen, eds. *Method and Order in Renaissance Philosophy of Nature: The Aristotle Commentary Tradition.* Aldershot: Ashgate, 1997.

Edwards, William F. "The Logic of Iacopo Zabarella (1533–1589)." PhD thesis, Columbia University, 1960.

Gilbert, Neal Ward. *Renaissance Concepts of Method.* New York: Columbia University Press, 1960.

Jardine, Nicholas. "Epistemology of the Sciences." In *The Cambridge History of Renaissance Philosophy*, 685–711, edited by Charles B. Schmitt, Quentin Skinner, Eckhard Kessler, and Jill Kraye. Cambridge: Cambridge University Press, 1988.

———. "Keeping Order in the School of Padua: Jacopo Zabarella and Francesco Piccolomini on the Offices of Philosophy." In Di Liscia et al., *Method and Order*, 183–210.

Maclean, Ian. "Mediations of Zabarella in Northern Europe, 1586–1623." Chap. 3 in *Learning and the Market Place.* Leiden: E. J. Brill, 2009.

Mikkeli, Heikki. *An Aristotelian Response to Renaissance Humanism. Jacopo Zabarella on the Nature of Arts and Sciences.* Helsinki: SHS, 1992. The only English-language monograph on Zabarella.

———. "Giacomo Zabarella." *The Stanford Encyclopedia of Philosophy* (Winter 2012 Edition). Edited by Edward N. Zalta. http://plato.stanford.edu/archives/win2012/entries/zabarella/.

Olivieri, Luigi, ed. *Aristotelismo veneto e scienza moderna: Atti del 250 anno academico del Centro per la storia della tradizione aristotelica nel Veneto.* 2 vols. Padua: Antenore, 1983.

Palmieri, Paolo. "Science and Authority in Giacomo Zabarella." *History of Science* 45 (2007): 404–27.

Piaia, Gregorio, ed. *La presenza dell'aristotelismo padovano nella filosofia della prima modernità: Atti del colloquio internazionale in memoria di Charles B. Schmitt*. Rome: Antenore, 2002.

Poppi, Antonino. *La dottrina della scienza in Giacomo Zabarella*. Padua: Antenore, 1972.

——. "Zabarella, or Aristotelianism as a Rigorous Science." In *The Impact of Aristotelianism on Modern Philosophy*, edited by Riccardo Pozzo, 35–63. Washington, D.C.: Catholic University of America Press, 2004.

Randall, John Herman. "The Development of Scientific Method in the School of Padua." *Journal of the History of Ideas* 1.2 (1940): 177–206. Reprinted in *The School of Padua and the Emergence of Modern Science*. Padua: Antenore, 1961.

Wallace, William A. "Circularity and the Paduan Regressus: From Pietro d'Abano to Galileo." *Vivarium* 33 (1995): 76–97.

——. "Randall redivivus: Galileo and the Paduan Aristotelians." *Journal of the History of Ideas* 49 (1988): 133–49.

Wear, Andrew, Roger Kenneth French, and Iain M. Lonie, eds. *The Medical Renaissance of the Sixteenth Century*. Cambridge: Cambridge University Press, 1985.

Cumulative Index to Volumes 1–2

ॐᢤᢤॐ

Publication of this volume has been made possible by

The Myron and Sheila Gilmore Publication Fund at I Tatti
The Robert Lehman Endowment Fund
The Jean-François Malle Scholarly Programs and Publications Fund
The Andrew W. Mellon Scholarly Publications Fund
The Craig and Barbara Smyth Fund
for Scholarly Programs and Publications
The Lila Wallace–Reader's Digest Endowment Fund
The Malcolm Wiener Fund for Scholarly Programs and Publications